The Design and Implementation of US Climate Policy

A National Bureau of
Economic Research
Conference Report

The Design and Implementation of US Climate Policy

Edited by **Don Fullerton and Catherine Wolfram**

The University of Chicago Press

Chicago and London

DON FULLERTON is the Gutgsell Professor in the Department of Finance at the University of Illinois at Urbana-Champaign, where he is also a faculty associate at both the Institute of Government and Public Affairs and the Center for Business and Public Policy. He is a research associate and director of the Program on Environmental and Energy Economics at the National Bureau of Economic Research. CATHERINE WOLFRAM is associate professor of business administration at the Haas School of Business, University of California, Berkeley, where she codirects the Energy Institute at Haas. She is a research associate of the National Bureau of Economic Research.

The University of Chicago Press, Chicago 60637
The University of Chicago Press, Ltd., London
© 2012 by the National Bureau of Economic Research
All rights reserved. Published 2012.
Printed in the United States of America

21 20 19 18 17 16 15 14 13 12   1 2 3 4 5
ISBN-13: 978-0-226-26914-6 (cloth)
ISBN-13: 978-0-226-92198-3 (e-book)
ISBN-10: 0-226-26914-0 (cloth)
ISBN-10: 0-226-92198-0 (e-book)

Library of Congress Cataloging-in-Publication Data

The design and implementation of US climate policy / edited by Don Fullerton and Catherine Wolfram.
    pages ; cm.—(National Bureau of Economic Research conference report) "This book contains the proceedings of an NBER conference held in Washington, DC, on May 13–14, 2010"—Page 3.
    Includes bibliographical references and index.
    ISBN-13: 978-0-226-26914-6 (cloth : alkaline paper)
    ISBN-10: 0-226-26914-0 (cloth : alkaline paper)
    ISBN-13: 978-0-226-92198-3 (e-book)
    ISBN-10: 0-226-92198-0 (e-book) 1. Climatic changes—Government policy—United States—Congresses. 2. Global warming—Government policy—United States—Congresses. 3. Climatic changes—Government policy—Economic aspects—United States—Congresses. 4. Global warming—Government policy—Economic aspects—United States—Congresses.
I. Fullerton, Don. II. Wolfram, Catherine D. III. Series: National Bureau of Economic Research conference report.
QC903.2.U6D47 2012
363.738'745610973—dc23
                                                        2011050604

Relation of the Directors to the
Work and Publications of the
National Bureau of Economic Research

1. The object of the NBER is to ascertain and present to the economics profession, and to the public more generally, important economic facts and their interpretation in a scientific manner without policy recommendations. The Board of Directors is charged with the responsibility of ensuring that the work of the NBER is carried on in strict conformity with this object.

2. The President shall establish an internal review process to ensure that book manuscripts proposed for publication DO NOT contain policy recommendations. This shall apply both to the proceedings of conferences and to manuscripts by a single author or by one or more co-authors but shall not apply to authors of comments at NBER conferences who are not NBER affiliates.

3. No book manuscript reporting research shall be published by the NBER until the President has sent to each member of the Board a notice that a manuscript is recommended for publication and that in the President's opinion it is suitable for publication in accordance with the above principles of the NBER. Such notification will include a table of contents and an abstract or summary of the manuscript's content, a list of contributors if applicable, and a response form for use by Directors who desire a copy of the manuscript for review. Each manuscript shall contain a summary drawing attention to the nature and treatment of the problem studied and the main conclusions reached.

4. No volume shall be published until forty-five days have elapsed from the above notification of intention to publish it. During this period a copy shall be sent to any Director requesting it, and if any Director objects to publication on the grounds that the manuscript contains policy recommendations, the objection will be presented to the author(s) or editor(s). In case of dispute, all members of the Board shall be notified, and the President shall appoint an ad hoc committee of the Board to decide the matter; thirty days additional shall be granted for this purpose.

5. The President shall present annually to the Board a report describing the internal manuscript review process, any objections made by Directors before publication or by anyone after publication, any disputes about such matters, and how they were handled.

6. Publications of the NBER issued for informational purposes concerning the work of the Bureau, or issued to inform the public of the activities at the Bureau, including but not limited to the NBER Digest and Reporter, shall be consistent with the object stated in paragraph 1. They shall contain a specific disclaimer noting that they have not passed through the review procedures required in this resolution. The Executive Committee of the Board is charged with the review of all such publications from time to time.

7. NBER working papers and manuscripts distributed on the Bureau's web site are not deemed to be publications for the purpose of this resolution, but they shall be consistent with the object stated in paragraph 1. Working papers shall contain a specific disclaimer noting that they have not passed through the review procedures required in this resolution. The NBER's web site shall contain a similar disclaimer. The President shall establish an internal review process to ensure that the working papers and the web site do not contain policy recommendations, and shall report annually to the Board on this process and any concerns raised in connection with it.

8. Unless otherwise determined by the Board or exempted by the terms of paragraphs 6 and 7, a copy of this resolution shall be printed in each NBER publication as described in paragraph 2 above.

Contents

Acknowledgments

This book contains the proceedings of an NBER conference held in Washington, DC, on May 13–14, 2010. Recent political developments in Washington might make adoption of comprehensive climate policy unlikely in the near term, but economics research must take a longer view. The conference was held in Washington, DC, because we hope that the various topics covered in this book will be useful to economists and policymakers contemplating future climate policy, whether that may include direct approaches such as a cap-and-trade system or indirect policies such as tax credits, clean energy subsidies, or renewable fuel standards.

We are very grateful for the generous financial support of the Smith Richardson Foundation, for intellectual encouragement from Jim Poterba, and for terrific organizational support from the wonderful people within NBER, including efficient conference organizing by Carl Beck and Rob Shannon, effective production of the book by Helena Fitz-Patrick, and timely budget assistance from Denis Healy, Alterra Milone, and Alison Oaxaca. We received excellent substantive suggestions from two anonymous reviewers. We also benefited from panel presentations at the conference by three Washington practitioners, including Joseph Aldy, Nathaniel Keohane, and Al McGartland. Several policymakers generously provided written comments and extended conversations that helped us refine the set of topics the book covered. These include Terry Dinan, Doug Elmendorf, Jud Jaffe, Jeff Liebman, and Billy Pizer. We also want to express our extreme gratitude to the chapter authors. They not only undertook new research on challenging topics and wrote the chapters, but each author also served as a discussant for one of the other presentations. This interchange was interesting and helpful enough that we decided to publish each discussant's comments along with the corresponding chapter in the book.

Introduction and Summary

Don Fullerton and Catherine Wolfram

Climate change is one of the most challenging issues facing policymakers today. Greenhouse gas emissions create externalities across the globe, which means that climate change mitigation requires internationally coordinated policy intervention. At the same time, every sector of the economy creates greenhouse gas emissions, some in large quantities. Therefore, climate change action, whenever it occurs, will be an expansive undertaking for any government.

The prospects for US federal climate change legislation have waxed and waned over the past several years. In 2007, the Senate Environment and Public Works Committee approved the Lieberman-Warner Climate Security Act. At the time, this was the farthest climate legislation had progressed in the US Congress. In 2009, the full House of Representatives passed the Waxman-Markey American Clean Energy and Security Act (H.R. 2454). Since the eventual failure of that Act, Congress has not considered any new climate change legislation.

We launched this book with the aim of engaging economic researchers to answer specific questions on climate policy implementation. When we began the project in early 2009, we hoped our contributors would provide

Don Fullerton is the Gutgsell Professor in the Department of Finance at the University of Illinois at Urbana-Champaign, where he is also a faculty associate at both the Institute of Government and Public Affairs and the Center for Business and Public Policy. He is a research associate and director of the Program on Environmental and Energy Economics at the National Bureau of Economic Research. Catherine Wolfram is associate professor of business administration at the Haas School of Business, University of California, Berkeley, where she codirects the Energy Institute at Haas. She is a research associate of the National Bureau of Economic Research.

For acknowledgments, sources of research support, and disclosure of the authors' material financial relationships, if any, please see http://www.nber.org/chapters/c12123.ack.

timely research on policy designs, but we proceeded with caution because the risk appeared real that the federal government would enact comprehensive climate legislation before our authors could submit their first drafts. As we write this introduction and summary in 2011, the opposite concern appears more relevant, since legislation on climate change now seems unlikely for at least several years. Nevertheless, we believed in 2009 and believe even more firmly today that economists have valuable expertise and insight to offer policymakers as they work through legislative and other approaches to mitigating climate change. Addressing climate change will be a massive undertaking, but we can draw on useful economic models as well as analogous experiences that economists have studied to help guide the policy process.

Early economic research on climate change has already contributed to our understanding of the scope of the damages associated with global warming as well as the costs of broadly defined strategies to reduce the emissions of carbon dioxide and other greenhouse gases. For example, researchers described the costs of global warming to various sectors of the economy, the potential savings from market-based incentives, and the major tradeoffs policymakers confront when deciding whether to use a price instrument like emissions taxes or a quantity instrument like tradable permits.

While economic models have proven useful to analyze these big picture issues, the next steps of the policy process require answers to a long list of more specific questions that bear on the actual design and implementation of US climate policy. If a cap-and-trade program is chosen, how will permits be allocated initially? Can permits be banked for use in a later period? If so, under what rules? Who will be allowed to sell offsets for the reduction of greenhouse gas (GHG) emissions or the sequestration of CO_2? How will those offsets be verified? What are the many distributional effects of these policies? How can any adverse distributional effects be ameliorated? What other environmental or nonenvironmental goals ought to be incorporated into the design of climate policy?

To get answers to these and other questions, we took a more prescriptive approach to developing this book than is conventional in economics. In particular, for most edited volumes in our field, editors select authors and give them some general guidance about the topic or topics they would like to see addressed. In contrast, we began by developing a detailed set of design and implementation questions that we thought needed answers. We next identified an academic economist whose expertise was relevant to each question. In almost all cases, we approached authors who had worked on related topics, but who would have to address topics that were new to them and new to the literature to write the chapter for our book. For example, Hilary Sigman has worked on enforcement and monitoring issues before, but not in the context of climate change.

To help induce our authors to take on new research topics, we asked for chapters that were shorter than the usual research paper. We advised authors to think hard about their assigned question, start an economic model to analyze it, collect whatever initial data could be used in that model, and suggest initial answers. We hoped that starting to work on the topic for our book would lead these authors into further research on each topic, which we have been delighted to see transpire in several cases (Bushnell and Mansur 2011; Sigman and Chang 2011).

The remainder of the introduction summarizes the chapters and relates them to each other. We do not attempt to review all of the other important literature in this field. Both the book edited by Guesnerie and Tulkens (2008), *The Design of Climate Policy,* and the review article by Aldy et al. (2010) in the *Journal of Economic Literature* called "Designing Climate Mitigation Policy" provide more comprehensive reviews than is possible here.

I.1   Climate Policy in the Broader Context

The first six chapters consider the possible effects of US climate policy on a range of economic outcomes, including household income, employment, innovation, greenhouse gas emissions outside the United States, emissions of non-greenhouse gas pollutants, and the natural carbon cycle. All six authors use economic theory, developed through simple, intuitive models, to identify different pathways by which each effect might operate. Several of the authors also use simulations or empirical estimates to bring data-driven evidence to bear on the questions they examine.

The first chapter—arguably the broadest in scope—quantifies the effects of climate policy on several different factors that impact household disposable income. Specifically, Gilbert Metcalf, Aparna Mathur, and Kevin Hassett simulate the impact of a CO_2 price of fifteen dollars per ton and analyze the burden absorbed by households at different deciles of the income distribution. By way of comparison, several of their scenarios also examine households at different points of the consumption distribution. Consumption is a more reliable indicator of lifetime income, as some households, such as students, have income that is temporarily very low. They disaggregate household income into capital and labor sources and model the impact of carbon pricing on both of these components. They also analyze changes in the prices of consumption bundles.

Estimates like these are central to political debates about carbon pricing, which is often seen as regressive, given the rough logic that low-income consumers spend a higher share of their income on electricity, natural gas, and gasoline. As Metcalf, Mathur, and Hassett point out, however, this rough logic is contradicted by the fact that higher income households are more likely to be hurt by reductions in employment or lower returns to capital caused by a CO_2 price. Their chapter certainly suggests that we need to develop more thorough analyses of the extent to which carbon pricing is likely to be forward shifted (i.e., lead to higher consumer prices) or backward shifted (i.e., reduce returns to capital and labor). Another possibility, which goes beyond the scope of the model in this chapter, is that the burden will be shifted abroad, for instance, to the Saudi government if climate policy causes oil prices to fall. While the authors use assumptions designed to cover a range of possibilities, it is important to continue to get concrete data that could inform which of their scenarios is most relevant.

In this spirit, chapter 2 by Olivier Deschênes takes an important step toward quantifying one of the backward-shifting mechanisms identified by Metcalf, Mathur, and Hassett—the effects of climate policy on labor markets. Conventional wisdom suggests that putting a price on carbon will reduce employment, but, as in the first chapter, Deschênes' economic model points out that this simple logic does not capture the full story. He begins by writing down a basic economic relationship that elucidates how a change in energy prices, such as one induced by a positive price on CO_2 emissions, might impact labor. Any cost-minimizing, profit-maximizing firm confronted with a price increase for one of their inputs faces two options, which are not necessarily mutually exclusive. They can use less of the more expensive input and substitute to other inputs, or they can make less of the good. If manufacturers reduce their output, all else equal, employment will unambiguously fall. As Deschênes points out, this is commonly called the scale effect. But, for a given level of output, it is not clear whether energy and labor are complements in the production process or substitutes, in which case employment might rise. Ultimately, the answer is empirical, and it may vary by skill-level of the job, industry, or region of the country.

To begin to get some insight into these questions, Deschênes estimates the empirical relationship between state-by-year variation in electricity prices and employment. He finds that a 4 percent increase in electricity prices, consistent with estimates of the impact of the Waxman-Markey legislation that passed the House in 2009, leads to approximately a 0.5 percent reduction in US employment. Whether one interprets these effects as big or small depends on one's perspective. A 0.5 percent reduction means a loss of several hundred thousand jobs, which is a large number, but, as Deschênes notes, the 2008 recession caused employment losses that were almost ten times larger. We hope that in future work, Deschênes and others will also separate the effects along different dimensions, such as industry sector, region of the country, or skill-level of the jobs (which would speak to the assumptions in the Metcalf, Mathur, and Hassett chapter on distributional implications). This will help inform policy discussions, not just about who will be the winners and losers, but also about how policies might be designed to mitigate the harm to those bearing the largest burden.

Chapter 3 addresses a related topic, as conventional wisdom often highlights the concern that jobs will be exported abroad if the United States unilaterally imposes a price on carbon. If jobs are exported abroad, emissions may go with them, which can undo the benefits of US-based efforts to limit carbon emissions (this is called "leakage"). Kala Krishna begins by describing some of the specific findings from work that relies on computable general equilibrium (CGE) models, and she highlights findings on the effectiveness of border tax adjustments for leakage mitigation. Noting that a CGE model can be a "black box," she provides a clear description of the mechanisms at play in these models, focusing on how border tax adjustments are represented.

Krishna goes on to point out how different conditions in product and factor markets will lead to different effects of policies. She makes an interesting point, for example, in the case where the United States restricts emissions in a way that would normally cause leakage. If the rest of the world has a generous, perhaps even nonbinding cap, then that emissions leakage will be mitigated, as it would cause the cap for the rest of the world to become binding. Any further pressure to increase emissions in the rest of the world will not result in more emissions, as it will only drive up the price of carbon abroad.

In chapter 4, Charles Kolstad takes on another important consideration for any climate change mitigation policy—how might it affect innovation designed to reduce greenhouse gas emissions? Achieving the types of greenhouse gas reductions required to thwart dangerous climate change will involve fundamental changes to the way society produces and consumes energy. It is critical to understand, therefore, how polices that the United States is likely to enact in the next several years will affect investments in activities that could bring about these types of transformative changes.

Kolstad's model focuses on the incentives of the innovator. Specifically, he models a single innovating firm that licenses its technology to multiple identical atomistic polluting firms. He shows that a social planner can set either a tax or an emissions cap to achieve the first-best levels of both abatement and investment in innovations that reduce the marginal cost of abatement. He shows that under a permit system, the innovator captures the entire surplus through a license to the polluting firms. Under a tax system, however, the innovator shares the gains with the polluters in the form of lower abatement costs. The intuition for this result is that under the cap-and-trade system the polluting firms are required to abate a certain amount, so their objective is to find the cheapest way to do it (strictly speaking, Kolstad is modeling a pure cap system, since his model has no trading between the identical firms). As long as the licensing fee plus the lower cost technology is epsilon cheaper than the preinnovation abatement technology, the polluting firms will choose it. In the case of a tax, however, the cost of abatement factors into the polluting firms' decisions about how much to abate, so the optimal licensing fee leaves some rents to the abating firm.

Kolstad's result suggests that cap-and-trade systems may provide stronger incentives for innovation. Going forward, it will be important to evaluate this result under different assumptions, for instance, to allow the innovating firm to use a multipart price structure for the innovative technology or to otherwise enrich the depiction of the relationship between the innovating and polluting firms.

In chapter 5, Stephen Holland describes, both theoretically and empirically, spillovers from CO_2 emissions regulations to other pollutants. This is an important point, and one that has received attention from an environmental justice community that fears GHG mitigation policies could lead to increased criteria pollutant concentrations in disadvantaged areas. The academic literature, at least to date, has largely overlooked the topic. It is important to consider, since reducing GHG emissions may lead to significant increases or reductions in other pollutants. Efficient climate policy design would consider spillovers, though the specific way to account for any costs or benefits depends critically on the nature (or lack) of regulatory treatment of the other pollutants. Spillovers may also factor into political and distributional considerations about climate policy.

Since the United States currently does not have a comprehensive climate change policy, obtaining empirical estimates of the extent of spillovers is not straightforward. Holland takes a clever approach to solving that problem and looks for evidence of spillovers to CO_2 emissions from NO_x regulations. Under relatively strong assumptions (i.e., unconstrained, profit-maximizing firms and only marginal changes in the prices of both CO_2 emissions and NO_x emissions), the response of CO_2 emissions to a change in the price of NO_x emissions is equal to the response of NO_x emissions to a change in the price of CO_2. Holland finds that CO_2 and NO_x emissions both fall when the price of NO_x emissions increases, and this is primarily driven by the output effect, as higher NO_x prices cause older plants to reduce operation. While Holland takes an electricity-generating plant as his unit of analysis, it will be important to extend this type of analysis to more aggregate units of analysis, such as the western electricity grid.

The final chapter in this section addresses spillovers from regulations of anthropogenic carbon emissions to the larger carbon cycle. Some of the basic facts Severin Borenstein lays out are quite sobering and provocative: annual anthropogenic carbon emissions are about 9 gigatons, while the natural carbon flux emits and absorbs 210 gigatons of carbon per year! Importantly, human activities can alter the natural carbon flux in many ways. So, if global governments succeed in enacting policies that reduce anthropogenic carbon emissions by half, which is a much larger reduction than contemplated by *any* near-term policies, all that work could be undone if the adjustments to achieve the reductions in anthropogenic emissions led to a mere 2 percent change in the natural carbon absorption. Borenstein goes on to discuss the implications of this fact for market-based climate policies.

I.2   Interactions with Other Policies

The effect of a US federal climate policy depends on climate change mitigation strategies pursued by states or other national governments. Chapter 7 by Lawrence Goulder and Robert Stavins considers the problem of interactions between state and federal policies, focusing on cap-and-trade programs or a carbon tax. Take as an example the effects of a subnational cap-and-trade system such as enacted already in ten northeastern states (the Regional Greenhouse Gas Initiative, called RGGI). With no other climate policy anywhere, then RGGI might succeed in reducing emissions in those states. Other jurisdictions, however, might increase production, which could drive up their emissions (i.e., leading to leakage). As a result, the overall cost of emission reduction is not minimized because marginal abatement costs are not equalized.

Suppose instead that the federal government has a carbon tax (or a permit system with a binding safety valve). Then the subnational policy has very similar effects to those just described: any binding subnational restriction may result in some leakage if other states increase production at their unchanged emissions price. On the other hand, consider a stringent subnational policy in the context of a federal permit system with a lower price (not at any safety valve ceiling price). In that case, Goulder and Stavins show that leakage will be complete—with no net emissions reductions whatever. The reason is that firms in that subnational regime must reduce emissions by some quantity, which makes exactly that quantity of national permits available to any firms outside that subnational regime. It effectively increases the supply of permits to others, and so reduces the nationwide price of federal permits.

Interestingly, it also implies a difference between a carbon tax and a cap-and-trade program, even with perfect certainty. With a US carbon tax, RGGI could reduce emissions further. With a federal cap-and-trade system, however, RGGI would have no effect on the environment, but would only reduce overall cost-effectiveness by introducing a difference between permit prices and, therefore, marginal costs of abatement. Goulder and Stavins consider other interesting cases and a variety of complications, some of which change the simple result we have described.

While Goulder and Stavins look at climate policy interactions between different jurisdictions, chapter 8 by Arik Levinson looks at interactions between different policies. To reduce carbon emissions, a single jurisdiction may choose to enact both a market policy (such as carbon tax or cap and trade) and traditional standards (such as a low-carbon fuel standard or an energy efficiency requirement). Levinson points out that having both kinds of policies can lead to one of three outcomes: the policies may be mutually reinforcing (like "belts and suspenders"), the binding policy may render the nonbinding policy irrelevant, or, if both policies are binding, then they may raise costs relative to one efficient policy designed to achieve the same abatement.

The cost-raising outcome occurs, for example, if a binding standard such as a low-carbon fuel standard means that more abatement takes place by that expensive means rather than by some other means—at the lower marginal abatement cost given by the common permit price elsewhere. In contrast, the irrelevant outcome occurs if the standard is not binding. Even if the standard alone would bind, a stringent carbon-pricing policy may induce firms to reduce the carbon content of fuel below the standard's requirement. Finally, the mutually reinforcing outcome may occur either because of some other market failure, or because of administrative complexity. For an example of the former, consider that if landlords' energy efficiency investments cannot be observed adequately, then renters may not be willing to pay for them. (Lucas Davis' chapter, described later, considers this possibility directly.) A carbon-pricing mechanism alone might then raise the cost of heating fuel paid by renters but still not be enough to induce landlords to pay for low-cost abatement via energy efficiency investments. It may require additional regulations such as building codes. For an example of administrative complexity, consider the difficulty of applying carbon pricing to all forms of carbon, especially ad hoc fuels used in developing countries. A simple ban on the most carbon-intensive fuels may be more enforceable than collecting a price on the carbon content of it.

In addition to interacting with each other, both mandatory carbon pricing and more traditional regulations may interact with purely voluntary programs. In chapter 9, Matthew Kotchen considers a particular voluntary program. Specifically, in 2005, the state of Connecticut started a "Clean Energy Options" program that allows individual households to pay extra for "green" electricity (produced by a mix of wind and small-scale hydro sources). In return, any municipality that enrolls at least a threshold share of the local households can qualify for the Connecticut Clean Energy Communities (CCEC) program that provides free solar panels to display prominently in public locations. Kotchen regards the free solar panels as a nudge, as they provide a low-cost mechanism to encourage voluntary household participation, yet are not a true quid pro quo of any substantial value. He finds that the merely symbolic CCEC reward induced a 39 percent increase in household participation in the Options program to pay for green electricity. That increase represents 7,000 households, 31 percent of all participating households statewide, and prevents an estimated 23,000 metric tons of carbon dioxide emissions.

Kotchen thus demonstrates that a voluntary program can have significant impact. An interesting follow-on question is how that voluntary program might interact with other mandatory programs. If the state or federal government introduced a mandatory carbon abatement policy or carbon-pricing policy, would households see their extra mandated costs as reasons *not* to incur any other costs voluntarily? In the language of other chapters just described, a binding cap-and-trade policy might make a nonbinding voluntary program irrelevant. If so, it reduces the net abatement achieved by the cap-and-trade program by the loss of abatement that otherwise would have been achieved with just the voluntary program.

These studies explain just a few of the examples of climate policy interactions. More generally, climate policy can interact with any tax or regulation at the federal, state, or local level. Clearly a federal climate policy interacts with state or regional climate policy, but it also might interact with federal or state tax policy or even nonenvironmental regulations. For example, a federal tax or price on carbon may compound the effects of a federal or state tax on energy, such as the gasoline excise tax. Therefore a careful analyst must simultaneously consider the relevant taxes or regulations at all levels.

A climate policy may also interact with international policies, such as those intended to address the competitiveness of US industry in trade with other countries. If US producers face a price on each ton of carbon dioxide emissions, the cost of producing US goods would rise, so climate policy might best be paired with other policies that restore US competitiveness in some manner. One of the proposed methods to address US competitiveness is to give some CO_2 permits to firms in proportion to their output. Chapter 10 by Meredith Fowlie studies this kind of output-based permit allocation (OBPA).

As she notes, a standard carbon tax or price minimizes the total abatement cost, because it works via two effects. First, the "substitution effect" induces firms to shift from carbon-intensive inputs toward other inputs, which reduces the carbon per unit of output. Second, the "output effect" raises the cost of production and thus reduces the number of units of output demanded. In an open economy, however, the latter effect may harm US competitiveness, move production overseas, and cause leakage.

Some US proposals would combat this competitiveness problem with an OBPA, which essentially rewards firms for producing more output. Fowlie points out that this implicit output subsidy has both pros and cons. The advantage is that it can offset some of the climate policy's effect on US output prices, which helps US firms compete and reduces leakage. The disadvantage is that it raises the overall cost of carbon abatement, by moving away from the cost-minimizing combination of abatement methods. A cap-and-trade program with OBPA still induces firms to shift toward less carbon-intensive production (the substitution effect), but it no longer induces consumers to reduce purchases (the output effect). With a fixed total number of permits and, therefore, a fixed requirement for total abatement, any attempt to protect one industry by OBPA means that more of the abatement must be undertaken by other industries. Those other industries presumably will need to undertake more expensive abatement strategies, as they move up their rising marginal cost of abatement schedule.

Moreover, the House Bill (H.R. 2454) specified that eligibility for this output subsidy would be based on some combination of the industry's energy intensity and trade intensity (that is, import penetration, or trade vulnerability). Industries with energy or emissions intensities above 20 percent are eligible regardless of trade intensity. But Fowlie shows that these are exactly the industries for which OBPA is most costly. Giving this output subsidy to energy-intensive industries means not reducing the output of energy-intensive industries. Instead, emissions must be reduced in industries that are not emission intensive, which can be very costly.

I.3   Design Features of Climate Policy

Many economists like to characterize a carbon tax in simple models as a rate, t, on all carbon emissions, implicitly assuming perfect administration, measurement, and enforcement. This section describes issues in the detailed design of a climate policy, which includes decisions about how to administer it, how to monitor actual emissions, and how to enforce rules. An eventual policy will apply to particular firms and not others, and it may include various exemptions, varied rates, and offsets.

One issue in the design of climate policy is whether to apply it "upstream" on the producers of fossil fuel (mines, oil wells, and importers) or "downstream" on the users of fossil fuel (drivers, electricity generators, and manufacturing plants with smokestacks). Chapter 11 by Erin Mansur points out that most pollutants are best regulated downstream, because the actual emitters may have means of reducing the emissions per unit of fuel. If those abatement methods are omitted, then overall cost of abatement is not minimized. In the case of carbon dioxide emissions, however, some have argued that those "end-of-pipe" methods are negligible or too expensive (such as carbon capture and sequestration [CCS]). The actual emissions may be based entirely on the carbon content of the fuel. Moreover, the tax or permit price could be collected from 150 refineries in the United States instead of from 105,000 gasoline service stations—or even worse, from drivers of 244 million motor vehicles. Measurement devices on all such vehicles would be prohibitively expensive.

Mansur develops a theory of cost-minimizing decisions about where to apply the tax on the vertical chain of production ("vertical targeting"). He models the tradeoffs explicitly, with choices both about fuel inputs and end-of-pipe abatement technology. He then adds transactions costs that depend on the number of firms under the policy, and shows how the additional costs of administering more downstream firms might offset any cost advantages from capturing end-of-pipe abatement technology downstream. He discusses how the choice might also be affected by leakage, which might be minimized by aiming at whatever part of the vertical chain has the least elastic foreign supply. He also notes problems with "offsets," which are essentially payments for end-of-pipe or postemission sequestration. Finally, he discusses how the analysis is changed by consideration of imperfect competition, price regulation by Public Utility Commissions that may or may not allow cost pass-through, and tax "salience" (where a more explicit payment of tax might affect actual behavioral reactions).

Chapter 12 by James Bushnell focuses on offsets. He begins by pointing out than an ideal carbon tax or cap would apply to *all* emissions. For a variety of reasons, however, actual climate policy is virtually bound to exclude certain firms, industries, or countries from the taxed or capped sector. First, monitoring and enforcement may be particularly difficult for some other greenhouse gases, or for small businesses, residences, and agriculture. Second, political pressures from certain sectors seeking an advantage may expand the definition of "small business" and other exemptions. Third, some jurisdictions might not participate in the carbon policy agreement. Fourth, the lowest cost mitigation might include activities that take carbon out of the atmosphere in the form of sequestration. In those cases, economic efficiency suggests that the policy not only place a positive price on emissions, but also provide a subsidy to sequestration activities that are *outside* the capping jurisdiction or capped sectors.

One way to achieve very low cost mitigation is to pay for sequestration though offsets, but the chapter by Bushnell points out a number of problems with those programs. First of all, any payments from firms in the capped jurisdiction to those in the uncapped jurisdiction inherently test the limits of interjurisdictional regulatory cooperation. Officials in the host nations must provide verification data or at least allow access to such data. Second, those host nations are often developing countries with weak regulatory or governance structures. Third, the system must set an emissions baseline against which to measure reductions. This step is literally impossible to do accurately, as it requires knowing the counterfactual emissions in the absence of the program. Firms may have better information than regulators about steps they would have taken in the absence of the program, which gives rise to problems of moral hazard and adverse selection.

If authorities correctly gauge each firm's true baseline (emissions without any offset policy), then no such problems arise. With imperfect information, the moral hazard problem suggests that firms will have the incentive to invest in high-carbon projects or to delay investments in abatement, so that regulators set a high baseline. That way, they can receive offset payments for undertaking more abatement than they would have pursued absent the program. The adverse selection problem arises not from changes in firm behavior, but because authorities do not know which firms have high or low actual baselines. The effects of offsets will then depend on whether the authorities are right *on average* about firms' baselines. If so, only firms with low actual baselines will opt into the offset program. Those with high actual baselines opt out and undertake no abatement. The result is more emissions and less overall abatement than anticipated. If authorities are wrong on average, then all baselines may be overestimated, and payments may be high. In this case the offset program does not inefficiently allocate abatement, but it may result in less total abatement than anticipated—and thus may require tighter controls in the capped sector.

This study has implications for actual carbon policy design and implementation, particularly suggestions that the problems with offsets be addressed by placing a ceiling on the total number of offsets or a devaluation of all offsets. The former does nothing to fix the problem of adverse selection when only some firms opt into the program, and the latter may inappropriately treat all offsets as equally nonadditional. More efficient responses might include overall program reviews, or randomized trials to collect better information.

Hilary Sigman provides a formal treatment of monitoring and enforcement issues in chapter 13. She assumes that the firm's compliance level depends on the cost of reducing emissions, the price of a carbon dioxide permit, the probability of detection for noncompliance, and the fine for noncompliance. She points out that both the fine and the probability of detection are low in existing permit programs in Europe and the United States, while observed compliance is high. This combination is somewhat puzzling, given the predictions of the model, but perhaps firms are concerned with public perceptions—the firm's image with customers, host communities, and potential employees. She also looks at the trend over time in the price of actual carbon dioxide permits in Europe, as opposed to the price of credits for reducing emissions elsewhere (offsets). Since the EU-ETS allows one-for-one trades between permits and offsets, we might expect these prices to be similar. Yet the difference in price is sometimes large, indicating that the offsets are not worth as much as permits to European firms. Again those firms may be concerned about the public relations problem of avoiding actual abatement in Europe, or they perceive a greater risk that offsets will be declared noncompliant.

With heterogeneous monitoring and enforcement costs among firms or emissions sources, Sigman notes that policymakers have a choice about how many to include within the emission cap. Regulators might want to exclude emission sources with very high monitoring and enforcement costs, where a firm might find cheating easier, but Sigman shows that extending the program to include more sources can bring down the price of a permit enough to discourage noncompliance generally. Thus policymakers might find more compliance with a broader program that includes more sources—even those that are more difficult to monitor.

In both the economic research and the policy spheres, most discussions have focused on mitigation—addressing climate change by restricting GHG emissions. Chapter 14 by Kerry Smith, by contrast, models adaptation to the warmer temperatures, reduced rainfall, and other changes associated with higher GHG concentrations. This can be a policy issue, as governments face the choice of either doing nothing (essentially waiting to see the degree of climate change before responding), or taking steps now to anticipate climate change and to facilitate adaptation.

Some goods represent substitutes for climate. If the climate gets hotter, we could substitute into more electricity for air conditioning. If climate change means reduced rainfall in some areas, one substitute good is increased storage of water in reservoirs. Many margins of substitution are possible, as residents could also substitute into goods that require less water. In any case, Smith's chapter points out that economic incentives can facilitate adaptation. If electricity or water is capacity constrained, for example, then policymakers can help allocate those scarce resources with pricing policies that take into account the scarcity at any particular time and place—perhaps using new metering technologies. Old technologies allow only one price per unit of water or electricity, so past analyses find the best single price and best single capacity that maximize expected social surplus given uncertain supply and demand. New technologies allow real-time pricing, however, which allows better allocation of the resource given any total availability within one period. Economic welfare then can exceed the level under current rules, where a drought leads to arbitrary decisions about water allocation (e.g., rules against certain uses of water, regardless of value).

In other words, efficient policy planning for adaptation should not focus only on building the right number or type of power plants, dams, and other infrastructure. The need for that infrastructure depends on how goods like water and electricity will be priced. The bottom line is that policymakers must make decisions about build capacity, pricing policies, and access to resources during times of shortage; these decisions are related to each other, and they all affect economic welfare.

A final design decision considered in this section is the question of whether or not to phase-in the provisions of climate policy, either by raising the carbon tax rate gradually over time or by reducing the number of permits over time. To address this question, chapter 15 by Roberton Williams builds a simple analytical, dynamic model with one sector that uses two inputs: emissions and one type of capital. Investment in new capital entails adjustment costs, providing a reason not to switch too rapidly away from emissions and into new capital. He then considers several different cases: a flow pollutant or stock pollutant, where marginal damages are either constant or rising with pollution.

For climate change, the relevant case is that of a stock pollutant, because damages depend on the concentration of greenhouse gases in the atmosphere, which depends on accumulated emissions. If damages are proportional to that stock, so that marginal damages are constant, then Williams shows that the optimal price of emissions is constant—no phase-in of a carbon tax. In this same case, however, the optimal emissions each year are falling. Thus the optimal permit policy is phased in, with a falling number of permits issued each year.

If marginal pollution damages increase with the stock of GHG, however, then an optimal policy that reduces the stock of pollution over time will result in marginal damages that also fall over time, and therefore a price of emissions that falls over time. Then the optimal price path for emissions is one that jumps immediately to a level above its long-run level. The optimal carbon tax then falls gradually, which is the *opposite* of the usual phase-in with a rising carbon tax.

Finally, Williams analyzes other considerations that may alter this optimal phase-in rule. If policymakers are concerned about the distribution of burdens, for example, then they may phase in a gradually increasing tax rate to limit the cost imposed on current owners of polluting capital. If authorities must take time to build capacity for monitoring or enforcement, then they may need to start with a subset of polluters and gradually expand the program to more firms. In any case, having dug into the topic, Williams concludes that these issues deserve more study.

I.4   Sector-Specific Issues

The remaining chapters consider climate-policy issues that are specific to four important areas: urban policy, plus the agricultural, automotive, and buildings sectors.

Much of Matthew Kahn's recent work, summarized in Kahn (2010), considers the interaction between cities and climate change. As temperatures rise, for example, which cities are likely to gain population and which will lose population? Will higher temperatures lead people to move from rural and suburban neighborhoods into city centers? If the answer to the second question is "yes," urban economic theory predicts that center-city residents will use less energy and therefore emit fewer greenhouse gases. This is because land prices are higher in cities, so residents will live in smaller spaces, own fewer cars (which require land to store) and use the ones they do own to drive fewer miles (as urban density makes alternatives like walking or public transportation better substitutes).

In chapter 16, Kahn sets out to evaluate this theory empirically. He uses three distinct data sets to evaluate whether center-city residents (a) drive fewer miles, (b) use public transportation more, and (c) use less electricity in their homes. He finds empirical support for the predictions of urban economic theory in all three cases, and the magnitude of the effects he measures is quite large. For instance, he finds that households living in census block groups at the twenty-fifth percentile of population density drive 25 percent more than households at the seventy-fifth percentile (and this distribution is taken over households that already live within thirty-five miles of a major city center). It is interesting to consider Kahn's estimates relative to the gasoline price elasticities estimated by Knittel and Sandler (in chapter 18, discussed later). This comparison suggests that the same change in driving would require gas prices to approximately double. Kahn's work forces us to consider the fact that urban policies, such as redevelopment or crime prevention programs, also may impact greenhouse gas emissions. As Chris Knittel's comments make clear, this chapter by Kahn is a first step, but has not fully addressed the possibility that the observed relationships reflect selection. For instance, if households that currently live in the suburbs were forced to relocate to the city center, they might make different choices than households currently choosing to live in dense, urban areas.

Chapter 17 by Michael Roberts and Wolfram Schlenker focuses on the agricultural sector, which is a small share of the US economy (less than 2 percent of the gross domestic product [GDP]), but which creates large consumer surplus both in the United States and abroad. They focus on corn and soybean yields, noting that together with wheat and rice, these crops account for about 75 percent of world caloric consumption. Their estimates, which are consistent with previous work, suggest that US crop yields fall dramatically in response to extreme temperatures. Specifically, annual yields decrease once average temperatures over any day exceed approximately 30°C, and the effects are predicted to be quite large (yields decrease by 5 percent for every twenty-four-hour period that the temperature averages 40°C). A natural question to ask is whether technological progress is likely to make crops more resilient to heat in the future. Roberts and Schlenker look to the past as a guide, first noting the tremendous progress over the last seventy to eighty years in efforts to increase yields, particularly for corn. This progress has largely been attributed to advances in new seed engineering and fertilizer use. As they document, however, increased yields have, if anything, come at the expense of heat resistance, as decade-by-decade estimates suggest that yields may be declining more during periods of extreme heat than they did at the beginning of the sample period. They conclude by discussing the extent to which private companies will have an incentive to invest in research and development on heat-resistant seeds, as well as any possible role for policy.

Chapter 18 by Christopher Knittel and Ryan Sandler considers the automotive sector. Noting that environmental policies to price carbon emissions are likely to lead to higher gas prices, they examine how consumers have responded to recent changes in gas prices to provide insight into how they would respond to carbon pricing. As the authors point out, consumers can adjust their behavior along a number of margins when faced by higher gasoline prices—driving less, buying more fuel-efficient new or used vehicles, scrapping fuel-inefficient vehicles, servicing their vehicle more frequently, or not driving too fast on the highway. While much of the previous literature has focused on the car purchase decision, they use a novel data source to consider both retirements (scrapping) and vehicle miles traveled. Specifically, they use information from California smog tests, which monitor every car older than six years at least once every two years.

They find large effects for scrapping decisions—vehicles in general are scrapped less when gas prices are high. This may reflect an income effect, whereby households that are paying more for gasoline are less likely to invest in a new vehicle and so keep their old one around longer. The more fuel inefficient cars, however, are more likely to be scrapped. Their results are provocative, yet the importance of the control variables suggests more room for further research. Also, while rich, the authors' data do not perfectly measure scrapping, so they must assume that vehicles that disappear from the data are scrapped. As mentioned earlier, they also find a large effect on vehicle-miles traveled.

The final chapter, by Lucas Davis, considers the buildings sector, which accounts for 40 percent of greenhouse gas emissions in the United States and has been singled out as a likely source of opportunities to reduce emissions at very low or even negative costs (McKinsey 2007). The remaining question to economists is why the people who live and work in buildings have not taken advantage of these opportunities already, particularly if they would reduce energy bills by more than they would cost. Davis considers one of the potential explanations for the so-called "energy efficiency gap." Specifically, he evaluates whether renters are less likely to have energy efficient appliances than homeowners. This pattern is consistent with a principal-agent problem whereby landlords purchase the inefficient appliances because tenants pay the bills, and tenants cannot observe or do not consider the energy efficiency of the appliances when deciding whether to live in a particular home. Using cross-sectional survey data, Davis finds this to be the case, and his results stand up to a very careful consideration of alternative explanations and functional forms. In terms of magnitudes, his results suggest that renters are between 1 and 10 percentage points less likely to have energy efficient appliances, which, relative to baseline penetration rates below 50 percent in all cases, accounts for a reasonable share of the variation between renters and homeowners.

Each chapter of this book makes an initial contribution to the economic analysis of an issue related to the design of US climate change policy. Many of the detailed issues that our authors analyze must be resolved before climate policy can be implemented, so the compilation of initial efforts amounts to a major step forward. We expect that the studies in this book will draw attention to new research areas of vital importance to any efforts to reduce future climate change. The work will also contribute to better policy regarding whether and how to mitigate damages from global warming, sea level rise, loss of coastal areas, increased storm severity, loss of biodiversity, and increased frequency and duration of droughts. We look forward to reading follow-on studies and hope that economists will continue to engage in future policy developments.

References

Aldy, Joseph E., Alan J. Krupnick, Richard G. Newell, Ian W. H. Parry, and William A. Pizer. 2010. "Designing Climate Mitigation Policy." *Journal of Economic Literature* 48 (4): 903–34.

Bushnell, James B., and Erin T. Mansur. 2011. "Vertical Targeting and Leakage in Carbon Policy." *American Economic Review Papers and Proceedings* 101 (3): 263–7.

Kahn, Matthew. 2010. *Climatopolis: How Our Cities Will Thrive in the Hotter Future.* New York: Basic Books.

Guesnerie, Roger, and Henry Tulkens, eds. 2008. *The Design of Climate Policy* (CESifo Seminar Series). Cambridge, MA: MIT Press.

McKinsey and Company. 2007. *Reducing US Greenhouse Gas Emissions: How Much at What Cost?* June. Available at http://www.mckinsey.com/Client_Service/Sustainability/Latest_thinking/Reducing_US_greenhouse_gas_er.

Sigman, Hilary, and Howard F. Chang. 2011. "The Effect of Allowing Pollution Offsets with Imperfect Enforcement." *American Economic Review Papers and Proceedings* 101 (3): 268–72.

I

Climate Policy in
the Broader Context

1

Distributional Impacts in a Comprehensive Climate Policy Package

Gilbert E. Metcalf, Aparna Mathur, and
Kevin A. Hassett

1.1 Introduction

Distributional considerations figure importantly in the design of comprehensive climate policy legislation. The allowance allocation in the American Clean Energy and Security Act of 2009 (H.R. 2454), popularly known as the Waxman-Markey bill, that was passed by the House of Representatives in June 2009, suggests the care and attention paid to distributional considerations in crafting the bill. Both the Kerry-Boxer bill and the Cantwell-Collins proposals in the Senate also paid close attention to distributional considerations.

This chapter uses data from the 2003 Consumer Expenditure Survey to allocate the burden of carbon pricing from possible cap-and-trade legislation under different assumptions about the relative importance of uses- and sources-side heterogeneity as well as differing assumptions about relative factor price changes. It builds on previous research using the Consumer Expenditure Survey by generalizing the incidence assumptions beyond the assumption of full-forward shifting of the carbon price. It also improves on the measurement of capital income burden allocation by using capi-

Gilbert E. Metcalf is deputy assistant secretary for environment and energy, US Department of the Treasury. He is on leave from the Department of Economics at Tufts University. Aparna Mathur is a resident scholar at the American Enterprise Institute for Public Policy Research. Kevin A. Hassett is a senior fellow and director of economic policy studies at the American Enterprise Institute for Public Policy Research. The views expressed are those of the authors and do not necessarily reflect those of the US Department of the Treasury.

We thank Don Fullerton, Hilary Sigman, and conference participants for useful suggestions. For acknowledgments, sources of research support, and disclosure of the authors' material financial relationships, if any, please see http://www.nber.org/chapters/c12136.ack.

tal income distribution data from the 2004 Survey of Consumer Finances (SCF) to augment the data in the Consumer Expenditure Survey.

The approach detailed in this chapter provides a method for carrying out a back-of-the-envelope calculation of the distributional impact of carbon pricing using readily available data that allows for sensitivity analysis of assumptions on sources- and uses-side incidence of carbon pricing. We find that accounting for sources-side impacts of carbon pricing yields less regressive impacts on households looking across the income distribution.

1.2   Background

Households differ on a number of dimensions that policymakers may care about. When designing a climate policy bill, policymakers have made it clear that many of these dimensions are important and affect the allocation of allowances as well as the mechanisms of allowance use. Households differ by income, regional location, primary heating source, and predominant mode of electricity generation among other things. We focus in this chapter on measuring the impact of carbon-pricing policies on households looking across the income distribution.

In carrying out distributional analyses, a number of considerations come into play. First is the question of how best to sort households to distinguish them by some measure of relative well-being. Income is often used for this ranking and this analysis sorts households by annual income. This brings a potential bias to the analysis to the extent that annual income is a poor proxy for lifetime well-being. As discussed elsewhere (see, for example, Fullerton and Metcalf [2002]) many low-income households are not poor in a lifetime sense. They may have transitorily low income or may be at a low income-earning stage of their careers. In both these cases consumption-to-income ratios may be unusually high and may provide a misleading picture of the distributional impact of consumption-related taxes (like energy taxes) or carbon-pricing policies. As a check for the importance of our income measurement we also provide results where we use current consumption as a proxy for lifetime income under the assumption that households engage in consumption smoothing.

A second issue is that the economic impact of carbon pricing depends importantly on how prices adjust to the new equilibrium with carbon pricing. This is particularly important for a policy that creates and distributes financial assets in excess of $100 billion by the middle of this decade (see Congressional Budget Office 2009). A number of computable general equilibrium economic analyses have argued that carbon pricing will predominantly be passed forward to consumers in the form of higher energy prices. See, for example, Bovenberg and Goulder (2001) and Metcalf et al. (2008).

Based on analyses focusing on uses-side incidence impacts of carbon pricing, a number of economists have carried out distributional analyses of carbon pricing using the Consumer Expenditure Survey, including Bull,

Hassett, and Metcalf (1994), Dinan and Rogers (2002), Metcalf (1999), Parry (2004), and Hassett, Mathur, and Metcalf (2009). The Consumer Expenditure Survey is particularly useful for this analysis given its high level of detailed disaggregation on household spending patterns. But these analyses are useful only to the degree that the assumption of full-forward shifting (e.g., impacts on uses side only) is correct.

In the following analysis we refer to forward shifting and backward shifting when we wish to analyze the distributional impacts of carbon pricing according to how households spend their income (uses side) or earn their income (sources side). The terminology of forward and backward shifting has a long-standing place in public economics, albeit an imprecise meaning. Whether a tax is shifted forward (leading to higher consumer prices) or shifted back (leading to lower factor returns) depends on the normalization employed in the general equilibrium framework. Since the normalization choice in a general equilibrium model has no real effects, forward or backward shifting cannot have real effects either (see Fullerton and Metcalf [2002] for more on this point). When we later refer to forward or backward shifting, we use this to refer to heterogeneous impacts of carbon pricing based on how different households spend or earn their income.

A recent study by Metcalf et al. (2008) found that for a given price normalization forward shifting of carbon pricing ranged widely depending on the fuel in question, the proposal under consideration, and the particular year of analysis. Carbon pricing on coal was nearly fully passed forward into higher prices, reflecting in large part the low Hotelling resource rents for coal. Shifting for natural gas ranged from a low of 14 percent to a high of over 200 percent. The latter occurs as demand rises for natural gas in the intermediate term as gas substitutes for coal in the production of electricity.[1] Finally, forward shifting for crude oil ranged from a low of 2 percent to a high of nearly 90 percent depending on the year and tax scenario.

If taxes are not passed forward to consumers in the form of higher product prices, then they are passed back to factors of production in the form of lower wages, returns to equity, and reduced resource rents. Changes in resource rents can also affect government revenues since much fossil fuel extraction in the United States occurs on publicly owned land (e.g., the Powder River Basin coal reserves in Wyoming and the Outer Continental Shelf oil and gas drilling). We ignore that complication in this analysis in part because the impact of taxes on government revenue from land-leasing activities is poorly understood.

This chapter uses burden-shifting insights from computable general equilibrium (CGE) models along with the Consumer Expenditure Survey to measure the burden of carbon pricing. A goal of the analysis is to demonstrate the ability to use the survey with a broader range of assumptions to

---

1. That natural gas prices may rise by over twice the tax rate indicates the complex price responses that can occur in general equilibrium.

obtain a rough-and-ready guide to the distributional impacts of carbon-pricing proposals without having to run full-blown CGE analyses.

1.3   Measuring Carbon Price Burdens

Our goal in this chapter is to provide a simple rough-and-ready measure of the burden impact of carbon pricing that builds on the insights of more complex economic analyses. This is in the tradition of a number of studies that use detailed data sets such as the Consumer Expenditure Survey (CEX) along with results and insights from sophisticated economic models to allocate the burden of government policies to different economic groups.

As noted earlier, previous studies using the CEX have assumed that carbon pricing is fully passed forward into higher consumer prices based on the carbon content of goods and services. Input-Output tables from the Bureau of Economic Analysis are used to trace through carbon content and thus carbon-pricing impacts. If carbon prices are passed back to factors of production, then we need to use income information in the CEX to distribute the carbon-pricing impacts. We distribute the burden of carbon pricing that falls on owners of capital in proportion to capital income shares as a proxy for capital ownership shares.[2]

Carbon-pricing burdens may also fall on owners of fossil fuel resources. To the extent these resources are privately owned, carbon pricing may lead to a reduction in returns to owning property with fossil fuel resources. Some of this property is held by sole proprietors and partnerships while other tracts are owned by corporations. Lacking detailed information on resource ownership, we assume that resource ownership is distributed across households in the same manner as capital.

Turning to allowances, we can allocate the value of allowances to households either according to consumption or income patterns depending on how allowances are distributed. The Waxman-Markey bill sets aside roughly 30 percent of allowances in the early years for distribution to customers of electricity and natural gas utilities to compensate them for higher electricity and gas prices. We allocate the value of those allowances to households based on their electricity and natural gas expenditures, respectively. Allocations to industry are assumed to benefit owners of capital. Allocations to households are distributed to households.

In general we follow the distribution approach of Rausch et al. (2010) for distributing the value of allowances. One place where we differ is in the allocation of allowances to the US government for deficit reduction. Under the assumption that reductions in the deficit reduce pressure to decrease government spending, we allocate the allowances for deficit reduction based

---

2. This follows from the result in Harberger (1962) that partial capital income taxes are borne by all owners of capital.

Table 1.1   **Incidence scenarios**

Scenario	Consumers (%)	Capital and resources (%)	Labor (%)
1	100	0	0
2	80	20	0
3	80	10	10
4	50	25	25

on government spending that would otherwise have to be cut. Our assumptions on the benefits of government spending across the income distribution are taken from the Tax Foundation (2007).

Rather than assume a particular burden-sharing outcome, we report results for four different scenarios to illustrate the importance of the burden-sharing assumption on distributional outcomes. The four scenarios we consider are reported in table 1.1.[3] The first scenario assumes full-forward shifting of carbon pricing to final consumers (i.e., burden is based on heterogeneity in household expenditure patterns). The next three scenarios allow for a greater role in sources-side effects with different assumptions about relative price changes between capital and labor. These approaches are based on a particular normalization (price of non-carbon-based consumption goods held fixed). As noted previously, forward and backward shifting is imprecise (and potentially misleading) terminology though long used in public finance. More precisely we focus on distributional impacts based on uses-side impacts and sources-side impacts. Scenario 1 focuses on uses-side heterogeneity only. The remaining three scenarios allow for greater amounts of sources-side heterogeneity and also allows for differential impacts on wage and capital (and resource) income.

1.4   Issues in Using the Consumer Expenditure Survey

The Consumer Expenditure Survey has been used by a number of researchers investigating the burden impacts of carbon pricing because of its rich detail on consumption patterns of US households. It also contains information on the demographic makeup of households as well as some income information. The CEX has a single capital income measure that researchers have used to allocate taxes to owners of capital in scenarios assuming some degree of backward shifting. The survey question for this data asks whether households received any regular income from dividends, trusts, estates, or royalties. A separate question asks about interest income from bank accounts, money market funds, CDs, or bonds. Researchers have used the dividend income amount (or dividends and interest) as a proxy for

---

3. This approach is in the spirit of the classic distributional analysis by Pechman (1985).

Table 1.2   Distribution of capital income across households

Annual income decile	Consumer expenditure survey	Survey of consumer finances
1	0.004	0.001
2	0.007	0.005
3	0.007	0.011
4	0.159	0.015
5	0.033	0.019
6	0.027	0.015
7	0.050	0.037
8	0.020	0.027
9	0.156	0.060
10	0.542	0.810

Source: Authors' calculations from 2003 CEX and 2004 SCF. Entries are capital income shares for each decile. Each column sums to one.

capital holdings under the assumption that capital income is proportional to capital holdings.

The problem with using CEX-reported capital income is that it may misrepresent capital holdings across income groups. There are two possible reasons. First, the CEX focuses primarily on spending and the income data quality may not be as high quality as the spending data. Second, if holdings of growth stocks are disproportionately held by higher income groups, then the CEX capital income measure will be biased toward more capital holdings in lower income groups. Table 1.2 suggests that the first problem is significant with the CEX showing more capital income in the lower income deciles than the SCF.[4]

Using data from the 2004 SCF, Wolfe (2010) estimates that 85 percent of net worth capital is held by households in the top quintile and 92 percent of nonhousehold wealth by this quintile. The CEX in contrast reports only 70 percent of capital income accruing to the top quintile. Using CEX capital income distributions will skew any carbon-pricing distribution toward greater progressivity to the extent that any of the burden is placed on owners of capital.

One advantage of using the SCF is that it disproportionately samples wealthy families. Each survey consists of a core representative sample combined with a high-income supplement, which is drawn from the Internal Revenue Service's Statistics of Income data file. Further, the survey questionnaire consists of detailed questions on different components of family wealth holdings. For these reasons, the SCF is widely acknowledged to be the best at capturing both the wealth at the top of the distribution and the complete wealth portfolio of households in the middle. Since the wealth dis-

4. Income cutoffs for the deciles are $10,304, 17,000, 24,000, 32,000, 40,200, 50,655, 65,032, 81,700, and 108,768.

Table 1.3   **Distribution of labor income across households**

Annual income decile	Consumer expenditure survey	Survey of consumer finances
1	0.003	0.003
2	0.012	0.011
3	0.025	0.023
4	0.042	0.039
5	0.063	0.054
6	0.083	0.073
7	0.114	0.088
8	0.143	0.126
9	0.185	0.178
10	0.331	0.403

Source: Authors' calculations from 2003 CEX and 2004 SCF. Entries are labor income shares for each decile. Each column sums to one.

tribution is highly skewed toward the top, most other surveys (like the CEX) that have poor data on high-income families tend to underreport measures of income and wealth.

The problem of distributional bias is not as significant for labor income as for capital income. Table 1.3 reports labor income shares across deciles from the CEX and SCF. The distributions are more closely aligned than those for capital income.

In this analysis we distribute the burden of carbon pricing that is shifted to owners of capital based on the distribution of capital income from the SCF (table 1.2).

1.5   Results

For purposes of our analysis, we consider the effect of a carbon tax set at a rate of fifteen dollars per metric ton of carbon dioxide. We trace the effect of this carbon tax on the prices of consumer goods produced by the industries through the use of Input-Output matrices available from the Bureau of Economic Analysis. Once we obtained the effect of the tax on prices of consumer goods, we used data from the Consumer Expenditure Survey (CEX) to compute carbon taxes paid by each household in the survey. For a detailed discussion of this methodology as well as the computed price increases, see Metcalf (1999) and more recently, Hassett, Mathur, and Metcalf (2009).

We extend the analysis in this chapter by considering the incidence on the sources-side as well. Using capital and labor income shares from the Survey of Consumer Finances (SCF), we are able to compute the carbon tax burdens on capital and labor income for households in the CEX. Hence the total burden on any household is computed as the sum of the burden on the consumption side, as well as on the income side.

The final step in the calculations shown in tables 1.4, 1.5, 1.6, and 1.7 is the allocation of the allowance revenues under the three proposals. Every proposal allows some level of rebates to households that are based on their energy use, their labor and capital income shares, or whether they are low income. The final burden is lowered by the level of rebates allowed under the three proposals.

As noted earlier, the distributional tables are based on a carbon-pricing policy that yields a carbon price of fifteen dollars per ton CO_2. This is consistent with permit price estimates in the 2015 to 2020 period for either H.R. 2454 (Waxman-Markey) or the Kerry-Boxer bill in the Senate. In the analyses in which allowance revenues are returned to households, we assume full return of revenue to households allocating permit value using the assumptions in Rausch et al. (2010).

Table 1.4 shows results for a cap-and-trade program in which we ignore the rebate of permit revenue to households. This scenario focuses on carbon pricing itself without the confounding effects of allowance allocations. The left panel of the table sorts households by annual income while the right panel sorts households by annual consumption, a proxy for lifetime income under the assumption that households engage in consumption smoothing.

We first discuss the results in which we sort households by annual income. The first scenario assumes carbon pricing is fully reflected in higher consumer prices. Carbon pricing is regressive in this scenario with the burden of higher consumer prices falling from 3.7 percent of household income in the lowest income decile to 0.8 percent of household income for the top decile.[5] The ratio of burdens between the top and bottom deciles is 4.6. If 20 percent of the burden of carbon pricing is shifted back to owners of capital and resources, the regressivity of carbon pricing is blunted somewhat with the ratio of burdens between the top and bottom deciles falling to 2.3. Shifting part of the burden from capital to labor (scenario 3) increases the regressivity slightly relative to scenario 2. Scenario 4 shows that the regressivity of carbon pricing is blunted as more of the burden is shifted back to factors of production—with the burden shifting to capital the most important. In this case the burden share in the lowest decile is only 20 percent higher than the burden in the top decile.

As discussed earlier, using annual income to rank households may overstate the regressivity of carbon pricing and so we also report results where we rank households by current consumption in the right-hand panel of table 1.4. Regressivity is significantly blunted when households are ranked by consumption.[6] Now when sources-side heterogeneity is sufficiently im-

5. The incidence numbers look marginally different from those in Hassett, Mathur, and Metcalf (2009), since we are not accounting for the differential impact on electricity prices across regions in this study.

6. This result is consistent with previous findings on the relative progressivity of energy and environmental taxes when comparing consumption to income-based household rankings. See Poterba (1989, 1991), Bull, Hassett, and Metcalf (1994), Lyon and Schwab (1995), Metcalf

Table 1.4   **Distribution of carbon pricing across households: No rebate**

Decile	Annual income deciles incidence assumptions				Annual consumption deciles incidence assumptions			
	1	2	3	4	1	2	3	4
1	3.70	2.99	3.02	2.01	1.45	1.18	1.25	0.96
2	3.05	2.48	2.51	1.71	1.41	1.21	1.26	1.03
3	2.31	1.93	1.97	1.46	1.31	1.15	1.19	1.01
4	2.03	1.71	1.76	1.36	1.29	1.04	1.14	0.92
5	1.75	1.47	1.54	1.23	1.24	1.12	1.16	1.04
6	1.51	1.26	1.35	1.09	1.17	1.03	1.09	0.97
7	1.30	1.13	1.20	1.03	1.16	1.03	1.10	1.02
8	1.24	1.04	1.14	0.98	1.07	0.91	0.99	0.89
9	1.02	0.91	0.99	0.96	1.01	1.10	1.07	1.17
10	0.82	1.29	1.15	1.64	0.90	1.12	1.03	1.23
Low/High ratio	4.51	2.32	2.63	1.23	1.61	1.05	1.21	0.78

Source: Authors' calculations. Table reports burden as a percentage of household income in annual income decile columns and as a percentage of current consumption in annual consumption decile columns. Last row reports ratio of burden for first decile relative to burden for top decile.

portant (scenario 4), carbon pricing looks proportional to modestly progressive. This finding is consistent with the finding of Rausch et al. (2010) who find that sources-side impacts lead to carbon pricing being progressive in their CGE analysis.

Table 1.4 considers the burden of carbon pricing with no consideration as to the distribution of carbon revenues. Considerable effort has been taken in the various cap-and-trade proposals in the House and Senate to allocate allowances (or allowance value) to offset the price impacts of carbon pricing. We now turn to a comparison of distributional results for the various burden-shifting scenarios identified in table 1.1. The allocation of allowances is based on the analysis of proposed cap-and-trade legislation carried out by Rausch et al. (2010). As these authors stress, the analysis only focuses on the allowance allocations in the bill and ignores all other aspects of the legislation. Thus, one should not view these distributions as representative of the actual distributions that will result from enactment of any of these bills.[7] To emphasize that we refer to the scenarios as Targeted Allowance

---

(1999), and Hassett, Mathur, and Metcalf (2009) among others. Fullerton and Heutel (2010), in contrast, find that uses-side impacts are more regressive when a consumption-based ranking of households is used instead of annual income. This appears to arise from their specification of incidence in which households are classified using annual income deciles, but the burden is reported relative to annual consumption.

7. Rausch et al. (2010) note other differences—in particular, the ability to use domestic and international offsets in the various proposals. Those considerations are not relevant for our analysis.

Allocation (TAA) scenarios TAA-1 for the Waxman-Markey approach, and TAA-2 for the Kerry-Boxer approach. We refer to a Household Dividend (HD) scenario for the Cantwell-Collins approach.

Table 1.5 reports results for the TAA-1 allocation approach. The bill has a complex allocation schedule for each of the years between 2012 and 2050. For this and the other two proposals we analyze, we consider the distributions in 2020.

Focusing first on the annual income analysis, the carbon-pricing reform (taking into account the burden of carbon pricing and the distribution of allowance value) is progressive regardless of the assumptions made about burden sharing between consumers and factors of production. Assuming full-forward shifting of the carbon price (incidence assumption 1) the burden of carbon pricing with allowance allocation in 2020 falls from –2.4 percent of household income for the lowest decile to –0.02 percent for the top decile. The bottom 40 percent of the income distribution get back more in allowance revenue (either directly or indirectly through allocations that reduce product prices for them) than they pay in higher prices of goods and services because of carbon pricing.

Assuming 20 percent of the burden is shifted from consumers to owners of capital and resources, the progressivity increases in 2020. The highest degree of progressivity occurs under incidence assumption 4 where half the burden is shifted back to factors of production with labor and capital equally sharing the burden. To draw parallels, it is interesting to note that the carbon-pricing burden with the rebate is marginally less than half the value of the Earned Income Tax Credit subsidy for the bottom decile. The

Table 1.5   Distribution for targeted allowance allocation 1

Decile	Annual income deciles incidence assumptions				Annual consumption deciles incidence assumptions			
	1	2	3	4	1	2	3	4
1	−2.38	−3.09	−3.06	−4.07	−1.12	−1.39	−1.31	−1.60
2	−1.27	−1.84	−1.80	−2.60	−0.74	−0.94	−0.89	−1.12
3	−0.62	−1.00	−0.96	−1.47	−0.44	−0.59	−0.56	−0.74
4	−0.98	−1.31	−1.26	−1.66	0.07	−0.18	−0.08	−0.30
5	0.56	0.29	0.35	0.04	0.14	0.02	0.06	−0.07
6	0.54	0.29	0.37	0.12	0.22	0.08	0.14	0.02
7	0.43	0.26	0.32	0.16	0.29	0.17	0.24	0.16
8	0.56	0.36	0.46	0.30	0.37	0.22	0.30	0.19
9	0.23	0.11	0.20	0.16	0.28	0.37	0.35	0.44
10	−0.02	0.46	0.31	0.81	0.03	0.24	0.16	0.36

Source: Authors' calculations for 2020 assuming permit distribution as described in text. Table reports burden as a percentage of household income in annual income decile columns and as a percentage of current consumption in annual consumption decile columns.

Table 1.6   Distribution for targeted allowance allocation 2

	Annual income deciles incidence assumptions				Annual consumption deciles incidence assumptions			
Decile	1	2	3	4	1	2	3	4
1	–2.86	–3.58	–3.54	–4.56	–1.32	–1.59	–1.52	–1.80
2	–1.50	–2.07	–2.04	–2.84	–0.84	–1.05	–1.00	–1.22
3	–0.67	–1.05	–1.01	–1.52	–0.50	–0.65	–0.62	–0.80
4	–0.92	–1.25	–1.19	–1.60	0.01	–0.24	–0.14	–0.36
5	0.49	0.21	0.28	–0.03	0.10	–0.01	0.02	–0.10
6	0.49	0.23	0.32	0.07	0.19	0.05	0.11	–0.01
7	0.41	0.24	0.30	0.14	0.28	0.16	0.22	0.14
8	0.54	0.34	0.44	0.28	0.35	0.20	0.28	0.17
9	0.23	0.12	0.20	0.17	0.29	0.38	0.35	0.45
10	0.03	0.50	0.36	0.85	0.08	0.30	0.21	0.41

Source: Authors' calculations for 2020 assuming permit distribution as described in text. Table reports burden as a percentage of household income in annual income decile columns and as a percentage of current consumption in annual consumption decile columns.

share of the EITC in total adjusted gross income for the bottom decile was approximately 18 percent in 2007.[8]

The analysis based on consumption as a proxy for lifetime income mutes but does not overturn the progressive result. As with annual income rankings, the more important sources-side heterogeneity, the more progressive the reform.

Table 1.6 presents results for the TAA-2 scheme. Results are very similar to those for TAA-1. Results for the HD scheme are quite different from either TAA-1 or TAA-2 due to a very different approach to allocation taken by this proposal (table 1.7). Whereas the former two proposals have a complex allocation scheme distributing allowances to industry and to gas and electricity local distribution companies, HD rebates three-quarters of the allowance revenue to households on an equal per capita basis. The remaining allowance revenue is used for various clean energy investments and regional programs.

The Household Dividend distribution approach is markedly more progressive than the previous programs. This follows primarily from the largely lump-sum nature of rebate approach taken in this bill. Assuming some of the tax is passed back to owners of capital and energy resources increases the progressivity of the program relative to the assumption of full-forward shifting. This holds true whether we rank households by annual income or consumption.

8. Available at http://www.irs.gov/taxstats/indtaxstats/article/0,,id=133414,00.html.

Table 1.7   **Distribution for household dividend**

Decile	Annual income deciles incidence assumptions				Annual consumption deciles incidence assumptions			
	1	2	3	4	1	2	3	4
1	−3.36	−4.07	−4.04	−5.05	−1.77	−2.04	−1.97	−2.26
2	−1.42	−1.99	−1.96	−2.76	−1.01	−1.21	−1.16	−1.38
3	−0.68	−1.05	−1.02	−1.52	−0.56	−0.72	−0.69	−0.87
4	−0.21	−0.54	−0.48	−0.88	−0.25	−0.50	−0.40	−0.62
5	0.01	−0.26	−0.19	−0.51	−0.05	−0.17	−0.13	−0.25
6	0.12	−0.14	−0.05	−0.30	0.06	−0.08	−0.01	−0.13
7	0.20	0.03	0.10	−0.07	0.20	0.08	0.15	0.06
8	0.36	0.16	0.25	0.09	0.26	0.11	0.19	0.08
9	0.31	0.20	0.28	0.25	0.36	0.45	0.42	0.52
10	0.36	0.83	0.69	1.18	0.46	0.67	0.59	0.79

Source: Authors' calculations for 2020 assuming permit distribution as described in text. Table reports burden as a percentage of household income in annual income decile columns and as a percentage of current consumption in annual consumption decile columns.

1.6   Conclusion

A perennial concern with proposals to put a price on carbon emissions either through a carbon tax or a cap-and-trade program is the perceived regressivity of the policy. We find that carbon pricing is indeed regressive when annual income is used to sort households, though the extent of the regressivity depends on the degree of backward shifting of the carbon price. The story changes, however, if households are ranked by a proxy for lifetime income. Now carbon pricing is at most mildly regressive, and may in fact be progressive depending on the relative importance of uses-side versus sources-side heterogeneity.

Once one allows for a distribution of some or all of the value of the allowances back to households—either directly or indirectly through grants to industry—the policy now looks progressive however one ranks households.[9] This is true for allocation schemes that are similar to the three leading cap-and-trade proposals currently under consideration by Congress.

This chapter provides a simple analytic approach for measuring the burden of carbon pricing that does not require sophisticated and numerically intensive economic models, but is not limited to restrictive assumptions that only uses-side heterogeneity can be taken into account when measuring the tax burden. We also show how to adjust for the capital income bias contained in the Consumer Expenditure Survey, a bias toward regressivity in carbon

---

9. This highlights the distinction between a green tax and a green tax reform made by Metcalf (1999).

pricing due to underreporting of capital income in higher income deciles in the CEX.

Once one allows for sources-side heterogeneity, carbon policies look more progressive than when attention is only on how households spend their income. Perhaps more important than the findings from any one scenario, our results on the progressivity of the leading cap-and-trade proposals are robust to the assumptions made on the relative importance of sources and uses-side effects for the burden of carbon pricing.

References

Bovenberg, A. Lans, and Laurence Goulder. 2001. "Neutralizing the Adverse Industry Impacts of CO2 Abatement Policies: What Does It Cost?" In *Distributional and Behavioral Effects of Environmental Policy,* edited by C. Carraro and G. E. Metcalf, 45–85. Chicago: University of Chicago Press.

Bull, Nicholas, Kevin A. Hassett, and Gilbert E. Metcalf. 1994. "Who Pays Broad-Based Energy Taxes? Computing Lifetime and Regional Incidence." *Energy Journal* 15 (3): 145–64.

Congressional Budget Office. 2009. *H.R. 2454 American Clean Energy and Security Act of 2009 Cost Estimate.* Washington, DC: CBO.

Dinan, Terry, and Diane Lim Rogers. 2002. "Distributional Effects of Carbon Allowance Trading: How Government Decisions Determine Winners and Losers." *National Tax Journal* 55 (2): 199–221.

Fullerton, Don, and Garth Heutel. 2010. "Analytical General Equilibrium Effects of Energy Policy on Output and Factor Prices." *B. E. Journal of Economic Analysis and Policy* 10 (2): Article 15.

Fullerton, Don, and Gilbert E. Metcalf. 2002. "Tax Incidence." In *Handbook of Public Economics,* edited by A. J. Auerbach and M. Feldstein, 1787–872. Amsterdam: Elsevier Science.

Harberger, Arnold C. 1962. "The Incidence of the Corporation Income Tax." *Journal of Political Economy* 70 (3): 215–40.

Hassett, Kevin A., Aparna Mathur, and Gilbert E. Metcalf. 2009. "The Incidence of a U.S. Carbon Tax: A Lifetime and Regional Analysis." *Energy Journal* 30 (2): 157–79.

Lyon, Andrew B., and Robert M. Schwab. 1995. "Consumption Taxes in a Life-Cycle Framework: Are Sin Taxes Regressive?" *Review of Economics and Statistics* 77 (3): 389–406.

Metcalf, Gilbert E. 1999. "A Distributional Analysis of Green Tax Reforms." *National Tax Journal* 52 (4): 655–81.

Metcalf, Gilbert E., Sergey Paltsev, John M. Reilly, Henry D. Jacoby, and Jennifer Holak. 2008. "Analysis of a Carbon Tax to Reduce U.S. Greenhouse Gas Emissions." MIT Joint Program on the Science and Policy of Global Change. Report no. 160. Available at: http://mit.edu/globalchange/www/MITJPSPGC_Rpt160.pdf.

Parry, Ian W. H. 2004. "Are Emissions Permits Regressive?" *Journal of Environmental Economics and Management* 47:364–87.

Pechman, J. 1985. *Who Paid the Taxes: 1966–85?* Washington, DC: Brookings Institution Press.

Poterba, James. 1989. "Lifetime Incidence and the Distributional Burden of Excise Taxes." *American Economic Review* 79 (2): 325–30.

———. 1991. "Is the Gasoline Tax Regressive?" In *Tax Policy and the Economy,* (vol. 5), edited by David Bradford, 145–64. Cambridge, MA: MIT Press.

Rausch, Sebastian, Gilbert E. Metcalf, John M. Reilly, and Sergey Paltsev. 2010. "Distributional Implications of Proposed U.S. Greenhouse Gas Control Measures." Massachusetts Institute of Technology, Joint Program on the Science and Policy of Global Change. Working Paper.

Tax Foundation. 2007. *Who Pays America's Tax Burden, and Who Gets the Most Government Spending?* Special Report no. 151. Available at: http://taxfoundation .org/files/sr151.pdf.

Wolfe, Edward N. 2010. "Recent Trends in Household Wealth in the United States: Rising Debt and the Middle-Class Squeeze—an Update to 2007." Bard College. Levy Economics Institute. Working Paper no. 589.

Comment   Hilary Sigman

Metcalf, Mathur, and Hassett's chapter (henceforth, MMH) significantly improves understanding of the effects of climate policy on households. Two advances relative to the previous literature stand out. First, MMH do not assume all carbon price effects are borne by consumers. In some of their scenarios, carbon prices may be partly shifted to capital in the form of lower returns or to labor in the form of lower wages. Second, they consider the distribution of the value of allowances in prominent policy proposals.

The MMH chapter has several key findings. First, relative to the standard assumption of full-forward shifting, all other distributions of the burden make a carbon price less regressive. Since full-forward shifting is unlikely, this result suggests a more positive picture of the progressivity of climate policy. Second, all the specific proposals considered (Waxman-Markey, Kerry-Boxer, and Cantwell-Collins) allocate allowances in ways that increase progressivity. Finally, lower income groups may gain quite a lot under these policies. The households in the lowest income decile may gain 3 to 4 percent of their income (or even 5 percent for some policies and scenarios). Gains often extend to the middle of the income distribution. Thus, gains are not restricted to households reporting very low income, who may have poor-quality income data, or be socioeconomically idiosyncratic. Instead, the policies seem to confer gains systematically to lower income households.

Hilary Sigman is professor of economics at Rutgers University and a research associate of the National Bureau of Economic Research.

For acknowledgments, sources of research support, and disclosure of the author's material financial relationships, if any, please see http://www.nber.org/chapters/c12137.ack.

Metcalf, Mathur, and Hassett's analysis is careful and gives the reader a good sense of its "moving parts," despite the sophisticated thinking and complex data work that underlie its results. For example, the authors carefully render the bewildering allowance allocations in Waxman-Markey into their incidence framework.

Several complications need to be kept in mind in interpreting these results. Some of these complications are mentioned in the chapter and result from the standard assumptions of incidence analysis. Many reflect intentional simplifications by the authors, who chose not to force readers to have faith in a specific general equilibrium model.

First, a focused policy intervention, such as a carbon price, may hit a few households especially hard. We may worry about uncommon but severe impacts more than about the average effects of the policy. Some analysis of this problem might be possible within the current framework. For example, the model might be used to determine how many households lose more than 10 percent of their income. However, the Consumer Expenditure Survey (CEX) may not have sufficient observations on the few households with extreme burdens. In addition, employment effects may loom larger than price increases, so chapter 2 by Deschênes in the current volume also addresses the question of concentrated burdens.

Second, MMH use annual income to measure relative well-being. As these authors and others have demonstrated in previous work, annual income may not reflect longer-run well-being. Some households have only temporarily low income or are at lower income stages of life. In the CEX, many households have expenditures well above income, suggesting a fairly substantial mismatch between reported annual income and actual well-being. This mismatch also raises some concerns about the quality of the CEX income data.

Metcalf, Mathur, and Hassett argue that a longer-run measure of well-being would make a carbon price even less regressive. If the carbon price is passed forward, this argument seems valid. However, it is unclear that this rule of thumb holds with backward incidence or when distributing the value of allowances.

Third, behavioral responses are limited in this analysis. Following standard tax incidence models, MMH assume no elasticity of demand. Changes from other aspects of climate policy are not considered, although they may be significant. For example, if public policy generated a large shift to electric vehicles, incidence might change substantially. The possibility of longer-term demand shifts makes the results far more convincing for the short run than the long run.

Finally, the burden of the carbon price always sums to the value of allowances, again following standard tax incidence analysis. However, the total burden may be higher than the value of allowances if we consider the losses from reduced output. It may also be less than the value of allowances, for example, if the benefits from reducing local pollution externalities or counteracting inefficient subsidies are large enough. Considering the benefits of reductions in climate change might also lower the net burden.

In traditional incidence analysis, one might counter that losses from reduced output just mean the level of burden is too low, but the slope of the income-burden relationship will be preserved. For the current analysis, however, these considerations may make it difficult to draw conclusions from the comparison across scenarios. The amount of burden may differ across MMH's scenarios because losses from reduced output depend on price responses.

Despite these caveats, MMH have produced compelling results that can help guide climate policy.

Climate Policy and Labor Markets

Olivier Deschênes

2.1 Introduction

An important component of the debate surrounding climate legislation in the United States is its potential impact on labor markets. A main concern is the displacement of jobs from the United States to countries without carbon pricing, especially for energy-intensive industries facing import pressure from nonregulated countries. These concerns are rooted in the long-standing debate on the effects of domestic environmental regulations on US industries, although the empirical evidence regarding those effects is mixed (see, for example, Jaffe et al. 1995; Berman and Bui 2001; Greenstone 2002).

While concerns that higher energy prices will depress labor demand have received much attention in this debate, theoretically the connection is ambiguous and depends on the sign of cross-elasticity of labor demand with respect to energy prices, which is a priori unknown.[1] Evidence from studies conducted in the 1970s and 1980s indicates that energy and labor are *p*-substitutes, albeit weakly, suggesting that increases in energy prices lead to small *increases* in labor demand (see, for example, Hamermesh [1993] and references therein).[2] Therefore, credible empirical estimates of the short-run

Olivier Deschênes is associate professor of economics at the University of California, Santa Barbara, and a research associate of the National Bureau of Economic Research.

I thank Matthew Kahn, Peter Kuhn, and Catherine Wolfram for their detailed comments on an earlier draft, as well as Lucas Davis and Michael Greenstone for helpful discussions. Allison Bauer provided excellent research assistance. For acknowledgments, sources of research support, and disclosure of the author's material financial relationships, if any, please see http://www.nber.org/chapters/c12150.ack.

1. This presumes firms use other inputs in additional to labor and energy.

2. Two inputs are said to be *p*-substitutes (*p*-complements) when their cross-partial elasticity of factor demand is positive (negative). So in the case of *p*-substitute inputs, an increase in the price of one input leads to an increase in the demand for the other.

and long-run cross-elasticities of labor demand with respect to energy prices are the key statistics required to assess the employment effects of climate policies that lead to increases in energy prices. This chapter provides some new evidence on this question.[3]

To date, most of the research on the potential effects of carbon pricing on employment has been conducted using computable general equilibrium models. The approach typically combines various aggregate data sets with sophisticated models of the US economy and simulates the short-run and long-run effects of setting a price on carbon. For example, Ho, Morgenstern, and Shih (2008) find that the employment effects of a ten dollars per ton carbon tax decline over time as the economy adjusts to the new energy prices. Taken as a whole, their analysis suggests employment effects ranging from –1 to –2 percent, although declines in some sectors are larger.

An alternative approach is to estimate the relationship between measures of economic activity (such as production and employment) and energy prices using historical data, and use these estimates to predict the impact of a carbon price. In this vein, Aldy and Pizer (2009) use annual industry-level data on output, employment, and electricity prices to assess the effects of a ten dollars per ton tax on carbon. The advantage of this approach is that it is more transparent and does not hinge on particular assumptions about intersectoral and intertemporal elasticities. Its main disadvantage is that it ignores general equilibrium effects. The findings of Aldy and Pizer suggest overall modest effects of this carbon tax, although some electricity-intensive manufacturing sectors are more severely affected.

This chapter provides new estimates of the relationship between real electricity prices and indicators of labor market activity using data for 1976 to 2007. While the prices of all energy sources are predicted to increase in proportion to their carbon content under carbon-pricing policy, in this short chapter I focus only on electricity because it is the largest energy expenditure in most sectors of the economy. For example, in the retail trade sector, electricity purchases correspond to roughly 2 percent of total production costs, but 80 percent of total energy costs. Thus in principle, a first-order impact channel of climate policy on the labor market will be through its effect on electricity prices.

The chapter contributes to the literature in two important ways. First, it relies primarily on within-state variation in electricity prices for the period 1976 to 2007. This extends the analysis of Aldy and Pizer (2009), who utilized aggregate electricity prices for the period 1986 to 1994. Second, I consider all sectors of the US economy (which I classify in twelve categories)

---

3. There is also a long-standing macroeconomic literature on the effect of energy, and especially oil prices on economic activity (see Hamilton [2008] and Killian [2008] for recent surveys).

rather than focusing only on the manufacturing sector. This distinction is important since the manufacturing sector now represents less than 20 percent of total employment in the United States. The resulting cross-sectional and time-series variation allows me to control for unrestricted year, state, and industry shocks, as well as allowing for differential time trends across states or industry. This modeling effort is made in an attempt to minimize the confounding effects of industry-specific or state-specific permanent and/or transitory shocks that may be correlated with electricity prices. It also implicitly controls for state-specific labor demand shocks (as long they evolve smoothly over time) or arbitrary year-specific shocks to labor demand (perhaps because of changes in determinants of international trade such as tariffs).

The main finding is that employment rates are negatively related to real electricity prices and that the relationship is relatively weak. The cross-elasticity of full-time equivalent (FTE) employment with respect to electricity prices ranges from –0.10 to –0.16 percent. By comparison, the average annual change in FTE employment (normalized by population) over the sample period is about 1.5 percent, so the fluctuations in employment caused by electricity price shocks are well within the range of the normal historical variation. The estimated elasticities are precise with confidence intervals that rule out large short-run declines in employment. Although not reported in detail here, an industry-level analysis also reveals that employment in some industries (agriculture, transportation, finance, insurance, and real estate) is more responsive to changes in electricity prices. Notably these industries only make up 15 percent of total employment.

I then interpret these estimates in the context of predicted increases in electricity prices that are consistent with H.R. 2454, the American Clean Energy and Security Act of 2009. To this end, I use the empirical estimates to simulate the short-run employment response to higher electricity prices. The preferred estimates in this chapter suggest that in the short run, an increase in electricity price of 4 percent would lead to a reduction in aggregate FTE employment of about 460,000 or 0.6 percent.

There are several caveats to this research and its results that need to be emphasized. First, since the analysis is based on annual variation in electricity prices, it is only relevant for evaluating the short-run employment effects of a possible carbon policy. These short-run effects will be important determinants of the initial transition costs associated with a climate policy. However, the short-run response to a permanent change in electricity price caused by a carbon-pricing policy will likely differ from the short-run response to transitory changes in electricity price that are measured in this chapter. In addition, the long-run effects will presumably be smaller in magnitude once all the adjustments to the capital stock are made and the sectoral reallocation of labor takes place. Second, estimates based on historical data are dependent on the set of events, institutions, and regulations that applied during the period observed. As such, these estimates may not be applicable to the new economic environment that would follow climate legislation. Third, the observed historical variation in electricity prices may not overlap with the higher energy prices caused by a specific carbon-pricing policy, and so prediction of its effects may depend on functional form projections. Finally, this analysis does not quantify the effect of the policy incentives that could increase employment in "green" sectors. In addition, many climate legislation proposals, such as H.R. 2454, contain provisions for job assistance programs aimed at workers displaced by the policy, and industry-specific subsidies designed to counter some of the added costs imposed by the policy. It is possible that such provisions will cause increases in labor demand in some sectors and this possibility is not accounted for in this analysis.

2.2   Conceptual Framework

A natural starting point to conceptualize the effect of energy prices on labor markets is the neoclassical theory of labor demand. In a model where labor and energy are factors of production (along with other factors), the cross-elasticity of labor demand with respect to energy prices is given by $\eta_{LE} = s_E \times [\sigma_{LE} - \rho/(\rho - \theta)]$ where s_E is the share of energy in total production costs, σ_{LE} is the partial elasticity of substitution between labor and energy, ρ is a measure of market power of the firm ($= 1$ if the firm is a price-taker in the product market, and > 1 if the firm is a price-maker), and θ measures the degree of homogeneity of the production function (see Cahuc and Zylberberg [2004] for derivations). The first term in the parentheses is the substitution effect (which may be positive or negative in this case) and the second term is the scale effect (which declines in magnitude as the degree of market power increases). This formula has two key implications: (a) the cross-elasticity of labor demand with respect to energy prices is likely to be small since S_E is small for most industries, and (b) the sign of η_{LE} will depend on whether the substitution or the scale effect dominates.

The previous expression also highlights three key sources of variation in the cross-elasticity of labor demand to energy price across industries. First, there are differences in energy intensity (i.e., S_E) across industries. Second, there may be differences in market power across industries that determine the degree to which firms in a sector can pass the extra costs associated with the policy to the buyers of their products (either as intermediary inputs, or as final demand). For example, sectors producing goods that face low import pressure are less likely to be affected by carbon pricing, at least in the short run. Finally, differences in the production technology (i.e., σ_{LE}) across sectors will also contribute to differences in the responsiveness of labor demand to shocks to energy prices.

2.3   Data Sources and Preliminary Analysis

2.3.1   Data

The primary data for this chapter are taken from the 1977 to 2008 March Current Population Surveys (CPS), and covers calendar years 1976 to 2007.[4] Importantly, the March CPS contains information about labor force outcomes (employment status, hours worked, weeks worked in the last year), as well as information on industry affiliation at the three-digit level. Starting in 1976, weeks of work are reported continuously, which explains the choice of the sample period. In addition, the March CPS contains demographic information including state of residence, age, gender, race, education, and so forth. The state of residence information will be used in conjunction with the survey year to link the CPS with the electricity price data.

The annual worker-level data are then combined with retail electricity prices from the State Energy Data System (SEDS) maintained by the Energy Information Administration. The SEDS data is detailed, and contains prices and expenditures for a dozen primary energy sources (i.e., coal, natural gas, etc.), as well as "transformed" energy sources, such as retail electricity and total energy at the state-year level. The retail electricity price data from SEDS are then merged with the microlevel CPS data by year and state of residence to construct the final samples used in the analysis.

2.3.2   Sample Construction and Key Variables

For the purpose of this analysis, I consider individuals aged sixteen to sixty-five, working for pay (i.e., not self-employed), and residing in the continental United States. I then use the micro data to derive the number of full-time equivalent (FTE) workers. The approach could be extended to other measures of labor supply, such as total hours worked, number of part-time workers, and so forth. In practice there is a tradeoff between a fine industry classification (which provides a better characterization of the production technology in which a worker is employed) and statistical precision (because of empty or small cells) and so for this chapter, I consider a twelve-industry classification.[5] Full-time equivalent employment is obtained by summing annual hours worked in each state-year-industry cell, and then dividing by 2,080 (40 hours per week * 52 weeks per year). In all cases, I use the CPS person weight (perwt) variable for these calculations.

4. These data were accessed through IPUMS (http://cps.ipums.org/cps/).
5. The industry classification are Agriculture & Natural Resources, Mining, Construction, Durable Goods, Non-Durable Goods, Transportation, Utilities, Wholesale Trade, Retail Trade, Finance, Insurance and Real Estate (FIRE), Services, and Public Administration.

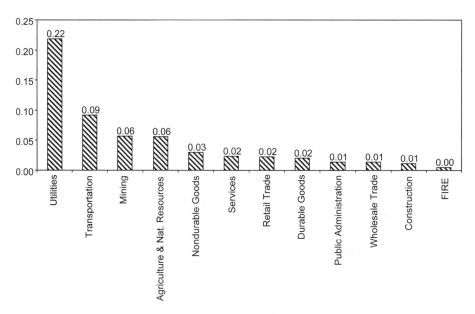

Fig. 2.1   Share of energy in total production costs, 2002

Notes: Tabulations from the Bureau of Economic Analysis "Industry Economic Accounts" for 2002. See the text for more details.

2.3.3   Preliminary Analysis

The formula for the cross-elasticity of labor demand with respect to energy price highlights that a mandated carbon price is likely to have differential effects across industries, reflecting in part differences in energy intensity. Unfortunately there are no comprehensive and comparable sources of data on electricity intensity available for each sector of the economy.[6] Instead, I report energy shares (defined as the ratio of the value of energy inputs over the value of all intermediate inputs and employee compensation) from the Bureau of Economic Analysis 2002 Industry Accounts data.

Figure 2.1 reports the energy shares for each of the twelve industry categories considered in the empirical analysis.[7] While there are evident differences in energy shares across sectors (ranging from less than 1 percent in the Finance, insurance, and real estate (FIRE) sector to 22 percent in the utility

6. For example, the Manufacturing Energy Consumption Survey (MECS) contains detailed information on electricity consumption in the manufacturing sector, but by definition this covers only roughly 20 percent of the US workforce. Similarly, the Survey of Business Expenses omits the agricultural, utilities, and public administration sectors.

7. The BEA data appears to slightly undercount energy inputs in some of the durable and nondurable manufacturing sectors. For these two sectors I use instead energy shares computed from the 2002 MECS data.

sector), for most sectors, and most of the employment, the energy share is 3 percent or less. In fact, the FTE weighted share across the twelve sectors is 2.6 percent. And since electricity is one of many possible sectoral energy inputs, these shares are upper bounds on the actual electricity shares (S_E). As such, this evidence, in connection with the previous theoretical formula, foreshadows that the cross-elasticity of employment with respect to electricity price is likely to be small.

Figure 2.2 presents a first look at the connection between real electricity prices and FTE employment over the period 1976 to 2007. The full line shows the yearly average of residuals from a regression of real electricity prices (in $2005 per kWh) on a quadratic time trend and unrestricted state effects. Similarly, the dashed line displays the yearly average of residuals from a regression of log FTE employment on a quadratic time trend and unrestricted state effects. The connection is remarkable: each period of higher than average electricity prices is accompanied by lower than average employment, especially in the early 1980s and late 1990s. In fact, the raw correlation between the two series is –0.77. This evidence clearly suggests the existence of a relationship between FTE employment and electricity prices. The following regression analysis will quantify and refine the magnitude of relationship by including more variables in order to control for unobserved shocks correlated with electricity price and labor demand.

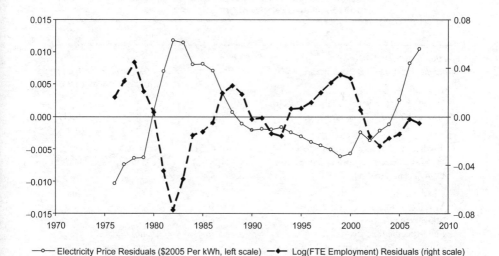

—○— Electricity Price Residuals ($2005 Per kWh, left scale)   —◆— Log(FTE Employment) Residuals (right scale)

Fig. 2.2   Residual relationship between real electricity prices and full-time equivalent (FTE) employment

Notes: Residuals from regressions based on 1,568 state * year observations. Each model controls for a quadratic in year and state fixed effects. Reported in the figure are the yearly averages of the residuals from the regressions. See the text for more details.

2.4   Regression Analysis

In order to estimate the cross-elasticity of labor demand with respect to electricity prices I consider group-level regression models of the form:

(1) $\text{Log}(Y_{st}) = \alpha_s + \alpha_t + \beta\text{Log}(P_{st}) + X_{st}\gamma + \varepsilon_{st},$

where Y_{st} represents employment in state s and observed in year t. The parameters α_s and α_t are fixed effects for state (s) and year (t). In some models, these fixed effects are also augmented by state-specific time trends. The key variable is P_{st}, the average retail electricity price in dollar per kWh in state s and year t (deflated to 2005 dollars). Variable β is the parameter of central interest in this chapter: it measures the percentage change in employment associated with a 1 percent change in real electricity prices. Table 2.1 reports estimates of this cross-elasticity for various specifications. The vector X_{st} contains the control variables, most importantly the size of the sixteen to sixty-five population in the relevant cell. In addition to the specification in equation (1), I also consider alternative models where the year effects are replaced with a quadratic time trend and where industry fixed effects and industry-specific time trends are included. The last term in equation (1), ε_{st}

Table 2.1   **Estimates of cross-elasticity of full-time equivalent (FTE) employment with respect to real electricity prices**

	(1)	(2)	(3)
A. Based on state * year cells			
cross-elasticity of FTE employment	–0.147	–0.096	–0.132
	(0.031)	(0.036)	(0.032)
B. Based on state * year * 12 industry cells			
cross-elasticity of FTE employment	–0.156	–0.097	–0.119
	(0.039)	(0.052)	(0.064)
C. Predicted FTE employment effect of 4%	–512,513	–334,702	–460,215
increase in electricity prices (based on	(108,081)	(125,513)	(111,567)
estimates in panel A).			
Quadratic in year	yes	no	no
Year fixed effects	no	yes	yes
State fixed effects	yes	yes	yes
State-specific time trends	no	no	yes
Industry fixed effects (panel B only)	no	no	yes
Industry-specific time trends (panel B only)	no	no	yes

Notes: Cross-elasticity estimates are from models based on 1,568 state * year cells (row A) and 18,471 state * year * industry cells (row B). Each model controls for the log of sixteen to sixty-five population in addition to the variables listed at the bottom of the table. Predicted FTE employment effects assume a 4 percent increase in electricity prices are evaluated at the sample average of aggregate FTE employment in the sample period (87,162,000). The standard errors in parentheses are corrected for within-state serial correlation. See the text for more details.

is an error term. Throughout the chapter the standard errors are corrected to allow for arbitrary within-state serial correlation.

Once the cross-elasticity of employment with respect to electricity price is estimated from equation (1), we can predict the impact of a particular climate policy on employment by multiplying the β coefficient by the predicted increase in electricity price. For example, the predicted change in FTE employment would be calculated as follows:

$$(2) \qquad \%\Delta\text{FTE} \approx \hat{\beta}_{\text{FTE}} \times \Delta P.$$

The credibility of this approach depends on the assumption that the estimation of equation (1) will produce unbiased estimates of the β parameter. The key assumption is that there are no residual labor demand shocks that are correlated with electricity price once we control for year, state, and industry fixed effects as well as industry-specific and state-specific time trends. This is a strong assumption; for example, it rules out state-specific labor demand shocks that do not evolve smoothly over time. Following, I further discuss the limitations of the empirical estimates produced by this analysis.

2.4.1 Cross-Elasticity of FTE Employment with Respect to Real Electricity Prices

Table 2.1 reports empirical estimates of the coefficient β in equation (1). In all models FTE employment and electricity prices are expressed in logs, so the reported coefficients correspond to the effect of a 1 percent change in electricity price on %FTE employment. Row A is based on state * year cells and ignores the variation in employment due to differences across states (and/or over time) in industry composition. Estimates in column (1) are based on models including a quadratic time trend and state fixed effects, column (2) replaces the quadratic time trend with year fixed effects and column (3) adds state-specific time trends to the specification. It is the more general model considered, and allows for differential shocks to labor demand in each state, provided that these shocks evolve smoothly enough. Estimates in row B are based on state * year * industry cells, but restrict the impact of electricity price on employment to be the same across industries. The specification of the models in columns (1), (2), and (3) of row B remains the same, with the exception that industry fixed effects are included in all specifications, and industry-specific time trends are added to the models in column (3).

The estimates are negative in all specifications and statistically significant in most. This indicates that increases in electricity prices lead to reductions in FTE employment and suggest that labor and electricity prices are p-complement. However, the cross-elasticities are relatively small: The largest point estimate in absolute magnitude is -0.156 and its 95 percent confidence interval ranges from -0.234 to -0.078. The preferred estimates in column (3) indicate that a 1 percent change in electricity price will lead to a –0.13 percent to –0.12 percent reduction in FTE employment. By comparison, the average annual change in FTE employment (normalized by population) over the sample period is about 1.5 percent, so the fluctuations in employment caused by electricity price shocks are well within the range of the normal historical variation.

Although not reported in table 2.1, I also estimated the impact of electricity prices on FTE employment separately for each of the twelve industry categories considered. With the caveat that this analysis lacks the statistical precision of table 2.1, it is notable that higher electricity prices lead to a reduction in FTE employment in most industries. The most affected industries are agriculture, transportation, and FIRE. The cross-elasticities for those three sectors are –0.426, –0.385, and –0.291, respectively, and are statistically significant at the conventional level. However these are smaller industries in terms of overall employment, representing about 15 percent of total employment in the United States over the sample period. There is a positive correlation between electricity prices and FTE employment in the mining and utilities sector, although the point estimates are not statistically significant.

Although not reported here, I have also considered alternative specifications of equation (1), notably to allow for nonlinearities and lagged effects of electricity prices on employment. In general, these considerations did not alter the main results significantly.[8] It is also worth noting that the analysis presented in table 2.1 could be extended to provide information about the "incidence" of electricity price shocks by examining responses specific to demographic groups or geographical areas.

2.4.2   Implication for Climate Policy

While the estimated cross-elasticities appear small, their implications on the possible aggregate employment effects of a climate policy may be more sizable. To put this in context, I evaluate the predicted aggregate employment effects associated with an increase in electricity price similar to the increase that would be caused by a climate policy like H.R. 2454, the American Clean Energy and Security Act of 2009.

To this end, I use the estimated cross-elasticities in row A of table 2.1 to simulate the short-run employment response to an increase in electricity price of 4 percent. This price increase is consistent with the projections from the Energy Information Administration (2009) about future electricity prices under H.R. 2454. The resulting predicted changes in FTE employment are reported in row C of table 2.1. Across the three specifications the estimates range from reductions of 510,000 to 335,000 in FTE employment. By comparison, the average aggregate FTE employment in the sample is about

---

8. Blanchard and Gali (2007) report that the effect of oil prices on aggregate employment has declined over time.

eighty-seven million. The preferred estimate in column (3) is −460,215 with a standard error of 111,567. It is worth noting that the predicted employment effects are a linear function of the estimated cross-elasticity and therefore could be implemented for alternative scenarios regarding future electricity prices under any specific climate policy.

2.4.3   Possible Sources of Bias in Cross-Elasticity Estimates

It is possible that the estimates reported in table 2.1 are biased if there are omitted factors in the regression models that are correlated with both electricity prices and labor demand. This bias would invalidate the results of this analysis, including the employment projections associated with specific climate policies.

A key issue is that within-state variation in electricity price provides the key identifying variation for the empirical analysis, and within-state electricity price changes are likely to be caused by many factors, including changes in regulator behavior, capacity constraints, changes in the relative price of primary energy inputs, and so forth. As such, these price shocks may be caused in part by factors related to labor demand in a way that is not controlled for by the year fixed effects and state-specific time trends included in the empirical models. This could occur if the electricity-pricing rule used by the utility regulators sets prices to equate average costs to average revenues. Since revenues depend on electricity sales, which may in turn depend on labor market conditions, this pricing rule may imply a reverse causality relationship from employment to electricity prices. As a consequence, this would lead to biased estimates of the cross-elasticity of employment with respect to electricity prices, and this bias is difficult to sign a priori.

A common solution to this problem is to rely on instrumental variables that are correlated with electricity prices but otherwise uncorrelated with labor demand. One possibility would be to use changes in relative prices of primary energy inputs used in producing electricity interacted with physical production capacity by fuel type in each state as instrumental variables for electricity prices. While a complete implementation is beyond the scope of this chapter, it is an approach I am undertaking in continuing work.

2.5   Implications and Concluding Remarks

Taken literally, the preferred estimates in this chapter suggest that in the short run, an increase in electricity price of 4 percent would lead to a reduction in aggregate FTE employment of about 460,000 or 0.6 percent. This estimate corresponds to the first-year response to higher electricity prices assuming firms did not anticipate the rise in electricity costs and that no production subsidies are given to sectors most affected by the introduction of a price on carbon. In reality, it is probable that a carbon-pricing policy will be phased in gradually and accompanied with subsidies to selected sectors. Such adjustment mechanisms should reduce some of the employment loss predicted by the approach in this chapter.

By comparison, the important recession that started in December of 2007 caused the number employed nationally to decline by 3.1 million between December 2007 and 2008.[9] Using this recent experience as a benchmark, it appears that climate policies that lead to increases to electricity price of 3 to 4 percent will lead to significant but not unprecedented employment loss.

There are many limitations to this research and its results need to be interpreted with caution. In my view the most significant limitation is that the approach taken here is only informative about the short-run effect of transitory shocks to electricity prices, and so ignores general equilibrium effects. Information about the differential dynamic adjustment paths across industries is essential to evaluate the full extent of the implications of climate legislation on labor markets. Insights into this question can be obtained by considering dynamic general equilibrium models.

References

Aldy, J., and W. Pizer. 2009. *The Competitiveness Impacts of Climate Change Mitigation Policies.* Pew Center on Global Climate Change. Available at http://www.cfr .org/climate-change/pew-center-competiveness-impacts-climate-change-mitigation-policies.

Berman, E., and L. Bui. 2001. "Environmental Regulation and Labor Demand: Evidence from the South Coast Air Basin." *Journal of Public Economics* 79: 265–95.

Blanchard, O. J., and J. Galí. 2007. "The Macroeconomic Effects of Oil Price Shocks: Why are the 2000s so different from the 1970s?" In *International Dimensions of Monetary Policy,* edited by Jordi Galí and Mark Gertler, 373–421. Chicago: University of Chicago Press.

Cahue, P., and A. Zylberberg. 2004. *Labor Economics.* Cambridge, MA: MIT Press.

Energy Information Administration. 2009. *Energy Market and Economic Impacts of H.R. 2454, the American Clean Energy and Security Act of 2009.* Washington, DC: EIA.

Greenstone, M. 2002. "The Impacts of Environmental Regulations on Industrial Activity: Evidence from the 1970 and 1977 Clean Air Act Amendments and the Census of Manufactures." *Journal of Political Economy* 110 (6): 1175–219.

Hamermesh, D. 1993. *Labor Demand.* Princeton, NJ: Princeton University Press.

Hamilton, J. 2008. "Oil and the Macroeconomy." In *The New Palgrave Dictionary of Economics,* 2nd ed., edited by S. Durlauf and L. Blume. New York: Palgrave Macmillan.

Ho, M., R. Morgenstern, and J. Shih. 2008. "Impact of Carbon Price Policies on U.S. Industry." Resources for the Future. Discussion Paper no. 08-37.

9. Laura A. Kelter (2009).

Jaffe, A., S. Peterson, P. Portney, and R. Stavins. 1995. "Environmental Regulation and the Competitiveness of U.S. Manufacturing: What Does the Evidence Tell Us?" *Journal of Economic Literature* 33 (1): 132–63.

Kelter, Laura A. 2009. "Substantial Job Losses in 2008: Weakness Broadens and Deepens across Industries." *Monthly Labor Review* 132 (3): 20–33.

Killian, L. 2008. "The Economic Effects of Energy Price Shocks." *Journal of Economic Literature* 46 (4): 871–909.

Comment   Matthew E. Kahn

This impressive chapter utilizes a state-level panel data set covering the years 1976 to 2007 to provide new estimates of the relationship between retail electricity prices and state employment activity. Based on an estimation strategy that controls for state and year fixed effects, this chapter exploits within-state variation in electricity prices. A key finding is that the electricity price elasticity is roughly –.12. Deschênes uses this estimate to predict the likely employment effects of a federal carbon mitigation policy. If such a policy would raise electricity prices by 4 percent, then he predicts that aggregate US employment would decline by 460,000. In absolute terms, this would appear to be a very large unintended regulatory effect, while relative to the nation's total workforce this effect is small.

In the summer of 2009, the House of Representatives barely passed the American Clean Energy and Security Act. In the summer of 2010, the Senate chose not to vote on that bill. The Congress' tepid efforts to battle climate change indicate that its members believe that such long-run regulation must have significant short-run costs. How do such senators know this? They are unlikely to have general equilibrium modelers on their staff. The Deschênes estimates offer credible evidence and represent a key "missing link" in public policy discussions. Combining state-specific predictions for how carbon regulation will affect state electricity prices with the Deschênes estimates would yield an expected job incidence measure that could help to predict congressional voting patterns on carbon mitigation legislation.

This chapter focuses on the short-run effects of electricity prices on employment. In the medium term, higher energy prices will induce some firms to innovate to economize on energy consumption (Popp 2002). Such nimble firms will be less likely to shut down or reduce employment when future electricity price increases take place. In contrast, there will be other

Matthew E. Kahn is professor in the Institute of the Environment, the Department of Economics, and the Department of Public Policy at the University of California, Los Angeles, and a research associate of the National Bureau of Economic Research.

For acknowledgments, sources of research support, and disclosure of the author's material financial relationships, if any, please see http://www.nber.org/chapters/c12151.ack.

firms in declining industries who are not making new investments to modernize their factories. Higher electricity prices may nudge these firms to shut down.

The Deschênes chapter implicitly assumes that employers are myopic and base their employment decisions on current electricity prices. Given that job creation represents an investment, firms should base such decisions on expected future input prices. A productive future line of research would be to resurrect some of the rational expectations efforts from labor economics (see Topel 1986) and apply them in this setting. For example, is state employment more sensitive to unanticipated increases in electricity prices or anticipated increases in electricity prices?

Within the same industry, rising electricity prices can have asymmetric effects on firms. Relative to new capital, older capital is likely to be much more energy inefficient and to be more likely to be rendered obsolete by increased energy prices. But, this chapter's state/industry/year aggregate analysis implicitly assumes that electricity price changes have symmetric effects on job creation and job destruction. This merits future research. Recent work by Bloom et al. (2010), using data from the United Kingdom, highlights that better managed firms are significantly less energy intensive. Such firms would be less likely to reduce their employment in the face of an unexpected increase in electricity prices.

This study reports disaggregated estimates by major industry. I am puzzled by some of the facts that emerge as reported in table 2.1's column (3). In particular, Deschênes cannot reject the hypothesis that there is no relationship between electricity prices and manufacturing employment. In fact, for nondurables manufacturing there is a positive but statistically insignificant correlation. This finding might surprise Rust Belt senators who are concerned about the continuing decline of this key sector in their states. Based on his table 2.1 findings, the sectors that are most affected by electricity prices include: agriculture, transportation, and FIRE (finance, insurance and real estate). This last fact surprises me. In states with rising electricity prices, firms in the FIRE sector can seek out office space in more energy efficient buildings or be charged lower rental rates as tenants. Eichholtz, Kok, and Quigley (2009) have documented that LEED certified and Energy Star commercial buildings command a price premium relative to the average building. Such premiums are likely to be larger in areas with higher electricity prices. In growing cities with rising electricity prices, the new buildings are likely to be built to economize on electricity consumption. Both of these facts would predict that FIRE employment would not be sensitive to increases in medium term changes in state electricity prices.

This chapter asks an important public policy question and utilizes a careful empirical strategy to generate new facts. I expect that a large amount of scholarship will build on this chapter's findings.

References

Bloom, Nicholas, Christos Genakos, Ralf Martin, and Raffaella Sadun. 2010. "Modern Management: Good for the Environment or Just Hot Air?" *Economic Journal* 120 (544): 551–72.

Eichholtz, Piet, Nils Kok, and John M. Quigley. 2009. "Doing Well By Doing Good? Green Office Buildings." *American Economic Review,* forthcoming.

Popp, David. 2002. "Induced Innovation and Energy Prices." *American Economic Review* 92 (1): 160–80.

Topel, Robert. 1986. "Local Labor Markets." *Journal of Political Economy* 94 (3): S111–S43.

Limiting Emissions and Trade
Some Basic Ideas

Kala Krishna

3.1 Introduction

On June 26, 2009, the American Clean Energy and Security Act (or the Waxman-Markey Bill after its authors) was approved by the US House of Representatives. It never cleared the Senate, however, and it now seems highly unlikely. Yet this event marked the first time either house approved a law meant to limit emissions to combat climate change and has resulted in a flurry of economics research in the area. The bill would have essentially created cap-and-trade programs for greenhouse gas emissions and specify reductions in total emissions of 17 percent starting from 2012. See Congressional Budget Office (CBO) (2009) for a good summary of the bill and its implications. News at the time indicated that the Senate version of the bill would have been weaker, with utilities being subject to caps by 2012 but with manufacturers being phased in only by 2016. Discussion at the time indicated floor and ceiling prices of ten dollars and thirty dollars per ton that will be adjusted for inflation.[1] It would have had product-specific import taxes based on the cost disadvantage created by such cap-and-trade measures on countries that do not limit their emissions (called border tax adjustments or BTAs for short). Such BTAs would, it was argued, both level the playing field for US firms and prevent leakage, where leakage is the change in foreign

Kala Krishna is the Liberal Arts Research Professor and professor of economics at the Pennsylvania State University and a research associate of the National Bureau of Economic Research.

For acknowledgments, sources of research support, and disclosure of the author's material financial relationships, if any, please see http://www.nber.org/chapters/c12154.ack.

1. The price ceiling would insure that businesses do not face too high a cost of permits as these are part of their costs. The floor protects them from the risk of investing in technology to reduce emissions only to find that it was not worth their while ex post.

emissions as a share of the domestic emissions reductions. They may also be legal under GATT/WTO; see Frankel (2009).

Existing studies suggest that the size of the BTAs would likely be quite small for most products. This is why, as drafted, US legislation envisioned BTAs mainly for producers in energy-intensive sectors. These include chemicals, paper, ferrous metals, nonferrous metals, and mineral products. However, there is considerable variation in the estimates of the effect of the kinds of emissions limits being discussed. Atkinson et al. (2010), which was a background paper for World Development Report 2010, uses a partial equilibrium model to estimate that if carbon is taxed at fifty dollars per ton of CO_2, Chinese exports to the United States would face an average tariff rate of 10.3 percent. Mattoo et al. (2009) employ a multicountry computable general equilibrium framework (the Environmental Impact and Sustainability Applied General Equilibrium Model, or ENVISAGE model). They compare outcomes under different scenarios for BTAs of a carbon tax that reduces emissions by 17 percent relative to 2005 by all Organization for Economic Cooperation and Development (OECD) countries.[2] Their work suggests some room for leakage, and that BTAs do little to reduce emissions. Most of the action in terms of growth of emissions comes from projected growth in the developing world. They calculate that a 17 percent reduction in emissions in energy-intensive goods only, as proposed by the United States, would lead to total emissions in 2025 relative to 2005 rising by about 54 percent (56.9 percent without BTAs). The 17 percent reduction in emissions by the OECD countries is more than undone by low- and middle-income countries raising their emissions by about 122.5 percent in the absence of BTAs (117.2 percent with BTAs based on foreign emissions).

The effect of BTAs on emissions and exports is also shown to be sensitive to who is reducing emissions. Boehringer, Fischer, and Rosendahl (2010) suggest that reducing emissions is significantly more costly in the European Union (EU) than the United States, mostly because EU emissions are already lower than comparable US ones. Moreover, because the EU is more open than the United States, leakage is greater from EU reductions than US ones. Full border tax adjustment policies, which include a tax on imports and a subsidy to exports, are quite effective in reducing leakage, with the import tariff being more important than the export subsidy.

The computable general equilibrium models used in the literature tend to be a bit of a black box. This chapter provides some intuition behind what goes on in general equilibrium by intuitively explaining what lies behind the demand for emissions. It traces out how a reduction in total emissions allowed in one country affects the general equilibrium and the determinants

2. Mattoo et al. (2009) and McKibbin and Wilcoxen (2009) among others, argue that whether developing country emissions or developed country ones are used as a basis for the BTA makes a substantial difference to developing country exports, leakage, and world emissions.

of the extent of leakage in the model. Finally, it concludes with some implications for policy.

3.2   Emissions in a General Equilibrium Setting

What is the easiest way of modeling emissions in production? A direct way
is to treat emissions permits as an input into production. In the absence of
emissions controls, this input is in unlimited supply and has a price of zero.
One way to think of emissions controls is that they make the supply of this
input (emissions permits) finite and binding; that is, their equilibrium price is
positive. Let us think of emission permits as being needed whenever carbon-
based fuels are used, with the number of permits needed being equal to a
constant multiple (e) of the fuel needed for production. Thus, using a unit
of fuel results in e units of emissions and so needs e emissions permits. This
makes the *effective* price of fuel used in production higher by the emissions
permit price times the emissions created by a unit of fuel. Assume that this
multiple is fixed for the time being, though of course, this is another margin
of adjustment as higher prices of emissions will create incentives to reduce
this multiple and economize on emissions permits. This is known in the
literature, see Copeland and Taylor (2003), as the technique effect.

Partial equilibrium and general equilibrium can give very different answers when analyzing the effect of emissions controls. In partial equilibrium
we would consider the demand and supply of emissions, keeping the prices
in other markets for good and factors constant. As the demand for emissions
is a derived demand, that is, it is derived from the demand for the goods that
use emissions to make them, we could write the demand for emissions as e
times the unit input requirement of fuel needed to make a unit of the good
in question times the domestic output of the good. If there is substitution
between inputs, an increase in the price of emissions will cause substitution
away from the use of fuel (and hence emissions) in production so that at a
given output of the goods produced, the demand for emissions (and fuel)
will fall. This will give a downward sloping demand for emissions permits
that comes solely from substitutability between inputs. If the supply of emissions is reduced, the price of emissions permits will rise. This increase in the
price of emissions will reduce emissions in partial equilibrium by making
firms economize on the use of emissions, which is what moves them up along
their demand curve for emissions.

Let us take this one step further. What is the effect on the demand for goods
made from using these permits? Well, under competition, an increase in the
cost of an input (emissions permits) will raise the marginal (and average)
cost of production, shifting the supply curve of these products (recall that
supply is just the marginal cost curve) inward and upward so that supply
and demand for these goods will intersect at a higher price. In this way, we
expect the price of goods that use emissions to rise as well when emissions

are targeted. However, this will have two effects: it will reduce our production and consumption of these goods (maybe we will substitute toward cleaner goods whose price has not risen) and this will shift the demand for emissions inward and reduce our emissions as well as our use of fuels (which should also reduce the world price of fuels). Second, to the extent that the higher prices for these pollution-creating final goods make the rest of the world want to produce more of them, and to the extent that lower fuel prices encourage the use of fuels (and hence emissions) in their production, the rest of the world's emissions will rise. To the extent that they use dirtier techniques of production, that is, their e is higher, this channel may even raise the total level of emissions in the world. This is the "leakage" that unilateral emissions controls will create. How large is this leakage?

Well, it turns out that the answer depends on the details of the model. It is well understood in trade that in the standard Heckscher-Ohlin model of trade and general equilibrium, where goods made in different countries are perfect substitutes for each other, if factors are reallocated between countries then, under certain conditions, there will be NO effect on world equilibrium prices.[3] All that will happen is that output will move from the country that has lost the input to the country that has gained it. Why? Well, if one country reduces its supply of emissions, say by ten units, and the other has NO emissions limits in place, then the latter can increase its emissions by the same amount (ten units) as the reduction by the former. But this is exactly like reallocating inputs between countries so that this classic trade result has immediate relevance. In this case, the leakage may be 100 percent: we may have no effect of emissions controls in one country on world emissions.[4]

To get around such issues, most computable models use settings where the goods made by different countries are imperfect substitutes for each other. As a result, unilateral emissions controls will, by raising the cost of production, make the emissions-controlling country's goods more expensive relative to those of other countries, causing substitution away from them toward the output of nonemissions controllers and consequent leakage. The extent of such leakage naturally depends on the *substitutability* between these goods in demand. If this substitutability is low, there may be little leakage, but if it is high, there may be a lot. A convenient way of modeling this substitutability between goods is to use the constant elasticity of substitution preference structure where a single parameter (or if a nested setup is used, a few parameters) define this substitutability. This is what makes the elasticity of substitution in the utility function a key parameter.

3. Technically, in the workhorse Heckscher-Ohlin model used in international trade, this is true if the world endowment point is inside what is called the Factor Price Equalization or FPE region. This region is large if there are enough goods relative to factors of production.
4. In the working paper version of this chapter, I lay out a Heckscher-Ohlin type model where this extreme result does not hold, yet remains reasonably tractable. I use the model to formally derive the effects of unilateral emissions controls.

3.3   Policy Implications

A point made in the literature is that unilateral emissions reductions will be at least partly undone by leakage. This leakage is the cause of much concern in the literature. Earlier, I argued that the extent of this leakage predicted by economic models will differ according to the model chosen (for example, goods being perfect substitutes across countries as in the HOS model or not) and the level chosen of certain key parameters (like the extent of substitution between goods made by different countries). The model and parameters used in simulating the effect of various policies will themselves yield the results. The computable general equilibrium literature tends to calibrate the chosen model to the data, and then run counterfactuals to help predict the effects of various policies. But this approach often does not give the reader insight into how alternative modeling assumptions would affect the outcomes of the policy simulations. In addition, the models themselves are often so complex that they are inaccessible to outsiders.

In addition to providing numbers from quantitative theorizing of this kind, general equilibrium models have a great deal of policy content in terms of simple insights. A good example might be the question of whether putting limits on the rest of the world's emissions, at levels that are not binding on them, is worth doing. After all, why bother if the controls are not yet binding, or only barely binding, where the price is very small? That would seem to make the controls on the rest of the world's emissions almost irrelevant. However, it is easy to see that the presence or absence of ANY controls can make a difference. If the rest of the world has caps on their own total emissions, *even if these caps are just binding,* then there will be no such leakage.

Why? Think of the standard Heckscher-Ohlin model of trade and general equilibrium, where goods made in different countries are perfect substitutes for each other. Now, as long as there are caps in all countries on emissions, emissions controls by one country is not equivalent to reallocating factors between countries, but to a reduction in world emissions permits. Thus, there will be an effect on the world equilibrium and the increasing scarcity of emissions permits in one country will result in a reallocation of supply of dirty goods toward the country with less stringent emissions limits until its emissions prices also rise! In this way, unilateral emissions reductions will have multilateral effects. With universal caps on emissions, a reduction in emissions permits in one country will raise the price of emissions everywhere and reduce world emissions one for one with the unilateral reductions undertaken in emissions. This insight has another important implication: the loss in competitiveness engendered by higher emissions prices in the country reducing emissions will be much less of an issue when all countries limit their emissions than when they do not and BTAs will also be less of an issue in terms of maintaining a level playing field.

The main point is that it is not just the level, but the *existence* of emissions controls in the rest of the world (ROW) that matters. Getting the rest of the world to commit to controls on emissions, even if the level of emissions they commit to is high, is a step in the right direction as it affects the nature of international transmission. If the ROW has no controls on emissions, then the price of emissions there is fixed at zero no matter what policy home enacts. Tighter emissions limits at home necessarily tilts the playing field in favor of rest of the world. But if the rest of the world has any limit on emissions, then tightening emissions at home will raise demand for emissions abroad and raise the price of emissions abroad, preventing leakage abroad, limiting the loss of competitiveness at home, and making the home country more willing to reduce its own emissions.

If emissions are controlled only in a subset of countries, there will inevitably be some leakage. How large might this leakage be? Trade theory has some further insights to offer here. First, if some factors are mobile, and in today's world they seem to be increasingly so, factor mobility can make emissions controls much less binding. It is well understood by now, that attempts to tax trade will be undone by the movement of capital (i.e., firm location) in certain situations a la Mundell (1957). In a similar vein, taxing emissions will result in firm relocation if factors are mobile. This relocation could be very large depending on the setting and model used. Babiker (2005) produces estimates for leakage of over 100 percent in an oligopolistic model with increasing returns to scale when relocation is explicitly allowed for.[5]

How large leakage would be is ultimately an empirical matter. Hanna (2010) shows that US multinationals increased their foreign assets by about 5.3 percent and foreign output by about 10 percent in response to the Clean Air Act Amendment, which dramatically strengthened US environmental regulations. Such responses even make things worse in terms of emissions if migrating firms use more polluting technologies abroad than at home.[6]

Recent work in trade, unrelated to the previous model, may also be germane. A concern in, for example, Mattoo et al. (2009) is that BTAs imposed in order to level the playing field may have large effects on the exports of non-emissions-controlling developing countries. While competitive models would suggest that lower exports to the United States when the United States has BTAs could be made up by larger exports elsewhere, in monopolistically competitive settings, the *opposite* prediction exists.

This point is made in Cherkashin et al. (2010). Suppose that the develop-

5. In related work, Cherkashin et al. (2010) show that in heterogeneous firm oligopolistic models, entry/exit by firms in response to trade policies are very large and account for most of the adjustment in output that occurs, suggesting that such settings might give large leakage effects in the emissions control context as well.

6. In contrast, while examining the EU's emissions trading program, Grubb and Neuhoff (2006) argue that the net value at stake is low for most sectors as the cost increases by emissions trading in the ten to thirty euro range are small for all but a few industries. However, if firms are very responsive to such differences, even small changes could have large effects.

ing world has no emissions limits in place and the United States does. Moreover, to prevent leakage, the United States also has BTAs. Cherkashin and colleagues argue that in a monopolistically competitive setting, the lower exports to the United States due to BTAs would be accompanied by *lower* exports to *all* other markets. The argument is elegant. The fall in expected profits from exports to the United States will make the expected profits of all existing firms negative. This will cause an exit of firms from these industries and this exit will raise the expected profits of all remaining firms. However, when firms exit, they exit from all their markets. As a result, developing country exports to *all* markets fall. Therefore the short-run effects of emissions limits, with entry held constant, are likely to be very different from the long-run ones. It would be unfortunate if the adverse effects on developing country exports of BTAs were underestimated.

Ultimately, the effects of unilateral emissions controls and the extent of leakage is an empirical issue. So why don't we ask what the data shows? After all, carbon taxes are levied in the EU.[7] Surprisingly enough, empirical work suggests that there is no effect of carbon taxes on international competitiveness for the EU. In a recently published paper, Kee, Ma, and Mani (2010) run a standard gravity equation for exports from country i to j of product k with the usual variables (exporter and importer fixed effects, product fixed effects, distance, a common border, a common currency, a common free trade area, distance, etc.) augmented by dummies that indicate whether both i and j have a carbon tax in place, only the exporter has a carbon tax in place, and only the importer has a carbon tax in place. They find that carbon taxes by exporting countries do not seem to matter! This is the exact opposite of what people might have expected.

However, the probable reason for this result is an institutional one. As part of the deal in imposing the emissions controls and consequent carbon taxes, firms are allocated emissions permits for free roughly equal to their emissions before the policy is instituted. These allocations are conditional on being in the industry. The allocations should have no real effects, merely being a transfer of rents from the government to the firms in question had the allocations been unconditional. Making them conditional prevents exit of firms due to the greater costs imposed by emissions restrictions, and this may minimize the effects on output and exports. Of course, to the extent that marginal costs are due to the need to have emissions permits, we should see a shift back in the supply curves and a reduction in exports coming from this. That we do not see this suggests that these marginal costs increases are small for most industries. There may be exceptions for the most energy-intensive

7. They may be joined by Australia, a country responsible for a disproportionate extent of GHG emissions and that is extremely vulnerable to climate change However, the proposal is modest: it calls for a low carbon price (of about twenty-three dollars a ton) and a low carbon emission reduction target of 5 percent by 2020.

ones. However, even in these industries the same results are obtained. This suggests that there may have been overcompensation of emissions permits allocated to firms, which raised entry into the industry and undid the expected fall in supply of existing firms. This suggests that incorporating the allocation of free emissions permits as a pure rent transfer as is commonly done may give misleading results.

Emissions permits are likely to be allocated to firms in practice for political economy reasons. Making this allocation conditional gives them a subsidy element that seems to raise exports in the limited evidence available. But then, why have BTAs? After all, if the allocation was fully tied to input use, it would not raise costs at all! In this event, additional border taxes would clearly be tilting the playing field in favor of domestic firms and not leveling it. Clearly, more work on exactly how permit allocation rules affect the behavior of firms in practice is required to better understand the role for BTAs.

References

Atkinson, Giles, Kirk Hamilton, Giovanni Ruta, and Dominique van der Mensbrugghe. 2010. "Trade in Virtual Carbon: Empirical Results and Implications for Policy." Policy Research Working Paper no. 5194. Washington, DC: The World Bank.

Babiker, Mustafa. 2005. "Climate Change Policy, Market Structure, and Carbon Leakage." *Journal of International Economics* 65:421–45.

Boehringer, Christoph, Carolyn Fischer, and Knut Einar Rosendahl. 2010. "The Global Effects of Subglobal Climate Policies." *B.E. Journal of Economic Analysis and Policy* 10 (2): 13.

Cherkashin, Ivan, Svetlana Demidova, H. L. Kee, and Kala Krishna. 2010. "Firm Heterogeneity and Costly Trade: A New Estimation Strategy and Policy Experiments." NBER Working Paper no. 16557. Cambridge, MA: National Bureau of Economic Research, November.

Congressional Budget Office. 2009. *The Economic Effects of Legislation to Reduce Greenhouse-Gas Emissions.* Washington, DC: CBO.

Copeland, Brian, and M. Scott Taylor. 2003. *Trade and the Environment: Theory and Evidence.* Princeton, NJ: Princeton University Press.

Frankel, Jeffrey A. 2009. "Addressing the Leakage/Competitiveness Issue in Climate Change Policy Proposals." *Brookings Trade Forum, 2008/2009,* 69–91. Doi: 10.1353/btf.0.0024.

Grubb, Michael, and Karsten Neuhoff. 2006. "Allocation and Competitiveness in the EU Emissions Trading Scheme: Policy Overview." *Climate Policy* 6:7–30.

Hanna, Rema. 2010. "US Environmental Regulation and FDI: Evidence from a Panel of U.S. Based Multinational Firms." *American Economic Journal: Applied Economics* 2 (3): 158–89.

Kee, Hiau Looi, Hong Ma, and Muthukumara Mani. 2010. "The Effects of Domestic Climate Change Measures on International Competitiveness." *World Economy* 33 (6): 820–9.

Mattoo, A., A. Subramanian, D. van der Mensbrugghe, and J. He. 2009. "Reconciling Climate Change and Trade Policy." Policy Research Working Paper no. WPS5123. Washington, DC: The World Bank.

McKibbin, Warwick, and Peter Wilcoxen. 2009. "The Economic and Environmental Effects of Border Tax Adjustments for Climate Change Policy." Center for Applied Macroeconomic Analysis (CAMA), Australian National University, Working Paper no. 9.

Mundell, Robert A. 1957. "International Trade and Factor Mobility." *American Economic Review* 47 (2): 321–35.

Comment   Meredith Fowlie

Policies designed to mitigate climate change are likely to have economy-wide impacts. Consequently, there is a strong case to be made for general equilibrium modeling that seeks to capture interactions between all sectors of the economy. A growing literature uses computable general equilibrium (CGE) models to quantify the economy-wide effects of greenhouse gas emissions regulations.

Kala Krishna begins her chapter with the observation that the CGE models commonly used in the literature tend to be nontransparent "black boxes." She provides a conceptual discussion of how greenhouse gas regulations imposed in one country can affect relative factor prices, trade flows, emissions, and emissions leakage in an open economy. The chapter provides useful insights into the inner workings of CGE models, emphasizing the value added vis-à-vis partial equilibrium approaches.

In this short comment, I first provide some context for Krishna's contribution. I then elaborate upon two of her key points. First, partial and general equilibrium models can yield very different predictions with respect to emissions leakage under incomplete climate change policy. Second, the extent of the emissions leakage predicted by CGE models will depend critically on the assumed structure of the model and the assumed values of some key model parameters.

Modeling Emissions and Emissions Leakage in an Open Economy

In her chapter, Krishna focuses primarily on general equilibrium modeling of emissions leakage. Leakage refers to any increase in emissions in one jurisdiction that occurs as a direct consequence of emissions regulation

Meredith Fowlie is assistant professor in the Department of Agricultural and Resource Economics at the University of California, Berkeley, and a faculty research fellow of the National Bureau of Economic Research.

For acknowledgments, sources of research support, and disclosure of the author's material financial relationships, if any, please see http://www.nber.org/chapters/c12155.ack.

imposed in another jurisdiction. The potential for emissions leakage has been a major obstacle to regional climate change policies.

There are at least three related channels through which leakage can occur. Consider the simple example of a home country imposing a binding cap on domestic emissions while the emissions in the rest of the world remain unregulated. The introduction of the emissions cap is likely to increase the operating costs of domestic producers relative to their unregulated rivals. In the short run, this may result in a shift of production activity and emissions to unregulated foreign producers (the first channel). Over the long term, the extent of this direct leakage can be exacerbated as firms relocate to jurisdictions with less stringent emissions control policies (the second channel). If demand for carbon intensive inputs in the home country is sufficiently large to affect world energy prices, indirect leakage can also occur. More precisely, as domestic demand for carbon-intensive fuels decreases, fuel prices fall, and producers in unregulated jurisdictions substitute toward these inputs (the third channel).

Ideally, an analysis of the potential for emissions leakage under a particular policy or program would account for all three channels. This is easier said than done! For the sake of tractability, partial equilibrium models typically hold factor prices constant. This shuts down the third channel. General equilibrium models can, in principle, capture both direct and indirect leakage effects. However, assumptions commonly invoked in CGE modeling limit the extent to which leakage can be realistically represented.

Theoretical Foundations of CGE Models

Theoretical general equilibrium models formalize the mutual interdependence of markets and serve as essential foundations for the CGE modeling of climate change policy impacts. In her chapter, Krishna offers a parsimonious and intuitive discussion of the underlying theory. She begins by describing a simple partial equilibrium modeling framework. Consider a competitive industry that produces an emissions-intensive good. If the home country imposes a price on emissions, this will increase the cost of domestic production which will lead, unambiguously, to an increase in the domestic equilibrium product price. Demand for imports from unregulated jurisdictions will increase, leading to leakage through on or both of the direct channels described earlier.

Whereas partial equilibrium models consider the policy impacts on one industry or sector in isolation, general equilibrium analysis considers impacts on the economy as a whole. How do the theoretical drivers of emissions leakage differ as we move from a partial to a general equilibrium framework? In short, it depends on the details of the model. In a recent paper, Karp (2011) explains how an increase in the emissions permit price need not decrease domestic production of an emissions-intensive good in a general equilibrium setting. If the production function for the dirty good is not separable in emissions and other inputs to production (e.g., capital and labor), the impact of the policy on regulated producers is mitigated by changes in relative factor prices. Thus, general equilibrium effects moderate the partial equilibrium effect of the emissions policy, such that a partial equilibrium model will overstate the magnitude of leakage. However, this result can be reversed under alternative general equilibrium modeling assumptions (Karp 2010).

From Theory to Applied Theory

The CGE modeling applies general equilibrium theory in empirically oriented analyses. A natural starting point, in terms of theoretical foundation, is the canonical Heckscher-Ohlin (HO) model. This model assumes perfect competition in all markets, constant returns to scale technology, and perfect substitution between goods produced in different jurisdictions. Unfortunately, this workhorse theory model can have limitations when used as a basis for applied analysis. The model can yield extreme results in realistic cases where the number of goods produced exceeds the number of factors of production. Sectoral production can be very sensitive to small changes in world prices (the so-called overspecialization problem) or indeterminate. Moreover, the HO model is inconsistent with some stylized empirical facts including frequent "cross hauling" (which occurs when a country imports and exports the same good) and price differences across trading partners that cannot be explained by transport costs.

Most of the CGE models used to analyze the effects of climate change policies avoid the aforementioned complications by assuming that imported goods and domestic goods are imperfect substitutes. So-called Armington elasticities specify the degree of substitution between domestically produced goods and goods produced in foreign countries. It is common and convenient to assume a constant elasticity of substitution. These assumptions greatly simplify the parameterization of the CGE model. But, as Krishna notes, they significantly affect the extent to which leakage occurs in a CGE model (see, for example, Babiker [2005]).

Conclusion

In this chapter, Kala Krishna presents an intuitive conceptual introduction to general equilibrium analysis of emissions regulations. The general equilibrium approaches she describes serve to highlight how sensitive emissions leakage can be to the linkages and intermarket interactions that partial equilibrium models ignore. To keep applied general equilibrium analysis tractable, many simplifying assumptions must be made. It is important to be aware of these modeling assumptions because they can significantly impact the extent to which emissions leakage manifests in CGE modeling and analysis.

References

Babiker, M. H. 2005. "Climate Change Policy, Market Structure, and Carbon Leakage." *Journal of International Economics* 65:421–45.
Karp, L. 2010. "Reflections on Carbon Leakage." Giannini Foundation. Working paper.
———. 2011. "The Environment and Trade: A Review." *Annual Review of Resource Economics,* forthcoming.

Regulatory Choice with
Pollution and Innovation

Charles D. Kolstad

4.1 Introduction

Probably the most fundamental issue in climate change is the role of innovation and invention in helping find a solution to the climate change problem. It is clear that many are depending on innovation to find cheaper ways to mitigate emissions and adapt to impacts. Governments around the world are trying to spur innovation. But nobody really knows how to efficiently induce innovation. No one knows what kinds of policies are effective in promoting the necessary amount of innovation. It is also unclear how the different approaches to regulating greenhouse gas emissions perform in inducing innovation. There is a sense that it is important to place a price on carbon, directly or indirectly, to send better signals to innovators. But how that carbon price translates into abatement-cost-reducing innovation is poorly understood.

In this chapter, we examine how environmental regulations work when there is an innovator with perfect property rights (perfect in the sense of a perfect patent with no spillovers). The innovator does not engage in pollution abatement but instead specializes in reducing the cost of pollution abatement, through innovation (which is then sold/licensed to polluters).

Charles D. Kolstad is professor of economics at the University of California, Santa Barbara, a university fellow of Resources for the Future, and a research associate of the National Bureau of Economic Research.

Research assistance from Valentin Shmidov is gratefully acknowledged. Comments from Don Fullerton, Sasha Golub, Rob Williams, Kerry Smith, Barry Nalebuff, Nat Keohane, and several anonymous referees have been appreciated. Research supported in part by the University of California Center for Energy and Environmental Economics (UCE³). For acknowledgments, sources of research support, and disclosure of the author's material financial relationships, if any, please see http://www.nber.org/chapters/c12152.ack.

The two questions we ask are: (a) do different types of environmental regulations perform differently in inducing innovation and abatement, and (b) do regulations differ in terms of how the gains from innovation are appropriated?

We develop a simple model, involving no uncertainty, in which we compare the performance of a cap-and-trade system (marketable permits) and an emissions tax system. Although other authors have examined this question, most authors use a highly simplistic representation of the innovation process. In this chapter, we focus more on the innovation process and less on other aspects of the economic environment.

As one might expect, given a lack of uncertainty, either regulatory policy is able to implement the first-best outcome. However, innovators clearly do better under a cap-and-trade system, capturing all of the rents from their innovation. Under a tax system, gains are split between the polluters and the innovators. Nevertheless, marginal conditions are such that efficiency is obtained.

4.2   Background

Innovation is at the core of dynamic economics. Hicks ([1932] 1966) put forward the idea that when relative prices of input factors shift, technical change will focus on saving the factor that has become relatively more expensive (the induced innovation hypothesis). One of the insights of the Solow model of growth is the so-called "Solow residual," which is the difference between growth in output and growth in input. It is attributable to technical change. This is a natural precursor to the more recent literature on endogenous growth (Romer 1994).

In the 1960s, a number of economists turned their attention to innovation, beginning with a seminal paper of Arrow (1962) and culminating in a host of papers including the classic papers by Scherer (1967) and Kamien and Schwartz (1968), the latter of which provides a theoretical model of induced innovation.

None of these papers deals with environmental externalities or regulation. That literature began to emerge in the 1970s, with a paper by Smith (1972). A common theme in the environmental literature is the comparative performance of different regulatory structures in terms of fostering innovation. Magat (1978) follows the common approach at that time of examining technical change through the lens of factor/output augmenting technical change (as did Kamien and Schwartz [1968]), within the context of optimal growth. He finds little difference between prices and quantities within this framework. Milliman and Prince (1989) compare a wide variety of environmental regulations (command-and-control, subsidies, taxes, free permits, auctioned permits) with a simple representation of regulation, cost-reducing innovation and diffusion, and then regulatory response to postinnovation costs. The focus is on who captures rents from innovation in a multiagent context, rather than on providing an explicit model of the innovation process. Fischer, Parry, and Pizer (2003) take this further by explicitly representing the process of innovation (making innovation endogenous). Abatement costs are $C(a, k)$ where a is the level of abatement and k is the level of technology, which results from research and development (R&D) at cost $F(k)$. The presence of the possibility of imitations of the innovated technology allows spillovers and thus diffusion to occur, which limits licensing fees. They do find differences among the different environmental regulations examined, though no clear regulatory approach dominates in terms of performance.

Denicolò (1999) focuses on innovation rather than diffusion and explicitly models the innovation process separately from the abatement process. He assumes the preinnovation emissions-output ratio is α (a constant) and the postinnovation ratio is β (a variable chosen by the innovator), with $\beta < \alpha$. The R&D cost of achieving that innovation is $C(\beta)$. The innovator licenses its innovation for a fee. With this simple structure of innovation, he shows that emission fees and marketable permits perform identically when the regulator moves first and commits to not change regulations postinnovation. When the regulator cannot so commit, the two instruments perform differently, though it is not possible to conclude that one regulatory approach dominates the other. Krysiak (2008) de-emphasizes the innovator as licensing a technology and focuses on how uncertainty might induce a preference for prices versus quantities, in the spirit of Weitzman's (1974) classic analysis. He concludes quantities are more efficient.

Scotchmer (2011) provides one of the most recent analyses of this issue, in the context of regulations for carbon emissions. Because of this, her model explicitly involves producing a good (energy) with an emissions-output ratio that can be reduced through innovation. Rather than focusing on the innovator's decision of how much innovation to undertake (with an explicit cost of innovation), she focuses on the returns to innovation from a specific reduction in the emissions-output ratio. She concludes that an emissions tax provides more innovation incentives than a cap-and-trade system.

The previous discussion concerns theoretical results on innovation. However, one of the key issues that has been of concern in the realm of climate policy and related empirical economics is how to empirically represent the extent of carbon-saving technological change (or, more generally, the rate of technical change for any factor). Although this literature is large, it is appropriate to mention two recent contributions by David Popp. Popp (2002) uses patent data to explicitly model the formation of the knowledge stock, using a perpetual inventory method (much as one would do using investment over time to estimate the capital stock). Using this approach he is able to disentangle the effect on energy consumption of prices as distinct from technological improvements. In Popp (2004) he carries this process further by modifying an optimal growth model commonly used for climate policy (Nordhaus' DICE model) to include endogenous technical change. One of the challenges is to represent private provision of R&D, acknowledging the inefficiencies of its provision, within a representation of the dynamics of economic activity and emissions.

4.3   A Model of Innovation and Abatement

We consider a situation with multiple atomistic firms in a polluting industry. Distinct from the polluting industry, there is one innovating firm, developing technologies to reduce the cost of abating in the polluting industry. The innovating firm conducts research, innovates, patents its innovation, and licenses the innovation to the polluting industry.

Our characterization of the polluting industry is straightforward. If the polluting industry chooses an aggregate amount of abatement a, then $C(a)$ is the cost of abatement incurred by firms in the industry (these costs are pre-innovation and exclude the costs of innovating). Furthermore, $B(a)$ is the environmental benefits from abatement, though those benefits do not accrue to the polluting industry. As is customary, C', C'', and $B' > 0$ and $B'' < 0$.

Our characterization of the innovating firm is also straightforward. Assume there is a firm that does not emit pollution but rather engages in innovation and licenses its abatement-cost-reducing innovations to the abating firms. The innovating firm undertakes R&D, which results in a technology that reduces the marginal cost of abatement. In particular, assume the innovator chooses the reduction in the marginal cost of abatement, σ. The cost of achieving this reduction in abatement costs is an R&D cost to the innovator of $R(\sigma)$, with $R' \geq 0$ and $R'' > 0$. Note that the unit of measurement for R is dollars whereas the unit of measurement for σ is dollars per ton (or dollars per unit of pollution abated). The R' is the change in R&D expenditures necessary to achieve a unit decrease in the marginal cost of abatement; R' thus maps dollars per ton into tons. Let the inverse of R' be given by the function S, which maps tons into dollars per ton. The innovating firm licenses its technology to the abating firms for a fee of φ per unit of abatement. This setup is shown in figure 4.1. The postinnovation social marginal cost of abatement is lower but the licensing fee offsets some or all of these cost reductions, from the perspective of the polluting firm. This model is similar to that of Denicolò (1999), though it differs in substantial ways, primarily in the representation of abatement and innovation.

The dynamics of this problem are as simple as possible: a three-period world. In the first period, the regulator acts, setting the level of the environmental regulation. In the second period, the R&D occurs and is licensed. In the third period, firms abate. This does not necessarily involve the actual passage of time but might be three stages to a single regulatory game.

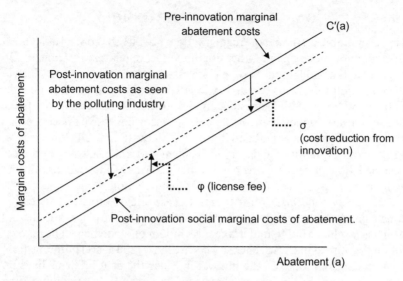

Fig. 4.1   The effect of innovation on marginal abatement costs

Superimposed on these market players is a regulator who is trying to maximize social welfare:

(1) $W(a, \sigma) = B(a) - [C(a) - \sigma a] - R(\sigma).$

Although it may seem like equation (1) is the obvious social welfare function, some ambiguity remains. Certainly the cost of R&D is a social cost. However, once the R&D is done, it becomes a sunk cost and abatement costs are forever lowered. Postinnovation, the regulator's objective is to balance $B(a)$ and $C(a)$-σa, without regard to the sunk cost $(R(\sigma))$. Recognizing this, a regulator may act in the first period to ignore $R(\sigma)$ in the social calculus. However, ignoring innovation involves viewing this problem through a different dynamic lens than is assumed here. In our simple structure, no further action occurs after innovation and abatement. This is equivalent to the regulator committing to not change the level of the regulation postinnovation.[1] It is clearly an interesting question as to what will prevail if a more realistic view of the dynamics of innovation is explored.

With exact control over abatement and innovation, the regulator can choose abatement and innovation to maximize welfare:

(2a) $B'(a^*) - C'(a^*) + \sigma^* = 0,$

and

1. The point about commitment to regulation and the distinction between the preinnovation and postinnovation period is clearly articulated by Denicolò (1999).

(2b) $$a^* - R'(\sigma^*) = 0 \qquad \Rightarrow \sigma^* = S(a^*).$$

However, we are assuming the regulator does not directly control abatement (a) and innovation (σ), but rather uses imperfect regulatory instruments. In particular, the regulator chooses a price instrument (t) or a quantity instrument (a). Polluting firms respond rationally and the innovating firm invests in the privately profit-maximizing amount of innovation and also sets the licensing fee (in dollars per ton abated), φ, accordingly. We are concerned about how much abatement and how much innovation result from an arbitrary price or quantity regulatory instrument and, further, when optimally designed, how these two instruments differ in terms of induced innovation, abatement, or distribution of rewards from innovation.

Quantity instruments. Consider first the case of a quantity instrument, $\hat{a}$, which mandates the amount of abatement that must take place. The abating firm has no choice but to undertake this amount of abatement. The innovator on the other hand, must choose a license fee, $\hat{\varphi}$, and a level of innovation, $\hat{\sigma}$, to maximize profits of the innovator. Since the abating firms have no ability to adjust the amount of abatement (it is mandated), the innovating firm can set the licensing fee to capture all of the rent, $\hat{\varphi} = \hat{\sigma}$. Profits for the innovating firms are then

(3) $$\Pi_I = \sigma a - R(\sigma),$$

which implies a resulting profit-maximizing level of innovation ($\hat{\sigma}$), as a function of the mandated abatement ($\hat{a}$), defined implicitly by the first-order conditions:

(4) $$d\Pi_I / d\sigma = \hat{a} - R'(\hat{\sigma}) = 0.$$

Equation (4) defines a condition for the amount of innovation that maximizes profit for the innovator: $\hat{\sigma}$ is set so that the marginal cost of reducing abatement costs is equal to the amount of abatement. By totally differentiating equation (4) one obtains an expression that shows how innovation changes as the abatement mandated increases:

(5) $$d\hat{\sigma} / d\hat{a} = 1/R''.$$

Because of curvature assumptions on R, this equation implies that as required abatement increases, the amount of innovation will also increase.

Price instruments. Now consider the more complex case of a price instrument. Compared to quantities, the price instrument sends a more indirect signal to both abaters and innovators. The regulator sets a price, t, for abatement (a payment for extra abatement is of course conceptually equivalent to charging a fee for unabated pollution). Profits for the polluting industry are given by

(6) $$\Pi_P = ta - C(a) + (\sigma - \varphi)a.$$

Profit maximization implicitly defines the abatement level, $\tilde{a}$, in response to a price $\tilde{t}$:

(7) $\tilde{a}: d\Pi_I / da = \tilde{t} - C'(\tilde{a}) + (\sigma - \varphi) = 0 \Leftrightarrow \tilde{t} = C'(\tilde{a}) - (\sigma - \varphi)$.

We now turn to the innovator's behavior. First, we totally differentiate equation (7), keeping t constant to determine how changes in σ and φ influence $\tilde{a}$:

(8a) $$0 = C''d\tilde{a} - d\sigma + d\varphi \Rightarrow,$$

(8b) $$d\tilde{a} / d\sigma = 1/C'',$$

and

(8c) $$d\tilde{a} / d\varphi = -1/C''.$$

The innovator's profit is

(9) $$\Pi_I = \varphi\tilde{a} - R(\sigma).$$

The innovator must choose both σ and φ to maximize profits in equation (9), resulting in first order conditions

(10a) $$\partial\Pi_I / \partial\sigma = \hat{\varphi}d\tilde{a} / d\sigma - R'(\tilde{\sigma}) = 0,$$

and

(10b) $$\partial\Pi_I / \partial\varphi = \hat{\varphi}d\tilde{a} / d\varphi + \tilde{a} = 0,$$

which implicitly define $\tilde{\sigma}$ and $\hat{\varphi}$ as functions of $\tilde{a}$ which in turn depends on $\tilde{t}$:

(11a) $$\tilde{\sigma}: R'(\tilde{\sigma}) = \tilde{a} \Rightarrow \tilde{\sigma} = S(\tilde{a}),$$

(11b) $$\hat{\varphi}: \hat{\varphi} = \tilde{a}C''(\tilde{a}).$$

In essence, the three equations, (7), (11a), and (11b) implicitly define $\tilde{a}$, $\tilde{\sigma}$ and $\hat{\varphi}$, as functions of t.

Socially optimal instruments. First-best levels of abatement (a^*) and innovation (σ^*) are defined by equation (2). If a quantity regulation is set such that $\hat{a} = a^*$, then innovation, $\hat{\sigma}$, will be set according to equation (4). Thus $\hat{\sigma} = \sigma^*$. A price regulation must be set (if possible) so that the same outcome prevails. In particular, set $\tilde{t}$ according to:

(12) $$\tilde{t} = C'(a^*) - S(a^*) + a^*C''(a^*).$$

It is easy to see that $\tilde{a} = a^*$ and $\tilde{\sigma} = \sigma^*$ satisfy equations (7) and (11), and thus a first-best outcome is supported by this level of the price instrument.

Equation (12) is intuitive, if somewhat more complicated than for the optimal quantity instrument. At an efficient level of abatement, a^*, and an efficient level of innovation, σ^*, the marginal costs will be reduced by $S(a^*)$ but then the license fee will increase the marginal cost seen by polluters by $a^* C''(a^*)$. This results in marginal costs equal to the right-hand side of equation (12). Setting the price instrument equal to that marginal cost, evaluated at a^*, supports the first-best outcome. Note that the optimal price instrument will be less than would prevail absent innovation. Similarly, the optimal quantity instrument will be more than would prevail absent innovation (since absent innovation, the σ^* would be missing from equation [2]).

This leads to the following result:

PROPOSITION 1. *Given the earlier structure and assumptions, price and quantity instruments are equivalent in implementing the first-best amount of abatement and innovation.*

Note, however, that the private return to the innovator from innovation differs for the two instruments. For the quantity instrument, all returns to innovation are captured by the innovator (the licensing fee is equal to the cost reduction from the innovation). In the case of the price instrument, only part of the marginal gains are captured by the innovator. As the licensing fee is raised from zero, direct revenue from the license obviously increases. However, an increased licensing fee increases the cost of abatement to the polluter and thus reduces abatement (see equation [8c]) and thus, indirectly, revenue to the innovator. So a tradeoff between raising the fee and lowering the fee implies there is some happy medium with the license fee strictly greater than 0 but strictly less than σ. Thus the polluter captures some of the gain from innovation in the form of reduced costs and the innovator also captures some of the gain. Of course, who captures the gain does not matter to efficiency in this case since the marginal conditions are such that innovation is the same with the two regulatory instruments.

4.4   Conclusion

Innovation is clearly a core issue for modern environmental regulation. Climate change is a case in point. Significantly regulating greenhouse gas emissions will be expensive, and innovation is the primary way of reducing costs (after regulatory efficiency gains have been exhausted). In fact, due to the long lag times of turning emissions reductions into temperature reductions, one of the primary reasons for implementing carbon regulation now is to spur innovation on reducing abatement costs in the future (when we get really serious about emissions). Thus the question of which environmental regulations tend to spur the most innovation is highly relevant.

A related question is how to represent the process of innovation, which is not well understood empirically. A better empirical understanding will help design better policies to encourage innovation and abatement.

This chapter provides a small step forward in terms of representing the process of innovation on abatement costs, though there is a considerable literature on this issue. One conclusion is that price instruments (e.g., a carbon tax) can be designed to induce the same amount of innovation and abatement as a quantity instrument (e.g., cap and trade). Although the two instruments can provide the same marginal incentives to innovators and abaters, the inframarginal rents from innovation differ in the two cases. In fact, the innovators appropriate all of the gains from innovation in the case of a quantity instrument, whereas innovators and abaters share the rents in the case of a price instrument.

The results reported here are suggestive more than definitive. In particular, most types of regulation lead to a first-best level of innovation, though different levels of rents to the innovators. What are the implications of this? Can a more realistic representation, perhaps with some uncertainty, lead to sharper distinctions between the two regulatory approaches? This chapter raises these issues but does not come close to resolving them.

References

Arrow, Kenneth J. 1962. "Economic Welfare and the Allocation of Resources for Invention." In *The Rate and Direction of Inventive Activity: Economic and Social Factors,* edited by R. Nelson. Princeton, NJ: Princeton University Press.

Denicolò, Vincenzo. 1999. "Pollution-Reducing Innovations under Taxes or Permits." *Oxford Economic Papers* 51:184–99.

Downing, P. B., and L. J. White. 1986. "Innovation in Pollution Control." *Journal of Environmental Economics and Management* 13:18–25.

Fischer, Carolyn, Ian W. H. Parry, and William A. Pizer. 2003. "Instrument Choice for Environmental Protection When Technological Innovation is Endogenous." *Journal of Environmental Economics and Management* 45:523–45.

Hicks, John R. [1932] 1966. *The Theory of Wages.* London: Macmillan.

Kamien, Mort, and Nancy Schwartz. 1968. "Optimal Induced Technical Change." *Econometrica* 36:1–17.

Krysiak, Frank C. 2008. "Prices vs. Quantities: The Effects on Technology Choice." *Journal of Public Economics* 92:1275–87.

Magat, Wesley A. 1978. "Pollution Control and Technological Advance: A Dynamic Model of the Firm." *Journal of Environmental Economics and Management* 5: 1–25.

Milliman, Scott, and Raymond Prince. 1989. "Firm Incentives to Promote Technological Change in Pollution Control." *Journal of Environmental Economics and Management* 17:247–65.

Montero, Juan-Pablo. 2002. "Permits, Standards, and Technology Innovation." *Journal of Environmental Economics and Management* 44:23–44.

Popp, David. 2002. "Induced Innovation and Energy Prices." *American Economic Review* 92:160–80.

———. 2004. "ENTICE: Endogenous Technological Change in the DICE Model of Global Warming." *Journal of Environmental Economics and Management* 48: 742–68.

Requate, Till. 2005. "Dynamic Incentives by Environmental Policy Instruments—A Survey." *Ecological Economics* 54:175–95.

Romer, Paul M. 1994. "The Origins of Endogenous Growth." *Journal of Economic Perspectives* 8 (1): 3–22.

Scherer, F. M. 1967. "Research and Development Resource Allocation under Rivalry." *Quarterly Journal of Economics* 81:359–94.

Scotchmer, Suzanne. 2011. "Cap-and-Trade, Emissions Taxes, and Innovation." *Innovation Policy and the Economy* 11:29–54.

Smith, V. Kerry. 1972. "The Implications of Common Property Resources for Technical Change." *European Economic Review* 3:469–79.

Weitzman, Martin B. 1974. "Prices vs. Quantities." *Review of Economic Studies* 61:477–91.

Comment   V. Kerry Smith

If we could rely on technological innovation to dramatically reduce the costs of mitigating greenhouse gases, then climate policy would be easy. All the analyses of the design and impacts of climate policy can agree with this point. Nordhaus (2008), for example, finds the present value of abatement costs would be about one-fourth that of his optimal approach if we could assume a low-cost backstop technology was available to replace fossil fuels when carbon's price reached five dollar a ton (in 2005 dollars). The Stern (2006) report makes exceptionally optimistic assumptions about technological advance, assuming abatement costs will decline by sixfold by 2050. Thus, the focus of Charles Kolstad's chapter is especially important. He notes that serious theoretical analysis to understand the effects of different climate policies on technical change needs to "unpack" the internal structure of the innovation process. He examines the interactions between three parties—the regulator, the firm facing environmental regulation and needing to control its emissions, and the firm offering new abatement technologies to reduce incremental abatement costs. In a stylized model that abstracts from uncertainty and the effects of regulatory policy in output markets, he finds that price and quantity instruments for regulating pollution can be made equivalent in terms of realizing the first-best (efficient) amount of abatement and innovation. However, the total return to innovation is not the same for these two instruments and the distribution of returns between the polluting firm and the innovating firm is also different. Innovators appropriate all the

V. Kerry Smith is the Regents' Professor, W. P. Carey Professor of Economics, and Distinguished Sustainability Scientist at the Global Institute of Sustainability at Arizona State University; a university fellow at Resources for the Future; and a research associate of the National Bureau of Economic Research.

For acknowledgments, sources of research support, and disclosure of the author's material financial relationships, if any, please see http://www.nber.org/chapters/c12153.ack.

gains from innovation with a quantity standard and share the gains with a price instrument.

Kolstad's recognition of the importance of the internal process of innovative activities is to be applauded. As with all good research, it helps to answer some questions and frames new ones. I consider three questions here: Does past experience with other pollution control policy suggest we should be optimistic about technical change reducing abatement costs? Is Kolstad's single pollutant focus limiting? and What lessons from other analyses of innovation are relevant for climate policy?

Past Experience

The signature example of an incentive-based environmental policy is the SO_2 permit trading program. Glowing accounts of its success can be found in the middle and late nineties.[1] It is difficult to disentangle all factors contributing to abatement costs and assess how much new innovations reduced control costs. Nonetheless, a simple comparison suggests that modest, not dramatic, unanticipated cost savings seems to be the most plausible conclusion that can be drawn from experience with this program. To arrive at this conclusion I extracted estimates of the long-run incremental costs of controlling SO_2 that were expected for 2010 from Burtraw's (1998) summary of what was known before the program was implemented. He suggested that the ICF (1990) study probably offered the best picture of expectations prior to the implementation of the SO_2 trading program because it included a detailed characterization of the ultimate design for the rule. This study estimated long-run marginal costs of controlling SO_2 in 2010 would be $579 to $760 per ton (in 1995 dollars). Using the consumer price index to convert these estimates to 2009 dollars, they are between $820 and $1,077 per ton.

The best estimates for the incremental control costs today would seem to be the spot price of SO_2 permits. Figure 4C.1, panel A reproduces a chart from Cantor Fitzgerald's trading records for the period 2003 to 2009, the last full year before the permits stopped trading in May 2010 (see figure

1. It is easy to get carried away with promises that technology will eliminate costs of pollution abatement. For example, Carol Browner noted when she was EPA administrator and was discussing the SO_2 program as EPA administrator that, "During the 1990 debate on the acid rain program, industry initially projected the cost of gas emission allowance to be $1,500 per ton of sulfur dioxide...Today these are selling for less than $100" (March 10, 1997). Unfortunately, today the story is very different. The US market for SO_2 emissions to control acid rain, for example, has had no trades since the Spring of 2010. One explanation is the uncertainty caused by court rulings and the development of recently finalized regulations to address SO_2 emissions that affect downwind ambient air quality. What is at issue is uncertainty over exactly how an SO_2 permit can be used. Recorded permit prices are effectively zero. Reestablishing the SO_2 permit market using new permits will require resolution of these sources of uncertainty. The recently finalized regulations attempt to do so by establishing a new, tighter, cap on SO_2 emissions from most, but not all, of the facilities subject to the acid rain SO_2 program. The prices of the new SO_2 permits under this program are expected to be positive, but those facilities not subject to the new program may now choose to emit more SO_2 given that the acid rain cap is no longer binding.

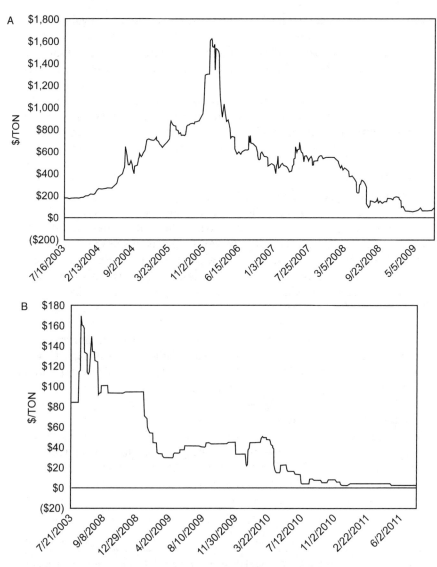

Fig. 4C.1   Spot prices for SO₂ permits: *A,* **To May 2009;** *B,* **A year later to May 2010**
Source: Cantor Fitzgerald website (accessed 5/21/2010).

4C.1, panel B). These results are in the dollars of the year of the exchange. There is not a smooth pattern. Spot prices for sulfur permits are, like other prices, influenced by a number of factors including expectations for what is to happen with other environmental policies.

Using a midrange of spot prices in the two years prior to the economic downturn and converting them to 2009 dollars yields about $635 a ton or a 23 percent decline from the low end of the expectations for incremental costs estimated in 1990. This would be about a 1.3 percent decline each year over these twenty years, certainly not negligible gains from unrecognized technologies, but also far short of the pace needed for a sixfold decline in costs in forty-five years, as assumed in the Stern report.

Pollutants Do Not Go Away

One of the earliest papers arguing that environmental externalities were pervasive, by Ayres and Kneese (1969), also emphasized the importance of an explicit recognition of materials and energy balances in modeling production and consumption. They argued we can change the form of pollution and where it is dispersed but the materials and heat comprising residuals do not go away. In the end we must decide where they are to go. Kolstad's model focuses on reducing abatement costs without dealing with the disposition of what is abated. This issue never comes up in his model and I believe in further refinements it should. Innovation may create a new problem. For example, suppose we are able to reduce the costs of controlling airborne emissions by passing them through a water mist instead of using a mechanical device. This innovation would create a watery sludge that captures the particles that would otherwise have been captured mechanically and removed as solid waste.

This point is important for several reasons. Changes in the regulations on a different pollutant—NO_x—could influence the costs of controlling CO_2. As Burtraw and Szambelan (2009) suggest, the interconnections between pollutants, reflecting Ayres and Kneese's warning to be sure economic analysis recognized the "physical realities" of production (and consumption), can be responsible for links between the markets for different tradeable pollution permits. A tangible example of such links, in the context of expectations about regulations, can be found in figure 4C.1, panel A. Court decisions about the implementation process for the Environmental Protection Agency (EPA)'s Clean Air Act Interstate Rule (CAIR) to control particulate matter and NO_x caused dramatic moves in the SO_2 allowance price (the spike in December 2005 in figure 4C.1, panel A) and ultimately the drop of prices to zero in 2010 (see figure 4C.1, panel B). In the United States, as of this writing (September 2011) macroeconomic policy is likely to be the major short-term source of uncertainty that can affect other environmental policies. In the future with economic recovery it will be climate policy. Such regulatory uncertainty has implications not only for controlling greenhouse gases but potentially for the consistency of the signals that permit prices provide for innovations in the abatement of all pollutants.

Lessons

At least two lessons from recent literature should be noted. First, Vernon Ruttan's (2006) last book argued that revolutionary departures from existing technological trajectories require new institutions. He was not optimistic that civil institutions could assemble the resources and create incentives that would lead to dramatic breakthroughs. Part of the reason for his relative pessimism was the inability of free societies to structure institutions that make these sustained commitments. In Kolstad's model commitments are given in the first stage of his regulatory game and remain consistent in his model. In the real world they change and may not be sufficiently consistent.

A more recent overview of innovation in different sectors reported in the Henderson and Newell (2011) volume suggests competition and not government is a more useful guide for innovation policy. This conclusion may well be right for private goods and services—but what about nonmarket services? Markets do not provide signals for them. The prices from permits are a start but we have seen how they can be undone by a court ruling and regulatory "fixes"—thus we must conclude that there are few areas where such markets are well established and reliable as a source of long-term signals. How do we avoid serious mistakes—creating with technological "cures" a set of problems that are worse than where we started? It is only by unpacking the details of the innovation process, as Kolstad has started, that we can hope to answer these questions.

References

Ayres, Robert U., and Allen V. Kneese. 1969. "Production, Consumption, and Externalities." *American Economic Review* 59:282–97.

Burtraw, Dallas. 1998. "Cost Savings, Market Performance, and Economic Benefits of the U.S. Acid Rain Program." Discussion paper no. RFF DP 98-28-REV. Washington, DC: Resources for the Future.

Burtraw, Dallas, and Sarah Jo Szambelan. 2009. "U.S. Emissions Trading Markets for SO_2 and NO_x." Discussion paper no. RFF DP 09-40. Washington, DC: Resources for the Future.

Henderson, Rebecca M., and Richard G. Newell, eds. 2011. *Accelerating Energy Innovation: Insights from Multiple Sectors.* Chicago: University of Chicago Press.

ICF Resources Incorporated (ICF). 1990. *Comparison of the Economic Impacts of the Acid Rain Provisions of the Senate Bill (S. 1630) and the House Bill (S. 1630).* Washington, DC: US Environmental Protection Agency.

Nordhaus, William. 2008. *A Question of Balance.* New Haven, CT: Yale University Press.

Ruttan, Vernon W. 2006. *Is War Necessary for Economic Growth.* Oxford: Oxford University Press.

Stern, Nicholas. 2006. *Stern Review on the Economics of Climate Change.* London: Her Majesty's Treasury.

Spillovers from Climate Policy to Other Pollutants

Stephen P. Holland

5.1 Introduction

Spillovers from climate policy (also known as ancillary benefits or ancillary costs) have important implications for policy design, modeling, and benefit-cost analysis. Spillovers arise since climate policy could lead, for example, to a reduction in particulate matter (PM) emissions as well as CO_2 emissions. In this case, the ancillary benefits of reduced PM emissions from the policy should be included in a benefit-cost analysis and may well lead the benefit-cost analysis to recommend more stringent climate policies. Unfortunately, spillovers can be either positive or negative since firms change production processes in response to climate policies, and these changes may lead either to an increase or decrease in emissions of other pollutants. After presenting a theoretical description of spillovers from climate policy, this chapter empirically tests for and decomposes climate policy spillovers in electric power generation.

Climate policy spillovers have received attention in the estimation of health benefits from reduced air pollution. This extensive literature, which is recently surveyed in Bell et al. (2008), varies considerably in its sophistication with regard to air quality modeling and the responses of polluters to climate

Stephen P. Holland is associate professor of economics at the University of North Carolina at Greensboro and a research associate of the National Bureau of Economic Research.

Special thanks to Severin Borenstein, James Bushnell, Chris Ruhm, and Catherine Wolfram for helpful discussions. Thanks to the University of California Energy Institute (UCEI) for generous research support during this project. Thanks also to seminar participants at the University of North Carolina at Greensboro and at UCEI and to Michael Mills for valuable research assistance. For acknowledgments, sources of research support, and disclosure of the author's material financial relationships, if any, please see http://www.nber.org/chapters/c12148.ack.

policy.[1] For example, Cifuentes et al. (2001) simply assumes climate policy uniformly reduces pollution across all spatial areas. Other studies use much more sophisticated air quality modeling to estimate the effects of emissions reductions. Bell and colleagues conclude that although the various studies are difficult to compare, the results provide "strong evidence" that the short-term ancillary benefits to public health of climate policy are "substantial."

Burtraw et al. (2003) focus on the responses to climate policy of electric power generators.[2] Using a sophisticated simulation model of electricity supply, the authors show that a carbon tax would have ancillary health benefits from reduced NO_x emissions of about eight dollars per metric ton of carbon. Since emissions of SO_2 are capped, they note that there are no ancillary health benefits from SO_2 emissions, but they estimate additional benefits from avoided future investment in emissions control equipment. Groosman, Muller, and O'Neill (2009) estimate similar effects with a sophisticated model of pollutant transport.[3]

Ancillary benefits from climate policy have also been studied in agriculture and forestry where climate policy could benefit soil quality, wildlife habitat, water quality, and landscape aesthetics.[4] Finally, ancillary benefits have been estimated to be substantial in developing countries where regulation of pollutants may be less stringent.[5]

5.2   The Theory of Spillovers from Climate Policy

Emissions are generally modeled using one of three equivalent approaches: as an input in the production process, as a joint product that is a "bad," or as abatement from some hypothetical level, for example, business as usual. The first approach has a number of advantages for modeling spillovers from climate policy since it is readily adaptable to modeling multiple pollutants and allows for a broad range of substitution possibilities. Moreover, it allows a simple way to model climate policies, for example, a carbon tax or cap and trade, as an increase in the price of CO_2 emissions (from a zero price).

In this framework, climate policy spillovers are shifts in input demands in response to an increase in the price of CO_2. Theory shows that input

1. See also European Environment Agency (2004).
2. Ancillary benefits have also been studied in transportation; see Walsh (2008) and Mazzi and Dowlatabadi (2007).
3. The more conservative estimates in Groosman and colleagues recognize that emissions of SO_2 are capped.
4. See Feng, Kling, and Gassman (2004), Plantinga and Wu (2003), and Pattanayak et al. (2002). Elbakidze and McCarl (2004) point out that ancillary benefits must be skeptically considered with agricultural offsets since offset emissions reductions from other sectors might also have ancillary benefits.
5. See Dudek, Golub, and Strukova (2003) for analysis of ancillary benefits in Russia; Dessus and O'Connor (2003) for analysis of Chile; and Joh et al. (2003) for analysis of Korea. Greenstone (2003) studies pollutant spillovers across media and finds little evidence that the CAAAs increase releases into waterways and the ground.

demand may either increase or decrease, depending on whether the input is a substitute or a complement to CO_2. Additionally, the effects of climate policy can be decomposed into two effects: an output effect, which generally decreases the demand for all inputs, and a substitution effect, which depends on whether the inputs are net substitutes or net complements for CO_2.[6] Importantly, demand for pollution inputs that are net substitutes can still fall with climate policy if the output effect outweighs the substitution effect.[7]

To illustrate these principles, consider electricity generation that leads to emissions of SO_2 and NO_x, as well as CO_2. Suppose climate policy caused dual fuel generating units to switch from fuel oil to natural gas. Since natural gas generally has lower sulphur content than fuel oil, SO_2 and CO_2 would be net complements: for a given amount of electricity emissions of SO_2 would be lower in response to climate policy. Since the output effect also serves to reduce SO_2 emissions, climate policy would have ancillary benefits from SO_2. Now suppose that climate policy caused natural gas-fired generating units to increase their combustion temperature, which reduces CO_2 emissions but increases NO_x emissions. In this case CO_2 and NO_x would be net substitutes. Note however, that since the output effect leads to a reduction in NO_x emissions, the overall effect may still be a reduction in NO_x emissions from climate policy if the output effect is stronger than the substitution effect. Thus, climate policy could have ancillary benefits or ancillary costs.

Spillovers are illustrated in figure 5.1 for the case of electricity production with emissions of CO_2 and NO_x. The first panel of figure 5.1 shows the input demand for CO_2. If marginal productivity is decreasing (the usual case) then the input demand (equivalently the value of the marginal product) is downward sloping. The firm would increase use of an input if the value of the marginal product were greater than the input cost. Thus at the optimum, the value of the marginal product equals the input cost. In the unregulated equilibrium, this marginal product would be zero and CO_2 emissions would be $e^0_{CO_2}$. If climate policy increases the price of CO_2 emissions to t_{CO_2}, for example through a carbon cap or tax, then CO_2 emissions would fall to $e^1_{CO_2}$.

Panel B of figure 5.1 illustrates the spillovers to NO_x emissions from climate policy. In the absence of climate policy, the NO_x input demand is illustrated by the downward sloping solid line and NO_x emissions are $e^0_{NO_x}$. The response of NO_x emissions to climate policy depends on two factors: (a) whether NO_x and CO_2 are substitutes or complements, and (b) regulations on NO_x emissions. In general NO_x and CO_2 can be substitutes or

6. These effects are equivalent to income and substitution effects from demand theory. Deschênes (chapter 2, this volume) develops an equivalent framework for labor demand with scale and substitution effects.

7. Decomposing responses into output and substitution effects is also useful since output effects may not be effective for reducing emissions if regulations are incomplete or firms have market power. See Holland (2012) for further discussion of output effects with incomplete regulation.

Panel A.   Demand for CO_2 emissions.

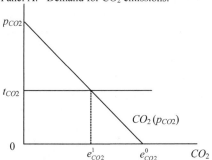

Panel B.   Demand for NO_x emissions.

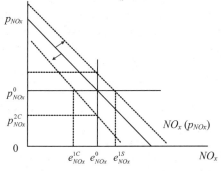

Panel C.   Supply and demand for electricity.

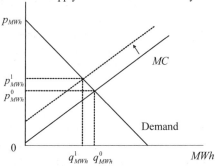

Fig. 5.1   Graphical model of spillovers from climate policy

complements. If NO_x and CO_2 are complements, then climate policy leads to an inward shift in the input demand for NO_x, that is, decreases the demand for NO_x emissions. On the other hand, if NO_x and CO_2 are substitutes, then climate policy increases the demand for NO_x emissions.

Whether or not climate policy changes, NO_x emissions depends crucially on the environmental regulation of the NO_x emissions. Two polar cases illustrate the effects: cap and trade in NO_x versus a NO_x tax. If NO_x is subject to an emissions cap (as in RECLAIM or in the NO_x Budget Program), then climate policy does not change NO_x emissions but changes the price of permits in the NO_x market. For example, if NO_x and CO_2 are complements, then climate policy decreases demand for NO_x emissions. Since emissions are capped, NO_x emissions remain at $e^0_{NO_x}$ and there are no spillover benefits, but the NO_x price falls from $p^0_{NO_x}$ to $p^{2C}_{NO_x}$.[8]

On the other hand, if NO_x emissions are subject to price regulation, then NO_x emissions change in response to climate policy. For example, if NO_x and CO_2 are complements, then climate policy would decrease demand for NO_x emissions and emissions would decrease from $e^0_{NO_x}$ to $e^{1C}_{NO_x}$. Alternatively, if NO_x and CO_2 are substitutes, then climate policy would *increase* NO_x emissions from $e^0_{NO_x}$ to $e^{1S}_{NO_x}$.

Panel C of figure 5.1 shows the effect of climate policy in the electricity market. Since climate policy increases the marginal cost of electricity production, the equilibrium price of electricity will rise from p^0_{MWh} to p^1_{MWh} and the equilibrium production will fall from q^0_{MWh} to q^1_{MWh}. This output effect will serve to reduce emissions of both CO_2 and NO_x. Note that the output effect makes it unlikely that NO_x and CO_2 would be gross substitutes since the substitution effect (which increases NO_x emissions) would need to outweigh the output effect (which decreases NO_x emissions).

Holland (2010) illustrates the proper valuation of climate policy spillovers for benefit-cost analysis. Two results are noteworthy. First, spillovers can affect the optimal carbon price. In particular, if there are ancillary benefits, then the optimal carbon price would be set *higher* than the marginal damages. Second, spillovers should be included in benefit-cost analysis just as other benefits or costs are included. In fact, from a theoretical standpoint, spillovers are indistinguishable from changes in any other input, such as labor. However, care must be taken to evaluate environmental spillovers according to their damages since market prices are not available.

Holland (2010) also extends the theoretical analysis in this section by deriving theoretical predictions. In particular, both the input demand and conditional input demand must be decreasing in the own price and output effects must be negative. These predictions will aid in the identification of empirical models.

5.3   Estimation Strategy

Spillovers resulting from responses to climate policy cannot be directly estimated in industries that are not yet subject to climate policy. Moreover, in industries currently subject to climate policy, it would be difficult to dis-

---

8. Burtraw et al. (2003) note that the falling NO_x price may have benefits from avoided future control equipment.

entangle the effects of climate policy from the effects of other environmental regulations.

To overcome these difficulties, I exploit the symmetry of input substitution and estimate the response of CO_2 emissions to the change in the price of NO_x emissions.[9] This has two advantages. First, NO_x emissions have been regulated extensively so it is possible to design an estimation strategy with variation in NO_x regulations. Second, CO_2 was not regulated, so there is no need to disentangle the effects of the NO_x regulation from CO_2 regulation. To proxy for changes in NO_x prices, I use changes in attainment status under the Clean Air Act Amendments (CAAAs). Regions that fail to achieve an ambient air quality standard are deemed to be in nonattainment. Designation as nonattainment under the CAAAs triggers additional regulations, which vary according to each state's implementation plan (SIP).[10] In this study, attainment status for one-hour ozone proxies for the price of NO_x, which is a primary ozone precursor. Since California had multiple changes into and out of attainment, the analysis focuses on California power plants.

The estimation strategy uses a fixed effects estimator. The basic estimating equation is:

$$(1) \qquad \ln(Emiss_{it}) = \beta Nonattain_{it} + f_i + g_i t + v_{jt} + \varepsilon_{it},$$

where $Emiss_{it}$ is emissions (of NO_x, CO_2, or SO_2) from generating unit i at time t; $Nonattain_{it}$ is a dummy variable indicating that unit i is in nonattainment for one-hour ozone at time t; f_i is a unit-specific fixed effect; $g_i t$ is a unit-specific linear trend; v_{jt} is a market-year-month fixed effect for market j; and ε_{it} is the error term. To correct for possible serial correlation, the error term, ε_{it}, is clustered at the generating unit.

The parameter of interest, β, indicates the response of emissions to a change in attainment status. Since the nonattainment dummy is a proxy for an increase in the price of NO_x emissions, the estimated coefficient captures the own price effect when NO_x emissions is the dependent variable. With CO_2 emissions as the dependent variable, the estimated coefficient captures the spillover. A positive (negative) coefficient indicates that NO_x and CO_2 are gross substitutes (complements). The own and spillover conditional (net) effects can be estimated by controlling for output in equation (1), and the output effect can be estimated directly when output is the dependent variable.[11]

9. Exploiting symmetry requires care since it only holds for marginal changes. See Holland (2010) appendix 2 for details on the symmetry of input substitution.

10. For detailed descriptions of the regulatory effects of nonattainment designation under the CAAAs, see Greenstone (2002).

11. By estimating $\ln(Emiss_{it}) = \beta^c Nonattain_{it} + \beta^{MWh} \ln(MWh_{it}) + f_i + g_i t + v_{jt} + \varepsilon_{it}$ and $\ln(MWh_{it}) = \beta' Nonattain_{it} + f_i + g_i t + v_{jt} + \varepsilon_{it}$ in addition to equation (1), all four derivatives in the Slutsky equation in Holland (2010) appendix 2 are estimated separately. However, the identity $\beta = \beta^{MWh} \beta' + \beta^c$ holds since the sample and all conditioning variables are identical.

Most of the potentially confounding variation is controlled for by the fixed effects. The unit-specific fixed effects capture any differences in emissions across units due to fuel-mix, generation technology, generator capacity, installed emissions control equipment, or any other time-invariant characteristics of the generating units. The unit-specific linear trends capture any trends at the unit level, for example, phasing out of old units. The market-year-month fixed effects are a vector of indicators for each month of each year for each market; for example, one indicator is for January 1999 for the northern California market (NP15) and another indicator is for January 1999 for the southern market. The market-year-month fixed effects capture all variation over time such as seasonal effects and changes in relative fuel prices, in labor costs, in capital costs, and in regulations affecting all generators as well as differences across the markets. This flexible set of fixed effects captures most of the potentially confounding effects.

Given this extensive set of nonparametric controls, model identification is based on variation in the attainment status of generating units over time in the sample. Intuitively, the generating units with unchanged attainment status would serve as controls for the generators with changed attainment status (the treated group).[12] The estimated effect would be biased if there were unobserved differential trends in emissions that were correlated with the change in attainment status. This threat to identification is addressed in two ways. First, the multiple changes into and out of attainment in California diminish the potential for bias from unobserved trends. Second, the model incorporates unit-specific linear trends to control for any unit-specific trends, which would not be captured by the market-year-month fixed effects.

The estimated spillover effect could also be incorrectly identified if regulatory authorities used the additional statutory authority to attempt to reduce emissions of other pollutants. In this case, changes in attainment status would indicate variations in the prices of both NO_x emissions and other pollutants, and the estimated effect would combine the direct and spillover effects. This potential confounding is limited by analyzing spillovers on CO_2 emissions. During the sample period, there was still substantive disagreement over whether CO_2 was a harmful pollutant and CO_2 was neither listed nor regulated by the Environmental Protection Agency (EPA) as a criteria pollutant. This lack of regulatory attention to CO_2 emissions suggests that the nonattainment indicator is not a proxy for an increase in the price of CO_2 emissions and that the spillover effect is properly identified.[13]

Identification is supported further by the testable predictions from theory. In particular, Holland (2010) shows that own price effects are nonpositive

---

12. With change at one time in attainment status, the estimator would be similar to the well-known difference-in-differences estimator.
13. This argument does not hold for SO_2 emissions.

for both factor demands and conditional factor demands and that output effects are nonpositive. A nonpositive estimate of β in equation (1) with NO_x emissions as the dependent variable is consistent with the theoretical predictions. With CO_2 emissions as the dependent variable, there are no additional testable implications since cross price effects can be either negative or positive.

5.4   Data

This analysis requires data on emissions, generation, attainment status, and other regulations. Availability of the emissions data limit the sample to the years 1997 to 2004. Emissions data come from the hourly US EPA continuous emissions monitoring systems (CEMS) for power plants. The data are very accurate, include all fossil fuel-fired generators meeting certain requirements, and have been used in a number of studies.[14] The hourly generating unit-level data are aggregated to the month for three reasons. First, a number of units report emissions in hours for which they report no output. Aggregation accurately captures emissions and output while incorporating any start-up emissions from generating units. Second, if regulations caused a unit to be run fewer hours, disaggregated data would not capture this reduction with the proportional (log) estimating equations. Aggregation captures the zero production hours. Finally, the data are highly serially correlated. Aggregation reduces the problem of serial correlation.

Since California had the most variation in attainment status, the primary analysis focuses on California. Of the twelve counties in California with changes in attainment status, only three counties have relevant power plants: Contra Costa, San Francisco, and San Diego. After dropping nonreports and data inconsistencies, the model identification is based on changes in attainment status at 29 of 178 generating units. The data are discussed further in Holland (2010).

5.5   Estimation Results

The results from estimating equation (1) are presented in table 5.1. Each column reports the results from one of seven regressions. Column (1) reports estimates where $\ln(NO_x)$ is the dependent variable, that is, the NO_x factor demand, and columns (3) and (5) capture the factor demands for CO_2 and SO_2. Columns (2), (4), and (6) estimate the conditional factor demands since they control for output, that is, $\ln(MWh)$. Column (7) reports estimates from regressing output on the same set of controls. Throughout, the unit fixed

14. For example, see Puller (2007), and Holland and Mansur (2008).

Table 5.1   **Main results: California results for NO_x, CO_2, and SO_2 emissions and megawatt hours**

	$\ln(NO_x)$		$\ln(CO_2)$		$\ln(SO_2)$		$\ln(MWh)$
	(1)	(2)	(3)	(4)	(5)	(6)	(7)
Nonattain	−0.516**	−0.221*	−0.326*	0.003	−0.371**	−0.037	−0.365*
	(0.203)	(0.131)	(0.190)	(0.030)	(0.170)	(0.132)	(0.200)
$\ln(MWh)$		0.809**		0.900**		0.897**	
		(0.016)		(0.010)		(0.013)	

Notes: There are 8,239 monthly observations for 178 generating units. (8,188 observations for the SO_2 regressions.) Dependent variable is log of emissions or log of MWh of generation. Regressions additionally control for market-year-month fixed effects, generating unit fixed effects, and generating unit linear trends. Standard errors clustered at the generating unit. Controls for other regulations (CO, NO_2, and eight-hour ozone nonattainment and ARP NO_x Early Election) are not jointly significant in six of the seven regressions.
**Significant at the 5 percent level.
*Significant at the 10 percent level.

effects, unit-specific linear trends, and market-year-month fixed effects are highly significant but are not reported.

The estimates of the three testable implications, in columns (1), (2), and (7), are all negative. Thus, the regression results are consistent with the theoretical predictions. Moreover these results show that approximately half of the estimated 40 percent reduction in NO_x emissions can be attributed to substitution effects with the remainder being attributable to output effects.

The pollutant spillover effects are reported in columns (3) through (6). For CO_2, the point estimate indicates that nonattainment designation reduced CO_2 emissions by 30 percent, suggesting gross complementarity. Controlling for output, the point estimate is very near zero. This suggests that almost all of the reduction in CO_2 emissions can be attributed to output effects. Similarly, the results for SO_2, columns (5) and (6), also indicate gross complementarity almost entirely due to output effects. The coefficient for the output effect in column (7) estimates a 30 percent reduction in output with nonattainment designation.

The coefficients on output in (2), (4), and (6) imply emissions elasticities for the three pollutants of 0.8 to 0.9. These estimates are statistically less than one implying that the emissions rates (emissions per MWh) are declining in output. However, the limited net effects suggest that the emissions rates do not vary substantially with changes in prices of other environmental inputs, that is, pollutant spillovers do not change emissions rates.[15]

Table 5.2 splits the sample into old and new plants based on the average age of the plants' units. These results show that the reductions in table 5.1

15. Holland (2010) presents additional specifications and robustness tests.

Table 5.2   **Old and new plants: California results for NO_x, CO_2, and SO_2 emissions and megawatt hours**

	ln(NO_x)		ln(CO_2)		ln(SO_2)		ln(MWh)
Panel A: Old plants (average start year before 1980). 5,566 observations with 89 units.							
Nonattain	−0.715**	−0.297*	−0.462**	−0.011	−0.325*	0.124	−0.511**
	(0.230)	(0.159)	(0.198)	(0.020)	(0.174)	(0.140)	(0.222)
ln(MWh)		0.817**		0.883**		0.887**	
		(0.018)		(0.012)		(0.013)	
Panel B: New plants (average start year after 1995). 2,673 observations with 89 units.							
Nonattain	0.154	0.090	0.044	−0.037	−0.536	−0.615*	0.085
	(0.445)	(0.279)	(0.507)	(0.053)	(0.517)	(0.364)	(0.495)
ln(MWh)		0.754**		0.957**		0.926**	
		(0.037)		(0.022)		(0.030)	

Note: Regressions additionally control for other regulations, for market-year-month fixed effects, for generating unit fixed effects, and for unit-specific linear trends.

come primarily from the reductions in output and emissions at older plants. Since newer plants are less polluting, they use the NO_x input more efficiently and thus did not reduce output in response to the change in attainment status.

The results are subject to three additional caveats. First, the power of the test is reduced since electric power generators were likely not the marginal polluter targeted by the change in attainment status. In particular, the state implementation plans (SIPs) for reducing NO_x emissions do not focus on electric power generation. Second, the estimates cannot control for local economic conditions that may have been correlated with changes in attainment status. Finally, the symmetry assumption requires care in interpreting the coefficients as spillovers from climate policy. Although the estimates are valid estimates of spillovers from ozone policy, they are only locally valid estimates of spillovers from climate policy.

5.6   Conclusion

Spillovers from climate policy are important for policy design, modeling, and benefit-cost analysis. This chapter shows that spillovers arise from output effects (which have ancillary benefits) and substitution effects (which may have ancillary benefits or ancillary costs). The ambiguous net effect highlights the importance of polluters' responses to climate policy.

The chapter then tests for ancillary benefits from climate policy in electricity power generation. The estimates are consistent with ancillary benefits from climate policy arising primarily from reductions in output (primarily at older plants) rather than from changes in emissions rates.

References

Bell, Michelle L., Devra L. Davis, Luis A. Cifuentes, Alan J. Krupnick, Richard D. Morgenstern, and George D. Thurston. 2008. "Ancillary Human Health Benefits of Improved Air Quality Resulting from Climate Change Mitigation." *Environmental Health* 7 (41). doi:10.1186/1476-069X-7-41.

Burtraw, Dallas, Alan Krupnick, Karen Palmer, Anthony Paul, Michael Toman, and Cary Bloyd. 2003. "Ancillary Benefits of Reduced Air Pollution in the US from Moderate Greenhouse Gas Mitigation Policies in the Electricity Sector." *Journal of Environmental Economics and Management* 45:650–73.

Cifuentes, Luis, Victor H. Borja-Aburto, Nelson Gouveia, George Thurston, and Devra Lee Davis. 2001. "Assessing the Health Benefits of Urban Air Pollution Reductions Associated with Climate Change Mitigation (2000–2020): Santiago, São Paulo, México City, and New York City." *Environmental Health Perspectives* 109 (Supplement 3): 419–25.

Dessus, Sebastien, and David O'Connor. 2003. "Climate Policy Without Tears: CGE-Based Ancillary Benefits Estimates for Chile." *Environmental and Resource Economics* 25:287–317.

Dudek, Dan, Alexander Golub, and Elena Strukova. 2003. "Ancillary Benefits of Reducing Greenhouse Gas Emissions in Transitional Economies." *World Development* 31 (10): 1759–69.

European Environment Agency. 2004. *Air Quality and Ancillary Benefits of Climate Change Policies.* EEA Technical report no 4. Available at: http://www.eea.europa.eu/publications/technical_report_2006_4.

Elbakidze, Levan, and Bruce A. McCarl. 2004. "Should We Consider the Co-Benefits of Agricultural GHG Offsets?" *Choices* Fall:25–6.

Feng, Hongli, Catherine L. Kling, and Philip W. Gassman. 2004. "Carbon Sequestration, Co-Benefits, and Conservation Programs." *Choices* Fall:19–24.

Greenstone, Michael. 2002. "The Impacts of Environmental Regulations on Industrial Activity: Evidence from the 1970 and 1977 Clean Air Act Amendments and the Census of Manufacturers." *Journal of Political Economy* 110: 1175–219.

———. 2003. "Estimating Regulation-Induced Substitution: The Effect of the Clean Air Act on Water and Ground Pollution." *American Economic Review Papers and Proceedings* 93 (2): 442–8.

Groosman, Britt, Nicholas Z. Muller, and Erin O'Neill. 2009. "The Ancillary Benefits from Climate Policy in the United States." Middleburg College, Department of Economics. Working Paper Series no. 0920.

Holland, Stephen P. 2012. "Emissions Taxes versus Intensity Standards: Second-Best Environmental Policies with Incomplete Regulation." *Journal of Environmental Economics and Management* 63(3): 375–87.

———. 2010. "Spillovers from Climate Policy." NBER Working Paper no. 16158. Cambridge, MA: National Bureau of Economic Research, July.

Holland, Stephen P., and Erin T. Mansur. 2008. "Is Real-Time Pricing Green? The Environmental Impacts of Electricity Demand Variance." *Review of Economics and Statistics* 90 (3): 550–61.

Joh, Seunghun, Yunmi Nam, Sanggyoo Shim, Joohon Sung, and Youngchul Shin. 2003. "Empirical Study on Environmental Ancillary Benefits Due to Greenhouse Gas Mitigation in Korea." *International Journal of Sustainable Development* 6 (3): 311–27.

Mazzi, Eric A., and Hadi Dowlatabadi. 2007. "Air Quality Impacts of Climate

Mitigation: UK Policy and Passenger Vehicle Choice." *Environmental Science and Technology* 41:387–92.

Pattanayak, Subhrendu K., Allan Sommer, Brian C. Murray, Timothy Bondelid, Bruce A. McCarl, and Dhazn Gillig. 2002. *Water Quality Co-Benefits of Greenhouse Gas Reduction Incentives in U.S. Agriculture: Final Report.* Available at: foragforum.rti.org/documents/Pattanayak-paper.pdf.

Plantinga, Andrew J., and JunJie Wu. 2003. "Co-Benefits from Carbon Sequestration in Forests: Evaluating Reductions in Agricultural Externalities from an Afforestation Policy in Wisconsin." *Land Economics* 79 (1): 74–85.

Puller, Steven L. 2007. "Pricing and Firm Conduct in California's Deregulated Electricity Market." *Review of Economics and Statistics* 89 (1): 75–87.

Walsh, Michael P. 2008. "Ancillary Benefits for Climate Change Mitigation and Air Pollution Control in the World's Motor Vehicle Fleets." *Annual Review of Public Health* 29:1–9.

Comment   Charles D. Kolstad

This chapter addresses an important question in the economics of environmental regulation—a question often given cursory lip service but rarely the subject of rigorous analysis. When an industry is subject to emission regulations for pollutant x, there may be changes in emissions of pollutant y, due either to changes in the technology of production or changes in the quantity of the underlying good produced. For instance, regulating carbon dioxide emissions can result in changes in emissions of particulate matter.

This is an important issue on many counts. A cost-benefit analysis of a proposed regulation should appropriately take into account the benefits/costs of such spillovers. Furthermore, environmental justice issues are often important in regulatory debates and environmental justice frequently involves changes in pollutants that are not the ones being directly regulated. That is the case with carbon emissions in California. Environmental justice proponents are concerned that regulating carbon emissions will result in increases in criteria air pollutants (e.g., particulates) in low income areas of cities.

Although Professor Holland's main contribution is his empirical analysis, he does discuss the theory behind the issue of spillovers. If NO_x emissions are efficiently regulated then marginal costs and benefits from NO_x are always balanced, both before instituting a CO_2 regulation and after. Thus nonzero spillovers from CO_2 regulation in part are due to inefficient NO_x regulations.

Charles D. Kolstad is professor of economics at the University of California, Santa Barbara, a university fellow of Resources for the Future, and a research associate of the National Bureau of Economic Research.

For acknowledgments, sources of research support, and disclosure of the author's material financial relationships, if any, please see http://www.nber.org/chapters/c12149.ack.

From an empirical point of view, a central issue in measuring the spillovers from carbon regulation is that there is no carbon regulation. To address this issue, Professor Holland estimates the change in CO_2 emissions from a change in the price of NO_x emissions, arguing that this price effect is the same as the change in NO_x emissions from a change in the price of CO_2. This is a very interesting way of dealing with this issue, exploiting the symmetry of the Hessian matrix of the profit function. Unfortunately, he does not have the price of NO_x either, so he proxies for this using NO_x attainment status of different regions. Basically, if a region is nonattainment, NO_x regulation is strict and if the region is in attainment, the regulations are weaker. Although this is not quite the same thing as a price of NO_x, he is able to conclude that NO_x and CO_2 are gross substitutes, implying that if CO_2 regulation is tightened, one would expect NO_x emissions to decline. However, when output is included, this substitution effect disappears, suggesting that most if not all of the effect of regulation on emissions is due to a change in output, not substitution.

The empirical analysis focuses on fossil-fueled power plants in California, which happen to be almost entirely natural gas fired. One justification for focusing on California is the large variation in local attainment with ambient air quality standards. It is unfortunate that there is not more heterogeneity in the California market, particularly considering that natural gas is not particularly carbon intensive in its emissions. It would seem that using a richer national data set on fossil fuel generation would yield much more general results. One would expect that the key variable, attainment status for NO_x, would vary considerably over the country (not just California). However, additional analyses by Professor Holland, not reported in the chapter, suggest that there is not much change in attainment status outside of California, at least given the short time frame of the analysis. It would be interesting to see if and how results change with a richer data set.

Another issue has to do with the technology of production. Simply focusing on natural gas generation will not pick up the substitution that we would expect to see from carbon regulation as generation moves from carbon intensive fuels (e.g., coal) to other fuels such as natural gas. Because the focus is on California, that effect will be underestimated in the model.

In conclusion, this chapter represents a very important step forward in measuring the ancillary benefits and costs associated with carbon regulation.

6

Markets for Anthropogenic Carbon within the Larger Carbon Cycle

Severin Borenstein

6.1 Introduction

Among climate scientists, there is a strong consensus that carbon emissions from human activity are increasing atmospheric CO_2 concentration and causing climate change. Among economists, there is a strong consensus that the most efficient way to reduce such anthropogenic greenhouse gas emissions is to price them, through either a tax or a tradable permit system. The CO_2 emissions from burning fossil fuels and deforestation, however, are small compared to the earth's natural carbon flux. These human activities produce about nine gigatons of carbon (GtC) emissions per year against the natural carbon flux backdrop—emission and uptake—of about 210 GtC per year, to and from oceans, vegetation, soils, and the atmosphere. Human activities, however, affect the natural carbon cycle in many ways that have not been incorporated in plans for pricing greenhouse gas emissions. This in no way suggests that human activity is not the primary cause of climate change, but it does suggest that establishing markets and property rights to control these emissions may be more challenging than standard models for tradable pollution permits imply.

In this chapter, I explore the implications for pricing carbon emissions

Severin Borenstein is the E. T. Grether Professor of Business Economics and Public Policy at the Haas School of Business, University of California, Berkeley; codirector of the Energy Institute at Haas; director of the University of California Energy Institute; and a research associate of the National Bureau of Economic Research.

My thanks to Max Auffhammer, Jim Bushnell, Lucas Davis, Meredith Fowlie, Don Fullerton, Richard Muller, Wolfram Schlenker, and Catherine Wolfram for helpful comments and discussions. I am particularly grateful to Margaret Torn for taking the time to explain to me some of the science of the carbon cycle. Any remaining errors are my responsibility alone. For acknowledgments, sources of research support, and disclosure of the author's material financial relationships, if any, please see http://www.nber.org/chapters/c12158.ack.

when human impacts on the natural carbon cycle are numerous, heterogeneous, and likely to be quantitatively significant beyond the direct greenhouse gas emissions from fossil fuel combustion and deforestation. Because the natural carbon flux between oceans, vegetation, soils, and the atmosphere is so large, even small anthropogenic perturbations in it can significantly alter the impact of human activity on climate. Nearly all of the available economic analysis has treated anthropogenic emissions as a separate and measurable process distinct from the natural carbon cycle. Under certain conditions, this may be a valid approach, but climate science suggests that these condition do not hold and may not even be a good approximation. Thus, it is useful to consider more explicitly the interaction between human activity and the natural carbon cycle, as well as implications for the appropriate boundaries of a market for greenhouse gas emissions.

6.2   A Very Brief Review of the Carbon Cycle

Prior to the mid-nineteenth century when large-scale anthropogenic CO_2 emissions began, the oceans, vegetation, and soils are estimated to have released about 210 GtC of carbon into the atmosphere in the form of CO_2 every year and absorbed the same amount on average. About 90 GtC was transferred to/from the ocean and 120 GtC is transfered to/from vegetation and soils.[1] Atmospheric levels of CO_2 remained in the range of 260 to 280 parts per million (ppm), equivalent to approximately 550 to 590 GtC in the atmosphere.[2]

Nearly all of these natural processes, however, are affected by changes in atmospheric carbon and the climate. For instance, increases in atmospheric CO_2 cause plants to grow faster and absorb more carbon, and cause ocean uptake of carbon to increase; higher average temperatures and other changes in climate alter the rate at which plants decompose and release CO_2; and changes in ocean temperature affect its uptake of carbon. Prior to the fossil fuels era, this seems to have been part of the natural resilience of the biosphere that maintained fairly stable atmospheric CO_2 concentrations for millenia.

Since the mid-nineteenth century, direct anthropogenic impact on the carbon cycle has steadily increased, primarily through fossil fuel combustion—averaging about 7.6 GtC per year during 2000 to 2006—but also through human-caused deforestation and changes in land use—estimated to be about 1.5 GtC per year during 2000 to 2006.[3] The deforestation and

1. My characterization of the carbon cycle is based on Houghton (2007), Canadell, Le Quère et al. (2007), and Sarmiento and Gruber (2002).
2. If it were absorbed entirely into the atmosphere, one GtC would raise the atmospheric level of CO_2 by slightly less than 0.5 ppm.
3. See Canadell, Le Quère et al. (2007), table 1. The CO_2 release attributed to fossil fuels includes the release from heating calcium carbonate in cement production. Non-CO_2 forms of

land-use change impacts are known with considerable less certainty than fossil fuel combustion, because the full process of carbon flux between vegetation/soils and the atmosphere is not understood nearly as well as the combustion of oil, coal, and natural gas.

Anthropogenic carbon emission must go somewhere. About 45 percent shows up as an increase in atmospheric concentration of CO_2. Scientists are confident that the residual carbon ends up in vegetation, soils, and the ocean, but attempts to measure these changes directly are imperfect. Carbon is mixed much less uniformly in the ocean than in the atmosphere, so its concentration is more difficult to measure. Concentration in vegetation and soils varies even more and is an even greater measurement challenge. The best estimates are based on widespread sampling of ocean waters to estimate ocean uptake, then attributing the residual to vegetation and soils. This approach suggests that ocean uptake accounts for about 24 percent of anthropogenic carbon emissions and 30 percent goes to vegetation and soils. However, the processes of ocean and vegetation/soils carbon uptake are not well understood. Estimates of these components—often referred to as the "residual flux," or, somewhat less accurately, the "unidentified sink"—total about five GtC per year.

Figure 6.1 is a simplified representation of the carbon cycle from the US Department of Energy, with estimates of the anthropogenic carbon emissions and terrestrial and ocean uptake updated, based on figures from Canadell, Le Quère et al. (2007) (in white boxes). While there is some disagreement about the estimates of carbon uptake of vegetation, soils, and the ocean, there is widespread agreement that these have been large net carbon sinks over the last two centuries, offsetting a considerable share of the direct anthropogenic carbon emissions.

There is some evidence that the carbon uptake share of nonatmospheric sinks is declining over time, but a larger proportion is remaining in the atmosphere.[4] This suggests that the nonatmospheric sinks, both identified and unidentified, may have started to become saturated. To date, climate change models have handled ocean and terrestrial sinks fairly mechanically, assuming that they will continue to absorb about the same share of anthropogenic carbon as has been estimated from residual sink calculations for recent years, or assuming that the share will change in some gradual and linear way. This is a source of significant uncertainty because both the carbon uptake capacities of these sinks and the impact of human activities on their capacities are not well understood.

---

carbon in the atmosphere, such as methane, play a significant role in climate change, but are a very small fraction of the carbon cycle. Atmospheric concentration of methane is approximately 1.8 ppm.

4. See Le Quère et al. (2009).

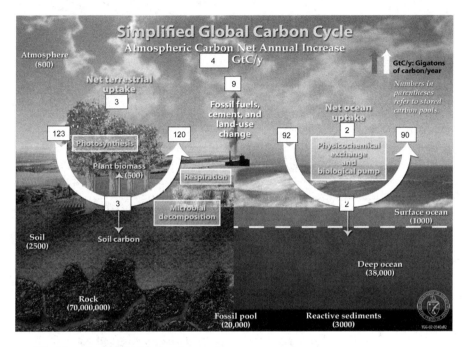

Fig. 6.1   Simplified global carbon cycle
Source: US Department of Energy and Canadell, Le Quère et al. (2007).

6.3   Markets for Carbon Emissions

If the human contribution to atmospheric CO_2 were completely distinct from the natural carbon cycle, setting and enforcing a cap on CO_2 released from fossil fuel combustion and deforestation would obviously address the carbon cycle imbalance. In that case, reduction of CO_2 emissions would translate one-for-one to reductions in atmospheric CO_2. From the description of the carbon cycle in the previous section, however, it is clear that this is not at all an accurate representation of the anthropogenic impact.

Apart from burning fossil fuels, most human activity that releases greenhouse gases is interacting with the natural carbon cycle on a short time scale. Cutting a virgin forest likely causes the trees to decompose and release carbon more quickly than would have occurred absent human interaction, in years rather than decades. Human-caused forest fires do so even faster. Agriculture raises many of the same issues, as tilling and crop management alters the soil release and uptake of CO_2. Livestock cultivation by humans also disrupts CO_2 uptake of soils and vegetation, as well as directly contributing significant quantities of methane. Nitrogen fertilizer, both at the location it is applied by humans and after it migrates through soils and water, interacts with CO_2 in complex ways to affect the growth of vegetation and its properties as a carbon sink.[5] Atmospheric anthropogenic nitrogen also seems likely to be significantly altering the carbon uptake of oceans as well as increasing emissions of nitrous oxide, potentially reducing the net carbon sink impact of oceans by more than half.[6] Man-made local air pollutants also interact with the natural carbon cycle: tropospheric ozone, a local pollutant created by the chemical interaction of man-made emissions and sunlight, disrupts the carbon sink effect of forests and other vegetation.[7]

Proposals for market mechanisms to control CO_2 emissions include restrictions on combustion of all types of fossil fuels, though usually with significant geographic and sectoral limits. Some proposals include limited applications to forestry and agriculture. Through offset programs, inclusion of some additional agriculture and livestock cultivation is often suggested, though it has played an extremely small role in the Clean Development Mechanism.[8] The impacts of nitrogen fertilization on vegetation, soils, and ocean uptake is invariably excluded, as is the impact of local air pollution. Many other ways in which human behavior impacts the natural carbon cycle to exacerbate or reduce atmospheric concentration of CO_2 are excluded from the functioning and proposed market mechanisms. The omissions are not because these are understood to be small factors. Some are estimated to be large, though none is estimated very precisely.

6.3.1 Climate Feedback Effects Are a Special Case of Interaction with the Natural Carbon Cycle

Market mechanisms do not explicitly incorporate aggregate interaction effects, known as feedback effects, in which the total planetary anthropogenic release of greenhouse gases causes changes in the nonanthropogenic carbon flux. Such effects are a function of aggregate anthropogenic emissions because CO_2 and other greenhouse gases mix nearly uniformly around the earth's atmosphere: increased atmospheric CO_2 concentration causes an increase in the carbon uptake of oceans, vegetation, and soils; it contributes directly to higher average temperatures and faster decomposing of dead vegetation, which releases more greenhouse gases; higher average temperatures cause faster melting of ice sheets, which then releases methane and also reduces the albedo of the earth. Warming also increases water evaporation and the concentration of atmospheric water vapor, which magnifies the greenhouse effect. Climate scientists attempt to account for these effects in modeling the relationship between atmospheric greenhouse gases and global temperature changes.

Conceptually these aggregate interactions are straightforward to handle within a market mechanism, though practical application faces substan-

---

5. See Reay et al. (2008).
6. See Duce et al. (2008).
7. See Canadell, Kirschbaumb et al. (2007).
8. See Grubb et al. (2010).

tial uncertainty about the magnitude of their climate impact. If the goal is to stabilize atmospheric carbon at a certain level, aggregate interaction effects would be incorporated into a cap-and-trade program by changing the total direct anthropogenic carbon emissions. The net effect of all aggregate interaction effects would determine a scale parameter, θ, that would change the cap on direct anthropogenic carbon emissions so as to meet the same level of atmospheric carbon as would be the target if $\theta = 1$ and there were no interaction effects. A $\theta < 1$ would indicate that the natural carbon cycle damps anthropogenic shocks, a net negative feedback effect, and a $\theta > 1$ would indicate that it exacerbates the shocks, a net positive feedback effect. The fact that about half of anthropogenic carbon is being absorbed by vegetation, soils, and the ocean suggests a θ well below one, but acceleration of vegetation decomposing and ice melting indicates the opposite. More importantly, a great deal of uncertainty remains about the longer run θ, though it seems likely to rise if the terrestrial and ocean sinks are becoming saturated and/or melting ice might accelerate the release of greenhouse gases and change the planet's albedo. Nonetheless, for any scientific model of these aggregate interaction effects, the cap on anthropogenic emissions can be adjusted in order to achieve (in expectation) any specified target for atmospheric carbon and climate change. Though the potential scientific impact of feedback effects is quite worrisome, they complicate market mechanisms much less than the idiosyncratic indirect impacts on which I have focused here.

6.4   From Incomplete Science to Incomplete Markets and Property Rights

Market mechanisms to address climate change have been aimed predominantly at reducing the greenhouse gas emissions from burning fossil fuels. Besides the enormous size of the fossil fuels industry, this focus is likely based on the fact that the scientific connection between fossil fuel combustion and greenhouse gas release is well established, and the fact that it is relatively easy to monitor fossil fuel consumption. While it is well understood that human behavior is affecting the natural carbon cycle, those effects are less direct, the relationship is less precisely established, and the emissions are more difficult to monitor. In the last decade, scientists have made important steps in understanding these relationships, but because the impacts are indirect and idiosyncratic it is likely that the links to greenhouse gas emissions will never be understood as precisely as the CO_2 release from burning a gallon of gasoline. For example, the greenhouse gas impact of nitrogen fertilizer appears to depend very much on where it is used, how it is applied, and how much escapes to neighboring soils and water.

Over time, the challenge of establishing scientific causality will transition to a challenge of establishing markets and property rights for the externalities created. Some empowered institution will have to determine a process for price setting and the initial allocation of the property rights. These appear to be particularly challenging tasks in the case of human impacts on the natural carbon cycle.

The heterogeneity and idiosyncrasy of these indirect impacts will pose a challenge for price setting. Of course, many government-regulated markets face a trade-off between precise cost-based pricing of each sale and the expense of implementing complex pricing schemes. The problem is present in congestion pricing of roads, differentiated time and locational impacts of criteria air pollutants, and time and location varying cost of supplying electricity.[9] In nearly all of these cases, prices vary much less than the underlying economic costs, usually based on appeals to equity and/or simplicity.

Such an outcome could be very inefficient in this case. While science does not yet provide complete answers, it seems likely that the variation in impact on the natural carbon cycle could be enormous for seemingly similar human activities. The impact of agricultural activities, for instance, depends not just on soil composition and alternative land use, but also on the quantities of fertilizers used and their ultimate disposition. Likewise, criteria air pollution has very different impacts on the natural carbon cycle depending on where the pollution is released. Due to the interaction with the natural carbon cycle, it seems quite possible that an activity could raise greenhouse gases if undertaken in some locations and lower it if the same activity is undertaken in other locations.

The idiosyncrasy of human impacts on the natural carbon cycle is also likely to greatly increase the complexity of allocating property rights and monitoring outcomes. Indirect impacts on the natural carbon cycle are likely to be difficult to monitor by their very nature, and large variation in impact from seemingly similar activities will make simplifying approaches less reliable—for example, a standard assumption about the carbon impact of releasing one pound of atmospheric nitrogen. Likewise, because property rights allocation will be concerned with distributional issues, difficulty in determining a participant's probable liability under a proposed price schedule could slow the political process and raise costs.

Scientific uncertainty is also likely to compound the difficulties of reaching agreements on property rights. Previous debates over the costs of environmental degradation—health impacts of criteria air pollutants, ozone depletion caused by chlorofluorocarbons (CFCs), and fossil fuels causing climate change—suggest that potential losers in the allocation of property rights will appeal to residual scientific uncertainty as a reason to postpone creation of the market. Indirect impacts on the natural carbon cycle seem likely to be particularly vulnerable to these delay strategies.[10]

---

9. See Tietenberg (1995).

10. Recent arguments over life cycle analyses of petroleum products and corn-based ethanol in California, including the impact of indirect land-use changes, are certainly consistent with this view. The parties that would have been harmed by recognizing indirect land-use effects argued that because considerable uncertainty about their magnitude existed, they should be counted as zero.

Ultimately, the value of incorporating human impact on the natural carbon cycle as part of carbon markets also depends on the potential for price incentives to change that interaction. In this dimension, it seems that the value is likely to be high. The human activities that science has already identified—including land management, use of nitrogen fertilizers, and control of criteria air pollutants—are generally thought to be responsive to economic incentives, certainly likely to be as responsive as energy demand. These are empirical questions, however, that remain to be addressed.

6.4.1   Can Carbon Offsets Better Address Interactions with the Natural Carbon Cycle?

The effects that I am discussing here are similar in practice to excluding a sector of the economy, or region of the world, under cap and trade. Carbon offsets are often presented as a way to reduce emissions from an excluded sector or region, as described by Bushnell (chapter 12, this volume). But the political, jurisdictional, and distributional concerns that give rise to sectoral or regional exclusion are not the primary impediments to incorporating interactions with the natural carbon cycle. Rather, uncertain science and costly monitoring of the human behavior that causes the interaction have led to the exclusion of these emissions from market mechanisms. Carbon offsets do not address either of these problems. If these barriers were remediated, policymakers still might run into the concerns that are addressed by carbon offsets depending on the location of the activity and people involved in it. There is, however, no obvious reason to think that the range of human activities that constitute interaction with the natural carbon cycle are more amenable to control through carbon offsets than through direct inclusion in a market mechanism such as cap and trade or a carbon tax.

6.5   Conclusion

Climate scientists have determined that many different human activities impact the levels of atmospheric greenhouse gases, not just burning fossil fuels. Many of these interactions are not well understood, but they are almost surely both heterogeneous and important in addressing climate change. Recent research suggests that human-caused air pollution, fertilizer dispersion, soil disruption, and other activities are having a significant effect on the net carbon uptake of vegetation, soils, and oceans. To date, market mechanisms for reducing greenhouse gases have largely ignored these interactions between human activity and the natural carbon flux.

My goal in this chapter is to argue that the scientific research on these interactions has matured to the point that it is time for economists and policymakers to take note, and to consider whether market mechanisms for greenhouse gases need to be extended to incorporate these complexities. Such extensions would be very challenging. The heterogeneity and idiosyncrasy of human impact on the natural carbon cycle would make appropriate pricing quite difficult, and the remaining scientific uncertainty about these interactions would likely impede efforts to assign property rights. Addressing some interactions would require determining property rights for a much broader range of activities than has ever before existed.

The costs of extending carbon markets in this direction must be weighed against the potential benefits. The benefits will depend on the magnitude of the interaction effects, which is the domain of natural scientists, and the price elasticities of the human activities that cause them, the determination of which should be economists' comparative advantage.

Finally, while I have focused here on market mechanisms—taxes or tradeable permits—the same concerns of heterogeneous and idiosyncratic interactions with the natural carbon cycle would apply to any attempt to address greenhouse gases with command and control regulation. Just as many more prices and property rights determinations are needed in a market setting due to indirect impacts on the natural carbon cycle, many more regulations would be needed under a command and control approach.

References

Canadell, Josep G., Miko U. F. Kirschbaumb, Werner A. Kurz, María-José Sanz, Bernhard Schlamadinger, and Yoshiki Yamagata. 2007. "Factoring Out Natural and Indirect Human Effects on Terrestrial Carbon Sources and Sinks." *Environmental Science and Policy* 10:370–84.

Canadell, Josep G., Corinne Le Quère, Michael R. Raupach, Christopher B. Field, Erik T. Buitenhuis, Philippe Ciais, Thomas J. Conway, Nathan P. Gillett, R. A. Houghton, and Gregg Marlandi. 2007. "Contributions to Accelerating Atmospheric CO_2 Growth from Economic Activity, Carbon Intensity, and Efficiency of Natural Sinks." *Proceedings of the National Academy of Sciences* 104 (47): 18866–70.

Duce, R. A., J. La Roche, K. Altieri, K. Arrigo, A. Baker, D. Capone, S. Cornell, et al. 2008. "Impacts of Atmospheric Anthropogenic Nitrogen on the Open Ocean." *Science* 320:893–7.

Grubb, Michael, Tim Laing, Thomas Counsell, and Catherine Willan. 2010. "Global Carbon Mechanisms: Lessons and Implications." *Climatic Change* 104 (3–4): 539–73.

Houghton, R. A. 2007. "Balancing the Global Carbon Budget." *Annual Review of Earth and Planetary Sciences* 35:313–47.

Le Quère, Corinne, Michael R. Raupach, Josep G. Canadell, Gregg Marland et al. 2009. "Trends in the Sources and Sinks of Carbon Dioxide." *Nature Geoscience* 2:831–6.

Reay, Dave S., Frank Dentener, Pete Smith, John Grace, and Richard A. Feely. 2008. "Global Nitrogen Deposition and Carbon Sinks." *Nature Geoscience* 1:430–7.

Sarmiento, Jorge L., and Nicolas Gruber. 2002. "Sinks for Anthropogenic Carbon." *Physics Today* 55 (8): 30–6.

Tietenberg, Tom. 1995. "Tradeable Permits for Pollution Control when Emission

Location Matters: What have We Learned?" *Environmental and Resource Economics* 5:95–113.

Comment   Wolfram Schlenker

The chapter by Severin Borenstein discusses how market-based approaches might have to be adjusted for human activities that impact the carbon flux differently across locations. The question is whether geographic variations in the carbon flux are large enough and measureable enough to usefully employ a spatially varied carbon-pricing scheme. Experience with markets for sulfur-dioxide emissions indicates that pollution-pricing mechanisms can greatly enhance efficiency, even if they do not perfectly account for geographic differences in the marginal costs and benefits of emissions.

Market-Based Regulation

The main appeal of market-based regulations (taxes or permits) is that they minimize abatement cost for a given reduction of pollution. Sulfur-dioxide (SO_2) emissions were subject to a permit trading system in the United States in the 1990s. Stavins (1998) argues that it resulted in annual cost savings of more than \$1 billion as firms with the lowest abatement cost reduced SO_2 emissions. It is celebrated as a big success story and has become a benchmark for modern environmental policy.

A tax or permit system can lead to suboptimal outcomes if it interacts with other distortions, or a unit of pollution causes damages depending on where it is emitted. For example, under the NO_x budget program, regulated or publicly owned utilities were more likely to install costly capital equipment (Fowlie 2010). These capital investments generated excess permits that could be sold to firms in deregulated and restructured electricity markets. Since NO_x is a local pollutant, it matters where a unit is emitted. As it turns out, deregulated markets are dirtier to begin with. Permit trading therefore shifted pollution toward areas where a unit of pollution is more damaging.

Market-based regulations have the potential to increase pollution damages if the location of the emission matters (a nonuniformly mixing pollutant). In such a case the tax rate would have to differ between locations, or permits would no longer be traded one-for-one. Instead, differentiated tax and trading ratios incorporate that marginal damages vary between regions.

While CO_2 perfectly mixes in the atmosphere, Borenstein emphasizes that

Wolfram Schlenker is assistant professor of economics at Columbia University and a faculty research fellow of the National Bureau of Economic Research.

For acknowledgments, sources of research support, and disclosure of the author's material financial relationships, if any, please see http://www.nber.org/chapters/c12159.ack.

a policy that reduces greenhouse emissions in one area might increase it in another area.

The Carbon Cycle

Anthropogenic emissions are a small part of the annual carbon cycle. Humans account for carbon emissions of roughly 9 GtC compared to a natural carbon flux of 210 GtC. About half of anthropogenic emissions stay in the atmosphere while the other half gets absorbed by oceans, vegetation, and soils. Each GtC of anthropogenic emissions therefore increases global CO_2 levels by roughly 2.25 ppm (parts per million).

Is this amount of anthropogenic emissions large? The CO_2 levels have fluctuated in the past. Reconstructions for the last 400,000 years by Hansen et al. (2008) show that CO_2 concentrations fluctuated between 200 ppm and 300 ppm, or roughly forty years worth of current emissions (holding current absorption rates constant). Reconstructed temperature records in the Arctic changed in phase by up to 10C. Higher latitudes show larger fluctuations in warming than the global average, and the authors estimate that a doubling of CO_2 concentrations would result in a +3C increase in temperature accounting only for fast feedback processes, and +6C when slow feedback processes are also included. Different models give various estimates of the climate sensitivity as there are large numbers of unresolved uncertainties.

There are at least several layers of uncertainty related to aggregate CO_2 emissions: (a) will sinks continue to absorb CO_2 or will they slow down in the future; (b) how do changes in greenhouse gas concentrations translate into changes in the global temperature; and (c) what are the feedback loops between a warming world and greenhouse gases releases; for example, will there be an extra release of methane when permafrost soils thaw? What is common to all these sources of uncertainty is that they relate to aggregate emissions. If the goal is to minimize compliance cost, a uniform tax rate around the world or a permit system where all countries participate would yield the optimal result. As uncertainty resolves, the regulating agency could adjust the tax rate or number of emissions permits. Several US regulations require periodic reauthorization, at which point these optimal levels could be revised.[1]

Another area with deep uncertainty depends on local anthropogenic interactions with the carbon cycle. Since the natural carbon flux is more than twenty times as large as anthropogenic emissions, even small local feedback loops with the natural carbon flux can in principle be an important component of anthropogenic emissions. Two examples can illustrate this point.

Agricultural policies can both directly impact the carbon flux (tilling

---

1. Weitzman (1974) demonstrates that the expected deadweight loss can vary under a tax or permit system when there is uncertainty about the marginal abatement cost curve. The optimal choice depends on the slopes of the marginal damage and marginal abatement cost curves.

releases carbon, and the use of fertilizer can release greenhouse gases) and indirectly as supply responses to changing prices occur predominantly on the extensive margin (Roberts and Schlenker 2010). If newly cultivated land comes from deforestation, large amounts of greenhouse gases can be released as 20 percent of carbon emissions are related to land-use change. If, on the other hand, fallow land is brought back into production, the land expansion might be a carbon sink. It crucially depends on where the expansion takes place.

Pollution control policies also affect plant growth and soil practices. Auffhammer, Ramanathan, and Vincent (2006) show that climate change due to brown clouds (air pollution) and greenhouse gases contributed to the slowdown in Indian harvest growth rates. A reduction in harvest growth again impacts world food prices and leads to expansions elsewhere. These local interactions have the potential to impact the carbon flux, and are not incorporated in current policy proposals.

Local feedback effects would require locally differentiated taxes or permit trading ratios. Incorporating local differences would increase economic efficiency. At the same time, the countries of the world have a difficult time agreeing to an overall limit; agreeing on local feedback loops might prove even more daunting: every country will have an incentive to argue that it is subject to a feedback loop that reduces the carbon flux to the atmosphere. In doing so it would obtain a more advantageous trading ratio.

Conclusion

There is considerable uncertainty about all the feedback effects between rising greenhouse gas concentrations in the atmosphere and changes in climate. Uncertainties related to aggregate emissions are easier to incorporate in a market-based system of taxes or permits as regulators only have to adjust the overall tax rate or pollution cap as more information becomes available. On the other hand, local feedback effects would require location-specific taxes or permit trading systems. Once uncertainties are resolved, the entire set of bilateral coefficients would have to be revised, which directly impact the cost of the regulation in each region. Individual countries have a strong incentive to play up feedback loops that reduce the carbon flux into the atmosphere while ignoring feedback loops that reduce the carbon flux. The potential gains from nonuniform regulations should be weighed against the possible implications they have for additional rent seeking and free-riding, which can also cause significant deadweight losses.

References

Auffhammer, Maximilian, V. Ramanathan, and Jeffrey R. Vincent. 2006. "Integrated Model Shows That Atmospheric Brown Clouds and Greenhouse Gases Have

Reduced Rice Harvests in India." *Proceedings of the National Academy of Sciences* 103:19668–72.

Fowlie, Meredith. 2010. "Emissions Trading, Electricity Restructuring, and Investment in Pollution Abatement." *American Economic Review* 100 (3): 837–69.

Hansen, James, Makiko Sato, Pushker Kharecha, David Beerling, Robert Berner, Valerie Masson-Delmotte, Mark Pagani, Maureen Raymo, Dana L. Royer, and James C. Zachos. 2008. "Target Atmospheric CO_2: Where Should Humanity Aim?" *Open Atmospheric Science Journal* 2:217–31.

Roberts, Michael J., and Wolfram Schlenker. 2010. "Identifying Supply and Demand Elasticities of Agricultural Commodities: Implications for the US Ethanol Mandate." NBER Working Paper no. 15921. Cambridge, MA: National Bureau of Economic Research, April.

Stavins, Robert N. 1998. "What Can We Learn from the Great Policy Experiment? Lessons From SO_2 Allowance Trading." *Journal of Economic Perspectives* 12 (3): 69–88.

Weitzman, Martin L. 1974. "Prices vs Quantities." *Review of Economic Studies* 41 (4): 477–91.

II

Interactions with Other Policies

Interactions between State and Federal Climate Change Policies

Lawrence H. Goulder and Robert N. Stavins

7.1 Introduction

Over the past five years, a series of climate bills with national cap-and-trade systems at their hearts have been introduced in the US Congress. But as of June 2010, only one bill—H.R. 2454, the American Clean Energy and Security Act of 2009—had been passed by a house of Congress, and no bill had been sent to the president for his signature. In this environment of relatively slow federal action, climate policy initiatives have emerged at the regional, state, and even local levels. In fact, state-level climate policies are being contemplated, developed, or implemented in more than half of the fifty states.[1]

Federal-level action may soon take place, however. This could come through congressional action or through greenhouse gas regulation by the US Environmental Protection Agency (EPA) under the Clean Air Act. In

Lawrence H. Goulder is the Shuzo Nishihara Professor in Environmental and Resource Economics and chair of the Economics Department at Stanford University, a university fellow of Resources for the Future, and a research associate of the National Bureau of Economic Research. Robert N. Stavins is the Albert Pratt Professor of Business and Government at the John F. Kennedy School of Government, Harvard University; a university fellow of Resources for the Future; and a research associate of the National Bureau of Economic Research.

The authors are grateful to Dallas Burtraw, Anthony Eggert, Arik Levinson, and Catherine Wolfram for helpful comments on a previous version of this chapter. For acknowledgments, sources of research support, and disclosure of the authors' material financial relationships, if any, please see http://www.nber.org/chapters/c12124.ack.

1. Most prominent among these are the Regional Greenhouse Gas Initiative (RGGI) in ten northeastern states, and AB 32, California's Global Warming Solutions Act of 2006. Throughout most of US history, state and local governments have had the primary responsibility for environmental protection (Revesz 2001). However, since the passage of the National Environmental Policy Act in 1969, the federal role has increased significantly. Federal laws for localized environmental problems generally leave room for states to exceed national standards.

the absence of congressional action, EPA action is called for as a result of the 2006 US Supreme Court decision in *Massachusetts v. EPA*, the Obama administration's subsequent "endangerment finding" that carbon dioxide (and other greenhouse gases) endangers public health and welfare, and the consequent designation in 2010 of carbon dioxide as a pollutant for regulatory purposes under the Clean Air Act both for stationary and mobile sources.[2]

No matter whether federal action comes through new legislation or via the EPA's authority under the Clean Air Act, important questions arise regarding the relationship of federal actions to ongoing state-level climate policy developments. In the presence of federal policies, to what extent will state efforts be cost-effective? How does the coexistence of state- and federal-level policies affect the ability of state efforts to achieve emissions reductions?

This chapter addresses these questions. We find that the coexistence of state and federal climate efforts can be mutually reinforcing or problematic, depending on the nature of the overlap between the two systems, the relative stringency of the efforts, and the types of policy instruments utilized. Problematic interactions arise when the federal policy involves restrictions on aggregate emissions quantities (as with a simple federal cap-and-trade program) or involves nationwide averaging of performance (as with fuel economy standards or renewable fuel standards). In these circumstances, the emission reductions accomplished by a subset of US states reduces pressure on the constraints posed by the federal policy, thereby freeing—indeed, encouraging—facilities or manufacturers to increase emissions in other states. This leads to "emissions leakage" and a loss of cost-effectiveness at the national level. In contrast, when the federal policy involves fixed prices for emissions (as under carbon taxes or under a cap-and-trade program with a binding "safety valve" or "price collar"), more aggressive climate policy in a subset of states does not lead to offsetting emissions elsewhere. Nationwide emissions are reduced, but the more aggressive state-level action generally leads to differing marginal abatement costs across states, implying that the same reduction could have been achieved at lower cost through an increase in the federally established price of emissions.

Even in situations where significant leakage is likely, there may be a case for state-level action to the extent that such action yields other, offsetting benefits. We articulate and evaluate a number of arguments that claim such benefits and are raised to support state-level climate policy in the presence of federal policies, despite the potential for leakage.

The chapter is organized as follows. Section 7.2 examines interactions between federal and state cap-and-trade programs, while section 7.3 examines interactions under other policies, including fuel economy standards and renewable fuel standards. In both of these sections, we highlight difficulties

2. See http://www.supremecourt.gov/opinions/06pdf/05-1120.pdf.

that stem from these interactions, and explore the extent to which avoiding these problems is consistent with the continuing presence of state programs. Section 7.4 evaluates several arguments claiming various benefits from state-level action that may offset the disadvantages (such as emissions leakage) identified earlier. Section 7.5 concludes.

7.2   National and Subnational Cap-and-Trade Systems

How would a federal cap-and-trade system interact with one or more state (or other subnational) cap-and-trade systems? Two key factors driving such interactions are the degree of overlap in coverage (scope of sources) between the federal and state systems, and the relative stringency of the two systems. We consider two important cases: programs with perfectly overlapping coverage, and programs with imperfectly overlapping coverage.[3]

7.2.1   Systems with Perfectly Overlapping Coverage

The simplest case is systems with perfectly overlapping coverage. This could include, for example, the case in which the federal and state cap-and-trade systems are broad in their sectoral coverage, as well as the case in which the federal and state systems both focus exclusively on the electricity generation sector.

Consider first the situation where the state program is *more stringent* than the national program in that it requires reductions from sources within the state that are greater than would be achieved under the national program alone. In this case, emissions sources must surrender both state and federal allowances to comply with the two jurisdictions. If a source only needed to surrender to one jurisdiction, it would choose abatement levels such that marginal abatement costs equaled the allowance price. If the source must offer allowances to two jurisdictions, it will equate marginal abatement costs with the *sum* of the two allowance prices.

Figure 7.1 depicts the impact of facing two allowance prices. The figure displays marginal benefits from emissions (corresponding to marginal costs of emissions abatement) for two groups of states; the "greener" states prefer more stringent cap-and-trade policy than the other states do. Suppose that initially the only cap-and-trade program is at the federal level. With allowance trading across all states, marginal abatement costs are equated across states, and a single allowance price of p_{FED} applies nationwide. Total emissions at the national level are e_{GS} plus e_{OS}, a total given by the federal policy's overall emissions cap.

Now suppose the greener states wish to impose a tighter cap-and-trade

3. Although our focus is on impacts of overlapping regulations across jurisdictions, the analysis has some formal similarities to the analysis of outcomes from overlapping regulations within a jurisdiction. Levinson (chapter 7, this volume) offers the latter analysis.

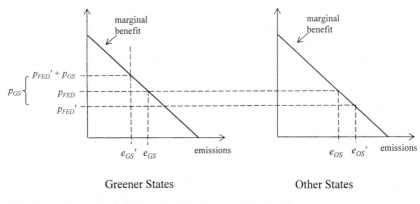

Fig. 7.1   Interaction of federal and state cap-and-trade programs

program within their own jurisdictions. They establish their own allowance cap of e_{GS}', allowing fewer emissions than their prior equilibrium emissions level e_{GS}. The tighter cap compels producers in the greener states to reduce their emissions further. This reduces demands for the federal-level allowances, causing the price of these allowances to fall, which leads to increased emissions in the other states. The new equilibrium price of federal allowances is p_{FED}'. The price of the green states' allowances is p_{GS}, determined such that the sum of the federal allowance price and the state allowance price equals the green states' marginal costs of abatement at e_{GS}'.

Importantly, the greener states' efforts do not lead to any reductions in national emissions beyond that mandated by the federal cap. These states face marginal abatement costs of p_{FED}' plus p_{GS}, higher than those (p_{FED}') in the other states. With marginal costs not equalized nationally, the country's overall abatement costs are greater than under the federal program alone.[4] Thus, the presence of the greener states' program compromises cost-effectiveness.

This is the likely outcome from the interaction of a stringent California cap-and-trade system implemented under Assembly Bill 32 (AB 32) and a less stringent federal system (assuming similar coverage). California's tighter cap would not achieve any further reductions in emissions. At the same time, it would add to the state's costs and to the nationwide costs of achieving the national target.

What would happen if the state program were *less stringent* than the federal program in the sense of requiring smaller reductions from sources within the state than would be achieved under the federal program alone? This would be a case where the greener states' cap is to the right of e_{GS}. In

4. Prior analyses by McGuinness and Ellerman (2008) and Burtraw and Shobe (2009) offered similar results.

this case, the federal allowance price would be sufficiently high to cause sources in the state to reduce emissions below the state cap; the state's cap is therefore not binding and the equilibrium price of state allowances is zero. Here the state program has no impact—it neither affects nationwide nor in-state emissions nor alters the cost-effectiveness of the federal program.

7.2.2   Systems with Imperfectly Overlapping Coverage

Now consider the case where the national and state programs involve imperfectly overlapping coverage. In this case, the nature of the interaction again depends upon which program is more comprehensive in its coverage of state sources. It also depends on which program is more stringent for the sources covered by both programs.

If the scope of the federal program envelops that of the state program (that is, includes all the sources in the state program plus others), then for those sectors covered by both the national and the state program, the results are the same as with perfectly overlapping coverage. If the state program is more stringent, the same leakage problems and losses of cost-effectiveness apply as previously discussed. If the state program is less stringent, then it has no impact. This is essentially the case with the Regional Greenhouse Gas Initiative (RGGI) in the Northeast, which covers only the electricity sector and is considerably less stringent than the major proposals for national economy-wide cap-and-trade systems.

If the state program is more comprehensive than the federal program, results again depend on relative stringency. If the state program is more stringent (for the common covered sectors) than the federal program, then the federal program becomes irrelevant regarding emissions within the state involved. In this case, the more stringent state program will loosen pressure on the federal cap, leading to a reduction in federal allowance prices and associated emissions leakage to other states. If the state program is less stringent for the common-covered sectors, reductions in common-covered sectors in the state will be governed by the federal program. In this case, the state policy has no direct impact on federal allowance prices and thus generates no leakage to other states. In both of these cases, the state can bring about reductions in nationwide emissions by causing reductions in sectors not covered by the federal program.

7.2.3   Other Design Features and Their Implications

How do other design features affect the nature of interaction of state and federal cap-and-trade systems?

Safety Valves

Some proposals for cap and trade include provisions for a "safety valve" or ceiling price on allowances. When a safety valve provision is included, the regulating authority prevents allowance prices from exceeding a given level by issuing additional emissions allowances as necessary. The effect of a safety valve depends on whether and how often it is triggered. Consider, for example, a scenario involving perfectly overlapping state and federal systems, and suppose that initially a safety valve in the federal system is active, so that allowance prices are at the ceiling price. Suppose that the level of emissions in a given state is e_1 in equilibrium in this situation. Now consider what happens if this state attempts to cap its own emissions below e_1. The given state's tighter cap will force additional abatement in that state, raising its marginal and total abatement costs.

The impact of the state's actions on nationwide emissions depends on whether the actions produce a large enough reduction in demand for federal allowances to disengage the safety valve. If the state's reduction is too small to do so, then the price of federal allowances will be unchanged: it will remain at the ceiling price. In this case, the state's actions will imply a reduction in the nation's overall emissions, since the state's own emissions reduction will not be accompanied by any increase in emissions in other states (the price of allowances to other states has not changed). On the other hand, if the given state's reduction yields a large enough reduction in demand for federal allowances to disengage the safety valve, then the price of federal allowances will fall, thereby inducing an offsetting increase in emissions from other states.

Allowance Allocation Methods

The nature of the allowance allocation in general has no affect on system interaction (although there are exceptions in the case of regulated industries). The interactions just described will be the same no matter how much one or both programs relies on auctioning or free allocation. Although the particular allowance allocation method has important distributional implications and can affect cost-effectiveness as well, it does not alter the general pattern of state-federal interactions just described.[5] This is in keeping with the fact that the allowance allocation method generally does not alter incentives at the margin (or allowance prices); it is the marginal incentives that determine emissions levels and cost-effectiveness.

7.2.4   Potential Resolutions

A Carve Out

There are ways to offset or avoid the leakage that would occur in the previous problematic cases. One is for the federal government to allow a state

---

5. Regulated firms generally face lower costs if they receive allowances free rather than need to purchase them in an auction. (See, for example, Bovenberg and Goulder [2001].) In addition, to the extent that allowances are auctioned and the proceeds are used to finance reductions in distortionary taxes, policy costs will be lower than in cases involving other uses of auction revenue or in the case of free allocation. (See, for example, Goulder, Parry, Williams, and Burtraw [1999].)

or group of states a "carve out" from the federal program if they implement or maintain a state program (or state programs) at least as stringent. In this case, two disjoint cap-and-trade programs emerge: the federal system applies only to states that do not carve out. The result is that there will be different allowance prices in some states and in the federal system, marginal abatement costs will not be equated, and so maximal cost-effectiveness will not be achieved.

Re-Denominating Federal Allowances

Another option is for a given state to require covered facilities with the state to submit more federal allowances per unit of emissions than would ordinarily be the case. This action by a greener state does not eliminate leakage, since it increases the effective price of reducing emissions in this state relative to the price in other states. However, in this case leakage is less than 100 percent: the state's action has the effect of tightening the national cap, since the given number of federal allowances in circulation now permits fewer nationwide emissions, assuming some emissions continue in the greener state.

Preemption

Another way to avoid problematic interactions is through federal rules preempting (that is, barring) state-level cap and trade in the presence of a federal program. Some consider this a useful method for preventing leakage and a loss of cost-effectiveness, as well as a way of assuring that private industry does not face multiple performance or technology standards. Others point out that to the extent that the greener state's actions raise costs, those costs are borne by that state alone; correspondingly, they oppose preemption on the grounds that states should have the freedom to decide whether to impose higher costs on themselves.

7.3   Interactions under Other Climate Policies

7.3.1   Fuel Economy Standards

Problematic interactions can also occur under policies involving automobile fuel efficiency standards or limits on automobiles' greenhouse gas emissions per mile. In response to the prospect of climate change, fourteen states moved to establish limits on greenhouse gases (GHGs) per mile from light-duty automobiles. These so-called "Pavley" standards require manufacturers to reduce per-mile GHG emissions by about 30 percent by 2016 and 45 percent by 2020 (California Air Resources Board 2008).[6]

Since CO_2 emissions and gasoline use are nearly proportional, the Pavley

---

6. The Pavley standards are named after California Assemblywoman Fran Pavley, who sponsored the California bill that launched this multistate effort.

limits effectively raise the fuel economy requirements for manufacturers in the states adopting such limits. These state-level actions can interact significantly with the existing federal Corporate Average Fuel Economy (CAFE) standards. Consider an auto manufacturer that prior to the imposition of the Pavley limits was just meeting the federal CAFE standard. Now it must meet the (tougher) Pavley requirement through its sales of cars registered in the adopting states. In meeting the tougher Pavley requirements, its overall US average fuel economy now exceeds the national requirement: the national constraint no longer binds. This means that the manufacturer is now able to change the composition of its sales outside of the Pavley states; specifically, it can shift its sales toward larger cars with lower fuel economy.

Indeed, if all manufacturers were initially constrained by the national CAFE standard, the introduction of the Pavley requirements would lead to emissions leakage of 100 percent at the margin, because the reductions within the Pavley states would be completely offset by emissions increases outside of those states. Using a numerical simulation model of the US automobile market, Goulder, Jacobsen, and van Benthem (2012) found that from 2009 through 2020 about 65 percent of the emissions reductions achieved in the new car market in the Pavley states would be offset by increased emissions in new car markets elsewhere.[7]

In May 2009 the Obama administration reached an agreement with the fourteen Pavley states, according to which the United States would tighten the federal fuel economy requirements in such a way as to achieve effective reductions in GHGs per mile consistent with the first-phase goals of the Pavley initiative. In return, the fourteen states agreed to abandon the first phase of the Pavley effort, which was no longer necessary, given the tightening of the federal standards. However, these states still intend to introduce further tightening of the greenhouse-gas-per-mile standards after 2016. This would imply fuel economy standards more stringent than those applying at the federal level. Hence the leakage issue remains alive.[8]

7.3.2   Renewable Fuel and Portfolio Standards

Renewable fuel standards require that the ratio of renewable to conventional fuels produced by refiners not fall below a given value. When these standards are imposed at both the state and federal levels, once again the

7. Another 5 percent of the emissions reduction is offset by increased emissions from used cars, as the Pavley effort leads to lower scrap rates of older, less fuel-efficient automobiles.

8. Despite the potential for leakage, the tougher state-level standards may conceivably accelerate the development of new technologies that auto manufacturers will eventually adopt throughout the nation, thereby leading to lower emissions and reduced fuel consumption. However, Goulder, Jacobsen, and van Benthem (2012) find that in the presence of the national CAFE standard, faster technological progress exacerbates the adverse fleet compositional impacts of state programs. As a result, in this context greater technological progress yields relatively little benefit in terms of reduced fuel consumption.

effort of individual states to exceed the federal standard could fail to bring about reduced emissions (or increased use of renewable fuels).[9]

This will be the case if, to meet the federal requirement, firms can apply a ratio based on overall (nationwide) use of renewable and conventional fuels. In this case the situation is perfectly analogous to that described earlier for fuel economy standards. If a firm's ratio of renewable-to-refined fuels was just high enough to meet the federal requirement, then when a given state imposes a higher ratio, the firm will more than meet the federal requirement. It is now able to utilize more conventional fuels in other states in which it operates. On other hand, if the federal rules require that each refinery operation—as opposed to each refinery company—meet the given ratio, the situation is different. In this case tighter requirements imposed by a given state will not free up firms to make compensating adjustments in other states.

The same interactions and pattern of outcomes would hold in the case of federal and state-level renewable portfolio standards, which require the electrical generators utilize renewable sources of energy (in particular, wind and solar) for a specific share of their annual generation. The federal systems contemplated in Washington would allow for national trading.

7.3.3 Interactions When the Federal and State Programs Involve Different Instruments

Significant interactions can also occur when the state and federal climate policy instruments differ. As mentioned in the introduction, federal climate policy might be undertaken by the US EPA under the auspices of the Clean Air Act. In this event, the EPA would probably make use of conventional regulatory approaches such as performance standards and technology mandates. Yet cap and trade is likely to continue in the Northeast under the Regional Greenhouse Gas Initiative, and many western states plan to implement cap and trade within the next few years. How would conventional regulation at the federal level interact with the state-level cap-and-trade programs?

Much depends on the particular instruments employed at the federal level and on the specific rules governing the use of these instruments. Consider the following plausible scenario. Suppose that the EPA imposes performance standards such as limits on emissions of certain greenhouse gases per unit of output. State- or regional-level cap-and-trade programs will induce changes in producer behavior, and in some cases these adjustments will cause particular facilities to exceed the federal performance standard. If the federal rules allow firms (or localities) to average their emissions-output ratios in

9. Apart from the leakage issue discussed here, some analyses indicate that a renewable fuel standard may have significant disadvantages relative to emissions pricing policies such as carbon taxes or cap and trade. Holland, Hughes, and Knittel (2009) show that the renewable fuel standard effectively subsidizes renewable fuels and that, as a result, it leads to more overall (renewable plus conventional) fuel use than is economically efficient. See also Wolak (2008).

determining whether they meet the federal standard, then the cap-and-trade initiatives at the state or regional level will precipitate offsetting adjustments in other states or regions. The same applies if the federal rules allow firms or localities to trade performance credits with one another.[10] Thus, the specifics of the federal rules are important.

7.3.4   Problematic Circumstances and Benign Cases

Thus, the potential for leakage and the associated loss of cost-effectiveness arise under a variety of circumstances. In general, problems result when both of the following two conditions apply: (a) the state-level efforts cause firms or facilities within the greener states to overcomply with the federal rules and, (b) the federal rules give firms or facilities the freedom to offset this overcompliance through various adjustments in other states.

We have already noted some cases where the two problematic conditions do not apply. One is when there is no overlap of the federal and state programs (condition [a] is not met). Another is when performance standards do not involve nationwide averaging (condition [b] is not met).

Another circumstance where problems are avoided is when the federal-level program sets prices. (This case was suggested by section 7.2.3's discussion of a safety valve.) Suppose, for example, a carbon tax were imposed at the federal level. If a state decided to impose new regulations requiring in-state reductions beyond what the federal tax would yield, the additional state-level reductions would not lead to offsetting increases elsewhere (apart from the usual "economic leakage"): the reductions in other states would remain governed by the federal carbon tax. Thus, price-based regulation at the federal level can avoid the problematic state-federal interactions. However, to the extent that the new state regulations imply differing marginal abatement costs across states, the potential exists for achieving the same further reduction in emissions at lower cost through a higher carbon tax.

7.4   Are There Other, Offsetting Benefits from State-Level Action?

Even in situations where significant leakage is likely, there may be a case for state-level action to the extent that such action yields other benefits. Here we assess a number of such arguments.

7.4.1   Stronger Arguments

We first present arguments that we regard as having some validity, although some require qualification.

States can contribute to cost-effectiveness by addressing market failures

---

10. Some instruments are more conducive to averaging or cross-facility trading than others. Trading or averaging is relatively straightforward with performance standards, but more difficult with technology mandates.

not addressed by federal climate policy. In addition to the environmental externality associated with climate change, there are some other market failures that merit attention. The presence of these other market failures would imply that getting relative prices right will not—on its own—yield the most efficient outcome. To the degree that federal climate policy disregards these other climate-related market failures, the potential exists for states to promote greater efficiency by addressing the neglected market failures.

States (and, for that matter, localities) may have an advantage over the federal government in addressing certain other market failures. For example, they may be most capable of dealing with the failure stemming from the principal-agent problem associated with renter-occupied buildings, according to which apartment renters have insufficient incentives to conserve electricity. States, counties, and cities can productively promote energy efficiency by addressing this market failure through building codes and zoning (Trisolini 2010). Note, however, that in some cases the additional market failure is most efficiently addressed through federal policy.

States can function as test-beds for alternative policy approaches not contained in an existing federal effort, thereby providing useful information for possible later adoption at the federal level. Clearly, experimentation has appeal, since experiments sometimes pay off handsomely (Ostrom 2009). Note that this argument seems to call for eventual implementation of the innovative policy approach at the federal level and a phasing-out of this effort at the state level after the benefits from given experiments are revealed. Note also that the question arises whether the experimentation is best carried out at the state, as opposed to federal, level.

State policies—particularly those that are more stringent than the federal policy—can exert pressure for more aggressive action at the federal level if the state efforts appear effective. To the extent that a state with more aggressive climate policy can demonstrate that greater reductions can be achieved at lower cost than previously thought, this can give impetus to stronger federal policy. Here again the state is functioning as a test-bed, providing new information. In the previous case, the new information comes from an experimental policy design; in this case it comes from the revealed impact of a more stringent policy.

When a given state imposes a tougher requirement than applies in other states, it can pressure manufacturers to adopt the tighter requirement nationwide rather than offer different technologies in different parts of the country. California's tighter auto pollution laws in the 1970s led to the tightening of the federal auto pollution standards—in part because auto manufacturers did not want to face two standards. Likewise, the Pavley effort initiated by California appears to have been instrumental in prompting the Obama administration's agreement to tighten federal fuel economy standards. Of course, in neither case does such causality imply that social welfare is maximized by the more stringent standard being adopted nationally.

7.4.2   Weaker Arguments

The following arguments seem to have considerably less merit.

States may face different costs of achieving greenhouse gas reductions, and may experience different benefits from avoided climate change (either because of different preferences or different physical outcomes). Differences of this sort exist and are important, but such differences do not provide a sound justification for state-level policy. Instead, they may justify compensation schemes and other elements that allow for differential net burdens across states, such as through the allocation across states of allowances or auction revenues from a federal cap-and-trade system.

States are more familiar with details related to in-state firms and institutions. With this better information, they may be most capable of exploiting low-cost opportunities for addressing climate change. Clearly, federal regulators—and state regulators as well—have limited information. Individual firms tend to have much better information about technological opportunities and abatement costs than do regulators. The information problem primarily provides a sound argument for market-based environmental policy—for policy approaches that give individual facilities or firms the flexibility to make best use of their (better) information. Market-based policies such as cap and trade or carbon taxes have this feature. Note that such policies can address the information problem effectively, even if the policies are introduced at the federal level. Thus, this information problem does not provide a good reason for state-level policy.

7.5   Conclusion

We have examined the nature and impacts of some important interactions between state and federal climate policy. Depending on the overlap and stringency of the state and federal policies, as well as the types of policy instruments employed, state efforts in the presence of a federal policy can be useful or counterproductive.

In general, problems result when both of the following two conditions apply: (a) the state-level efforts cause firms or facilities within the greener states to overcomply with the federal rules, and (b) the federal rules give firms or facilities the freedom to offset this overcompliance through various adjustments in other states. In these circumstances, state-level efforts do not succeed in reducing greenhouse gas emissions nationally, and they reduce the cost-effectiveness of the overall national effort.

We find that there is more potential for these difficulties when the federal policy sets limits on aggregate emissions quantities, or allows manufacturers or facilities to average performance across states. In contrast, the difficulties are usually avoided when the policies have little overlap or when the federal policy sets prices for emissions.

Even in circumstances involving problematic interactions, there may be offsetting attractions of state-level climate policy. We evaluated a number of arguments that have been made to support state-level climate policy in the presence of federal policies, even when problematic interactions arise, and found some arguments to be compelling and others much less so.

References

Bovenberg, A. Lans, and Lawrence H. Goulder. 2001. "Neutralizing the Adverse Industry Impacts of CO2 Abatement Policies: What Does It Cost?" In *Behavioral and Distributional Effects of Environmental Policy,* edited by C. Carraro and G. Metcalf. Chicago: University of Chicago Press.

Burtraw, Dallas, and Bill Shobe. 2009. "State and Local Climate Policy under a National Emissions Floor." Resources for the Future. RFF Discussion Paper no. 09-54.

California Air Resources Board. 2006. *The California Global Warming Solutions Act of 2006.* Text of Assembly Bill 32, August 31. Available at: http://www.arb.ca.gov/cc/docs/ab32text.pdf.

———. 2008. *Comparison of Greenhouse Gas Reductions for the United States and Canada under U.S. CAFE Standards and California Air Resources Board Regulations: An Enhanced Technical Assessment.* Available at: http://www.climatechange.ca.gov/publications/arb.html.

Goulder, Lawrence H., Ian W. H. Parry, Roberton C. Williams III, and Dallas Burtraw. 1999. "The Cost-Effectiveness of Alternative Instruments for Environmental Protection in a Second-Best Setting." *Journal of Public Economics* 72 (3): 329–60.

Goulder, Lawrence H., Mark R. Jacobsen, and Arthur van Benthem. 2012. "Unintended Consequences from Nested State and Federal Regulations: The Case of the Pavley Greenhouse-Gas-per-Mile Limits." *Journal of Environmental Economics and Management* 63 (2): 187–207.

Holland, Stephen P., Jonathan E. Hughes, and Christopher R. Knittel. 2009. "Greenhouse Gas Reductions under Low Carbon Fuel Standards?" *American Economic Journal: Economic Policy* 1 (1): 106–46.

McGuinness, Meghan, and A. Denny Ellerman. 2008. "The Effects of Interactions between Federal and State Climate Policies." Cambridge, MA: Massachusetts Institute of Technology, Center for Energy and Environmental Policy Research. Working Paper no. 08-004.

Ostrom, Elinor. 2009. "A Polycentric Approach for Coping with Climate Change." Washington, DC: The World Bank. Working Paper no. 5095.

Revesz, Richard L. 2001. "Federalism and Environmental Regulation: A Public Choice Analysis." *Harvard Law Review* 111:553–641.

Trisolini, Katherine A. 2010. "All Hands on Deck: Local Governments and the Potential for Bidirectional Climate Change Regulation." *Stanford Law Review* 62 (3) 669–746.

Wolak, Frank. 2008. *Low-Carbon Fuel Standards: Do They Really Work?* Stanford Institute for Economic Policy Research Policy Brief. Available at: http://www.stanford.edu/group/fwolak/cgi-bin.

Comment   Arik Levinson

Goulder and Stavins have provided a clear and useful framework for thinking about the complex interactions between comprehensive climate bills under consideration by the US Congress and existing state regulations already in place, planned, or contemplated. In this note I make four brief points, some new, some adding emphasis to points in their chapter: (a) the core of their analysis lends itself to a simple, two-by-two diagrammatic exposition; (b) their analysis is more general than their chapter suggests; (c) the justifications they explore for continued coexistence of overlapping state and federal regulations are exceptions that prove the rule; and (d) as they note, many of the problems caused by those overlapping regulations would be avoided by a federal pollution tax in lieu of cap and trade.

A Two-By-Two Diagram

Goulder and Stavins identify the two key criteria for whether and how state and federal climate laws would interact: how much abatement is required (stringency) and how many polluting sectors are covered by the legislation (comprehensiveness). That yields four possible outcomes, depicted in figure 7C.1.[1]

The upper left-hand corner (box [A]) of figure 7C.1 depicts the simplest case, where the federal policy covers more of the economy with more stringent legislation. For example, the northeastern states' RGGI requires a 10 percent emissions reduction by 2018 from the utility sector alone, while the Waxman-Markey bill that passed the US House of Representatives in 2009 would require a 17 percent reduction by 2020 from numerous sources including utilities, large manufacturers, refiners, and natural gas sales. The federal law, if enacted, would cover more sources more stringently than RGGI. The state-level regulation's environmental effects would effectively be made irrelevant by the federal law.

Box (B) of figure 7C.1 depicts the hypothetical case where the federal law covers more sectors, but the state law is more stringent. Suppose, for example, that a version of Waxman-Markey passed into law but required less than a 10 percent reduction. In that case, utilities in the Northeast could use the greater abatement mandated by state law to sell federal allowances to other states or sectors. This interaction between state and federal laws can be

Arik Levinson is professor of economics at Georgetown University and a research associate of the National Bureau of Economic Research.

For acknowledgments, sources of research support, and disclosure of the author's material financial relationships, if any, please see http://www.nber.org/chapters/c12125.ack.

1. Stavins, in comments on this note, pointed out that the state and federal policies could also be equally stringent or equally comprehensive, leading to a 3-by-3 diagram. For simplicity and brevity, I have left that unexplored here.

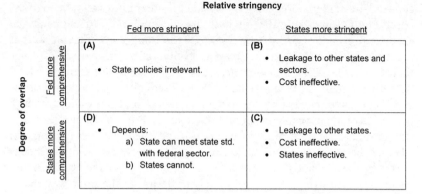

Fig. 7C.1   **Goulder and Stavins in a diagram**

seen in two ways: the federal law enables leakage of GHG emissions from the Northeast utility sector to other states and sectors, and the state law distorts the cost-effectiveness of the federal cap-and-trade system.

Box (C) depicts the hypothetical case where the state law is both more stringent *and* more comprehensive. Imagine a weak federal law covering only the utility sector, and a strict state law covering multiple sectors. Here the state's utilities could sell federal emissions allowances they accumulate as a consequence of meeting the strict state standard to sources in other states, but not to other sectors within the state. Like box (B), there is leakage here, but only across state lines, not across sectors within the state. Also like box (B), the interaction can be seen in two ways: the federal law enables leakage of GHG emissions from the utility sector, and the state law distorts the cost-effectiveness of the federal cap-and-trade system.

The most complex case is depicted in box (D), where the federal law is more stringent, but the state law is more comprehensive. Imagine a strict federal law governing only utilities, combined with a weaker state law covering more sectors. If we presume that the state standard cannot be met entirely by abatement within the federal sector (utilities), then this case reverses the outcome in the other boxes. The federal policy undermines the cost-effectiveness of the broader state cap-and-trade policy. And the state policy undermines the emissions reductions mandated by the federal policy, by enabling leakage across sectors within the state.

Although all of this is hypothetical given the current state of climate legislation in the US Congress, it illustrates how complex the potential interactions can be.

Goulder and Stavins Generalized

The analysis in Goulder and Stavins is in some ways more general than they describe. In chapter 8 of this volume, I discuss interactions between cap-and-trade climate legislation and other more traditional, non-market-based regulations that either predate the cap-and-trade rules or may be enacted alongside them. Retitle figure 7.1 in Goulder and Stavins so that the left graph is labeled "Renewable Energy Standards" instead of "Greener States," and the right graph is labeled "Energy Efficiency Standards" instead of "Other States." A cap-and-trade system will equalize marginal abatement costs between the two sources of abatement, renewable energy and efficiency, just as it would between greener and other states. But if a renewable energy standard coexists alongside the cap and trade, then its effect depends on whether the standard mandates less renewable energy than would be incentivized by the cap-and-trade permit price, or more. If the renewable energy standard is less stringent, it is effectively irrelevant in the same way that a less stringent, less comprehensive state regulation is irrelevant. Given the cap-and-trade permit price, utilities will opt to exceed the renewable standard. On the other hand, if the renewable energy standard is more stringent, it raises abatement from renewable energy, allowing allowances to be sold to energy efficiency sources (leakage), and raises the cost of abating GHG emissions without generating any more abatement—similar to the effects of a more stringent state regulation.

Justifications for Coexisting Federal and State Laws

The overarching conclusion of Goulder and Stavins's chapter and my chapter in this volume is that the coexistence of the two sets of regulations (federal and state, cap and trade and traditional mandate) is either irrelevant or costly. As I do in my chapter, Goulder and Stavins also devote space to identifying cases where that coexistence may be justified. My own impression of those justifications is that in both cases they appear more as exceptions that prove the rule rather than general reasons to enact both types of policies.

Goulder and Stavins provide three general justifications. First, states may address other market failures, such as the fact that landlords and tenants have incomplete incentives to conserve energy. If states have a local-knowledge advantage, regulations addressing building construction or appliance standards may be best set and administered by the states rather than the federal government. Of course, as they note, that argument does not apply to large-scale GHG abatement programs such as RGGI. Second, states are often described as laboratories of regulatory experimentation. Perhaps state-level experimentation will eventually lead to a better-designed federal climate policy. And third, state policies like RGGI and California's AB 32 may provide the political pressure that leads to comprehensive federal policy. Again, as Goulder and Stavins note, these justifications provide reasons for state policies eventually to be replaced by federal policy, not to coexist.

Conclusion and an Advantage of Emissions Taxes over Cap and Trade

Finally, the discussions in Goulder and Stavins and in my chapter illustrate an important advantage a GHG emissions tax would have over a cap-and-trade system. Economists have long argued that social problems like GHG emissions can most cost-effectively be solved by internalizing the externalities—placing a "price on carbon" in common parlance. That price can arise from two possible mechanisms: an emissions tax or a cap-and-trade system. The two share common advantages. Both would internalize externalities. Both would motivate research and development into alternative energy, conservation, and carbon sequestration. And most importantly, both would level the playing field across potential sources of GHG reduction, ensuring that market forces determine that whatever reduction occurs comes at the lowest possible total cost.

One important difference, however, between an emissions tax and cap and trade involves the logistical difficulty of introducing the policy in the first place. Policymakers considering a new, comprehensive, federal GHG cap-and-trade system face a dilemma with respect to sources already covered by other regulations—state regulations or other traditional regulatory mandates. If the federal policy excludes those sources, they lose the cost-effectiveness—the level playing field—of the comprehensive cap and trade. But federal policy covers those already-abating sources, and unless 100 percent of the allowances are auctioned, policymakers must decide how much credit to give sources for abatement that has already occurred, whether voluntary or mandated, raising issues of fairness with respect to sources that may have postponed abatement knowing the comprehensive federal system was coming.

As Goulder and Stavins note, an emissions tax would avoid some of this dilemma. A federal GHG tax could in theory be levied without concern about preexisting state or federal regulations. Those sources that have already abated GHG emissions would simply have an early lead on reacting to the new tax. Where the other state or federal policies result in more abatement than the federal tax would have generated, that excess abatement cannot leak to other states or sectors, because all sources must pay the federal emissions tax rate.[2]

In sum, Goulder and Stavins have cleverly and clearly framed the key issues in thinking about how proposed federal climate legislation may interact with existing state regulations, and that framework illustrates one of the key advantages held by emissions taxes over cap and trade in the contest to become America's preferred greenhouse gas regulatory instrument.

2. Of course, a comprehensive federal tax could interact in problematic ways with state and local or sector-specific emissions taxes.

8

Belts and Suspenders
Interactions among
Climate Policy Regulations

Arik Levinson

8.1 Introduction

Climate policy, if it is to be successful, will be large. Aldy and Pizer (2008) put the cost to the United States as comparable to the "total cost of all existing environmental regulation." Unfortunately, economists' models work best at the margins, predicting the consequences of small incremental changes in policy affecting isolated sectors of the economy. Models work less well for large discrete shifts in policy affecting many sectors simultaneously, the type of regulation likely to be necessary to reduce greenhouse gas (GHG) emissions. The difficulty inherent in assessing large policy changes is that their general equilibrium effects can be vast—even bigger than their direct effects.

Another word for general equilibrium effects, broadly speaking, is "interactions." The size and scope of proposed climate legislation means there will be important interactions with most of the economy, including government tax revenues, other environmental problems aside from climate change, labor markets, terms of trade effects, and other government regulations.

To define a reasonably limited area of attention, I focus on the simplest and most direct form of interaction—those between the tradable GHG emissions permit systems (cap and trade) that are part of many proposed and enacted new climate bills around the world, and the more traditional command-and-control regulatory standards. For climate regulations that

Arik Levinson is professor of economics at Georgetown University and a research associate of the National Bureau of Economic Research.

Gilbert Metcalf, Don Fullerton, and participants at the May 13–14 Design and Implementation of US Climate Policy Conference provided helpful feedback on an early draft. For acknowledgments, sources of research support, and disclosure of the author's material financial relationships, if any, please see http://www.nber.org/chapters/c12138.ack.

have already been passed, mostly in Europe, and for the climate regulations that have been proposed in the United States, the coexistence of these multiple instruments is "the norm, rather than the exception" (Bennear and Stavins 2007). In part, that coexistence has emerged because the cap-and-trade climate laws have been laid down on top of decades of traditional standards. But the coexistence is also written into the language of climate bills that typically include both tradable permits and traditional standards. Either way, we need to think about interactions between the two types of regulatory instruments.

The coupling of tradable permits with traditional standards has been called a "belt-and-suspenders" approach (Pearlstein 2009). In this case, however, it is not clear whether the belt and suspenders are mutually reinforcing, redundant but harmless, or working at cross-purposes. All three viewpoints have appeared in print. Krugman (2010) articulates the mutually reinforcing viewpoint: "I would advocate supplementing market-based disincentives with direct controls." Sijm (2005) makes the case for redundancy: "the coexistence of [tradable permits] and policies affecting fossil fuel use by participating sectors is hard to justify and, hence, these policies could be considered to be redundant and ready to be abolished." And the US Congressional Budget Office (2009b) sees the two as sometimes conflicting: "regulatory standards combined with market-based approaches often will increase the cost of meeting an environmental goal."

Which viewpoint is correct? The answer can be seen in a simple reinterpretation of the textbook partial-equilibrium model illustrating the cost-effectiveness of tradable permit schemes. And that answer depends on whether the price of the tradable GHG emissions permits, and hence the marginal cost of compliance with the cap-and-trade legislation, is higher or lower than the marginal cost of compliance with the traditional regulatory standard. Intuitively, if the permit price exceeds a firm's regulatory compliance costs, that firm would abate beyond the regulatory standard anyway, in response to the cap-and-trade incentives, and the regulatory standard would be irrelevant for that firm. By contrast, if the permit price falls below the regulatory compliance costs for a firm, the firm would meet the regulatory standard exactly and either sell excess permits or buy fewer than it would under cap-and-trade alone. The regulatory standard raises the firm's cost of abating emissions without any resulting increase in overall abatement. Are there economic reasons to pair a tradable permit system with traditional regulatory standards? If there are other market failures aside from the GHG externality, or there are administrative complications in directly targeting GHG emissions, then there may be rationales for combining the two policies, though here we must be careful not to extrapolate from logic that applies to local pollutants but not to greenhouse gases. And finally, economists' demonstrated experience forecasting regulatory costs suggest we are more likely to overstate the costs of meeting a cap-and-trade regulation than a traditional standard, and that therefore where the two instruments are paired, they are likely to increase costs without accompanying abatement benefits.

Before turning to focus on interactions between cap-and-trade and traditional standards, it is worth recognizing a few of the many important interactions the simple textbook model omits.

8.2   Other Interactions—An Aside

United States climate policy will interact with a long list of other important considerations. For example, analysts have long recognized that policies aimed at reducing one pollutant may result in more or less emissions of another (Sigman 1996). For another, an enormous literature exists on spillover effects across countries, either because environmental regulations in one country move polluting industry to less stringent countries (Brunnermeier and Levinson 2004), or because, more subtly, environmental regulations have terms-of-trade effects (Bohringer, Fischer, and Rosendahl 2010). Another vast literature looks at interactions between pollution taxes and other government taxes (Goulder 2002) and expenditures (Metcalf 2008).

The focus here, broadly speaking, is about how environmental regulations targeted at the same pollutant interact with one another. Economists have begun to recognize the importance of these interactions, as policies have begun to pile up and interact in complex ways (Oikonomou and Jepma 2008; Sorrell and Sijm 2003; Eichner and Pethig 2009). This work tends to provide semantic taxonomies of interactions, elaborate charts of interactions, or models with features designed to study specific but very complex parts of the European Union (EU)'s existing tradable permit system. And, to my knowledge, there has been no empirical work that would shed light on the extent of the possible interactions or their consequent effects.

8.3   The Textbook Model

For a long time, economists have focused on persuading policymakers to use market-based instruments—emissions taxes or cap and trade—*instead of* traditional regulatory standards rather than in *addition to* traditional standards. Some version of figure 8.1 appears in most undergraduate environmental economics texts, as a means of illustrating the cost-effectiveness of a tradable permit system compared with a regulatory standard. The bottom axis displays the total uncontrolled pollution from two sources. The sources could be two factories, two industries, two different control strategies, and so forth. Source one, for example, could be carbon mitigation from utilities using renewable energy portfolios, and source two could represent carbon mitigation from increased energy efficiency. Each source has a marginal abatement cost curve (MAC). Regulatory standards mandate a certain amount of abatement from each source. Figure 8.1 depicts two such

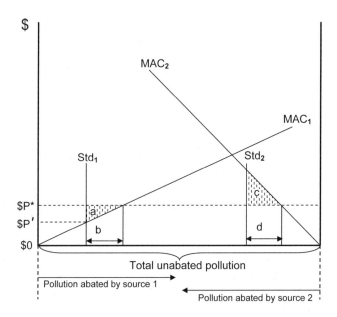

Fig. 8.1   Standards combined with tradable permits are either irrelevant or inefficient

standards, where the standard imposed on source one (Std_1) leads to lower marginal abatement costs than the standard imposed on source two (Std_2). The point of tradable permits is to allow source one to do more abatement and source two to do less abatement, until the MACs are equal (to **P***) and no further gains are possible. The cost savings are areas $c + d - b$, or equivalently the shaded areas $a + c$. These cost savings provide the justification for replacing standards with tradable permits.

In practice, however, US climate legislation will likely contain a tradable permit scheme along with regulatory standards, either because the standards predate the newer tradable permit scheme, or because the new legislation has both parts. For example, Title III of H.R. 2454, the bill the US House of Representatives passed in 2009, would impose a tradable cap on GHG emissions, while Title I of the same bill requires electric utilities to generate up to 25 percent of their output from renewable sources.

First suppose that the standards on the two sources, Std_1 and Std_2 in figure 8.1, are designed to achieve the same total abatement as the permit system acting alone, where the permit-only policy would result in the permit price **P***. Initially, suppose that Std_1 is in effect, that source two faces no standard, and that the permit policy is added on top of the single standard Std_1—a belt and suspenders approach. In this case, the marginal cost to source one of meeting Std_1 is **P'**, which is less than the permit price **P***. For source one, the regulatory standard is effectively irrelevant. Polluters in this situation would choose to do more abatement than required by the standard, even if the standard did not exist. There may be some regulatory costs associated with administering the standard (monitoring, compliance paperwork, etc.), but other than that, the standard has no economic costs.

On the other hand, suppose the single standard is like Std_2 in figure 8.1, combined with the same permit policy with price **P***. Here the marginal cost of meeting the standard exceeds the marginal costs of meeting the tradable emissions cap. By forcing more abatement via source two, Std_2 standard lowers the market price of the tradable permits from **P*** to **P′**, reducing the incentive for polluters to abate via source one (down to the same level as if they had faced only Std_1). In this simple two-source model, the efficiency costs from combining standard two with a cap-and-trade permit policy— belts and suspenders—are the shaded areas, $a + c$, the same as the total efficiency cost of imposing both standards with no tradable permits. The cost savings from the tradable permit scheme are eliminated by the imposition of standard two alone.[1]

Setting aside for a moment the possibility that the standard and permit schemes are mutually reinforcing in some way not described by figure 8.1, how can we tell if the standard is irrelevant like standard one, or costly like standard two? The key distinction is whether the marginal compliance costs for meeting the standard are lower or higher than the cap-and-trade permit price. If the costs from the standard are lower, the standard is largely irrelevant; if the costs from the standard are higher, it imposes real costs.

The CBO (2009a) estimates that the renewable portfolio standards in Title I of H.R. 2454 are like standard one in figure 8.1—largely irrelevant economically because the estimated cost of meeting the standard will fall short of the estimated tradable GHG emissions permit price. By contrast, Abrell and Weigt (2008) examine the European Union's Emissions Trading System, in conjunction with the renewable portfolio standards in Germany. They find the German renewable portfolio standard to be much more costly than the price of GHG permits, and that the renewable standards push the carbon price to zero. In other words, all of the abatement necessary will come from the one source—renewables, despite the fact that other sources are less costly.[2]

This finding is typical. Fullerton, McDermott, and Caulkins (1997) find that forcing electric utilities to abate carbon with scrubbers, rather than by purchasing sulfur dioxide (SO_2) emission permits, increases abatement costs by a multiple of five. Gonzalez (2007) surveys a number of papers that

---

1. Fischer and Preonas (2010) formalize this line of reasoning where a tradable permit system interacts with policies promoting renewable sources of electricity.

2. In fact, if Abrell and Weigt are correct, the cost-inefficiency of Germany's renewable portfolio may have a silver lining. The standards would lead to an excess supply of permits, meaning that they reduce GHG emissions by more than the total required by the carbon cap. In other words, renewables alone as a source of abatement reduce GHG by more than would be reduced by all sources combined under the tradable cap.

examine this tradeoff between tradable emissions permits and renewable electricity standards. The studies he examines find that the coexistence of the two instruments is generally costly, because renewable electricity sources are not typically the least-cost means of abating GHG emissions. For example, Unger and Ahlgren (2005) examine tradable GHG permits for the Nordic countries, and find that a renewable electricity standard of 10 percent reduces carbon emissions at a cost seven times higher than a pure cap-and-trade system.

All of these studies make predictions about whether the nonmarket regulations will be inframarginal, inducing less compliance than predicted by response to cap and trade, or binding, inducing more compliance. This turns out to be a tricky forecast, because the whole rationale for cap and trade is that compliance costs are difficult to predict. In fact, Harrington, Morgenstern, and Nelson (2000) compare ex ante and ex post assessments of US regulations issued by the Environmental Protection Agency (EPA) and the Occupational Safety and Health Administration (OSHA), and find that the ex ante forecasts of costs are typically too high.

> Of the rules initially examined, 14 projected inflated total costs, while pre-regulation estimates were too low for only 3 rules. These exaggerated adjustment costs are often attributable to underestimates of the potential that technological change could minimize pollution abatement costs.

Moreover, the largest overestimates occurred in the case of the market-based policies—taxes and tradable permit schemes, which makes sense because those rules leave polluters the most scope for flexible technological responses. This in turn means that we are more likely to overestimate the costs of a cap-and-trade component of any new climate bill, and less likely to overestimate the costs of any preexisting or accompanying traditional regulatory standards, leaving those standards more likely to interact badly with the permit trading mechanism, reducing its cost effectiveness. Even if we predict that the renewable portfolio standards will be inframarginal, as the CBO (2009a) predicts for the renewable portfolio standards in Title I of H.R. 2434, experience suggests that prediction is likely to overstate the carbon permit prices relative to renewable portfolio standards, and therefore to understate the degree to which the cost-effectiveness of carbon trading is undermined.

In an important sense, the problem here is worse than the usual comparison between standards and tradable permits. In the standard case, highlighted famously in a table in Tietenberg (1990) documenting the efficiency gains from moving to a market-based policy, there is a hidden benefit of traditional regulatory standards. Under standards, some sources of pollution overabate. For example, Atkinson and Lewis's (1974) study of particulates in St. Louis found that a market-based system that equated marginal abatement costs would meet the ambient standards at only one-sixth the cost of existing regulatory standards. But Oates, Portney, and McGartland (1989) point out that one of the reasons the regulatory standard's costs are high is that they overregulate some sources in order to meet the ambient pollution standard everywhere. An ideally designed market-based system would just meet the constraint at every locale, and hence yield more pollution in some places than would the nonmarket standard. If we take into account the *net* benefits of the market-based standard (net of those excess abatement benefits), the difference between market-based and nonmarket regulations is smaller. The key, however, to the Oates and colleagues result is the spatial heterogeneity of pollution. By imposing the same regulatory standard on all locations, some areas inevitably exceed the local ambient standard. A market-based solution that allows overcomplying areas to sell emissions permits until they just meet the local ambient pollution standard would comply with the regulation at lower cost, but impose some new environmental costs on those permit-selling regions. Oates, Portney, and McGartland account for that loss of environmental quality when they tally up the *net* benefits of market-based policies.

For greenhouse gases, however, there would be no such net adjustment, because there are no geographic differences, or "hot spots" in climate change. If a regulatory standard induces overabatement by once source, that depresses the permit price for all sources, reducing abatement by other sources so as to completely offset the overabatement in the first place. In the Oates and colleagues example, the regulatory standard reduces pollution in some locales, without a corresponding increase elsewhere, because all regions must meet the minimum ambient standard. With greenhouse gases, permit trading allows reduced emissions in some locales or by some sources to be completely offset by increased emissions elsewhere. The silver lining of nonmarket policies described by Oates and colleagues does not apply in the case of this global pollutant.

8.4   Rationales for Multiple Policies: Other Market
   Failures and Administrative Complexity

Figure 8.1 and the accompanying text describe two possible results of interacting tradable permit schemes and traditional regulatory standards: the standards could be irrelevant, or they could increase compliance costs with no associated benefits. But there is a third possibility. There could be an economically sound rationale for enacting a tradable permit regulation in combination with a traditional regulatory standard—the belt and suspenders combination could work better than either policy alone. These rationales fall into two broad categories: (a) other market failures, and (b) administrative complexity. While these rationales have been used to justify combining permits and traditional regulations for local air pollutants, such as the criteria air pollutants that have been regulated by the Clean Air Act since the

1970s, not all of the rationales turn out to be applicable to greenhouse gases and global climate change.

Start with the first category: other market failures. The main market failure is, of course, the pollution externality. The GHG emitters do not take into account damages they may impose on others or on future generations. That, however, is unlikely to be the only departure from perfectly competitive assumptions relating to GHG emissions. One additional market failure involves research and development (R&D) in new GHG-abating technologies. If one firm invests in R&D and invents a new abatement technology, or a new energy efficiency technology that by coincidence abates GHG emissions, some benefits from that invention will spill over to other firms, because they either imitate the technology or build upon it with further R&D. Consequently, firms will likely underinvest in R&D, relative to what would be optimal. Jaffe, Newell, and Stavins (2005, 166–67) nicely summarize the interactions between these two market failures: "Pollution creates a negative externality, and so the invisible hand allows too much of it. Technology creates positive externalities, and so the invisible hand produces too little of it."

In theory, however, R&D market failures can work in the opposite direction, and lead to overinvestment relative to the optimum. Competitive firms may duplicate each other's R&D efforts, resulting in wasteful investment by some firms. Similarly, firms may invest in rent-seeking R&D aimed at slight innovations that would replace existing technologies with new ones that are only marginally better, but would capture market rents.[3] On balance, empirical studies find that the industry-wide return to R&D is approximately two to four times as high as the returns to any one firm, suggesting underinvestment in R&D (Jones and Williams 1998).

To correct this underinvestment in R&D, we might consider pairing a tradable permit scheme to address the first market failure with an R&D subsidy to address the second, where the R&D subsidy induces GHG abatements like one of the two regulatory standards in figure 8.1. However, unless there is something else at work here, nothing about the R&D market failure is particular to the environment, and there is no reason a sensible R&D policy shouldn't be economy wide, rather than targeted at GHG-reducing technologies.

In fact, however, there are other factors at work that may justify targeting R&D subsidies at GHG technologies. One such justification involves the seeming insensitivity of consumers and businesses to energy price signals. Hausman (1979) showed that implausibly high discount rates would be needed to justify the choices consumers were making among room air conditioners with varying energy efficiency and prices. This "energy para-

---

3. Jones and Williams (1998) name this spillover aspect of R&D the "standing on shoulders" effect, and the socially wasteful duplication the "stepping on toes" effect.

dox" has been documented many times since then, and has been explained in various ways. Levinson and Niemann (2004) note that for apartment tenants, either the landlords pay for the utility bills and tenants therefore have no incentive to conserve energy on a daily basis, or tenants pay for the utility bills and landlords therefore have no incentive to invest in energy efficient appliances or construction. Any price signals from a tradable permit system would be weakened because either tenants or landlords do not face the true marginal cost of their energy decisions. This might provide a justification for combining a tradable permit policy with an R&D subsidy targeted at energy efficiency.[4] But more likely, it justifies pairing tradable permits with energy efficiency standards and building codes for appliances and construction.[5] Either way, some form of regulatory standard could complement a GHG emissions permit system.[6]

The second broad rationale for pairing traditional regulatory standards with tradable permit schemes involves administrative complexity—difficulty attaching a market price to emissions. One such source of complexity that has been used to justify pairing tradable permits with regulatory standards in analogous contexts, but which would *not* apply to GHG emissions, involves the spatial heterogeneity of damages. Unlike GHGs, the damages from most pollutants vary depending on where they are emitted. This makes organizing and administering a tradable permit scheme difficult. One could imagine, for example, a matrix of pollution transfer coefficients mapping pollution from each location of emission and to each location of deposition (McGartland and Oates 1985). To avoid this, designers of the US SO_2 trading program intentionally simplified the system. One ton of SO_2 is treated the same whether it is emitted in the Midwest and falls on New England, or emitted on the Atlantic coast and drifts out to sea. This spatial heterogeneity means that locations with high abatement costs risk becoming large net purchasers of SO_2 emissions permits and emitters of SO_2, and therefore having high ambient SO_2 concentrations. Some states responded to this by enacting command-and-control regulations on top of the SO_2 trading program, or by prohibiting trades. Wisconsin prevented some local utilities from buying SO_2 permits, and Illinois mandated scrubber installation (Johnstone 2003). These constraints, coupling tradable permit and traditional regulations, can be seen as a costly response to the complexity of regulating heterogeneous sources. But they are *not* relevant to GHG emissions because their justifica-

4. Another explanation for the energy paradox comes from Hassett and Metcalf (1993), who point out that energy prices are uncertain, but that energy-saving investments are irreversible, leading to rational unwillingness to invest. In that case there is no other market failure, and no economic rationale for a second policy instrument.

5. Another might be product labeling, which has been shown to be effective in combination with energy price increases (Newell, Jaffe, and Stavins 1999).

6. Acemoglu et al. (2009) model this formally in an optimal growth model with endogenous technical change and an environmental externality. They show that the optimal policy can involve both a (dynamic) pollution tax and an R&D subsidy directed at the polluting sector.

tion is based on eliminating hot spots of excess pollution, and for climate change no such heterogeneity of damages exists.

A second complexity justification involves uncertainty in predicted abatement costs. Since Weitzman (1974), economists have recognized that uncertainty in marginal pollution abatement costs means there is an important distinction between quantity regulations (cap and trade) and price regulations (pollution taxes). Cap and trade leads to certainty about the quantity of pollution, but uncertainty about the costs imposed on polluters. Pollution taxes yield certain costs, but uncertain pollution quantities. Roberts and Spence (1976) proposed pairing the two, so that a set amount of pollution permits are traded, but polluters can exceed their permitted quantities by paying a pollution tax. The tax puts a known ceiling on the otherwise uncertain permit price. One could also imagine a price floor where the government would agree to purchase all permits at some set price (Pizer 2002). This type of price collar is contained in both the CLEAR Act proposed in 2009 by Senators Cantwell and Collins and the American Power Act proposed in 2010 by Senators Kerry and Lieberman.

These price collars, however, are not the type of multiple instrument setup imagined in figure 8.1, in that both the tradable permits and the price collar are related market-based instruments. Price collars are more accurately described as slightly more elaborate versions of tradable permit policies, a single instrument. Moreover, in several cases where the tradable permit schemes have included price caps, those caps have never been reached and were therefore irrelevant—much as a low-cost standard would be. The Danish CO_2 trading mechanism had a price cap at forty DKK per ton, which was never reached (Johnstone 2003). Similarly, the US Acid Rain program set an initial SO_2 permit price cap at $1,500 per ton. Permit prices mostly traded between $100 and $300, and the price cap was scrapped. In both cases, it seems the existence of the price cap may have appeased worries about extremely high costs and eased passage of the legislation politically, but imposed no economic consequences.

A source of administrative complexity possibly more relevant to climate change involves difficulty monitoring emissions directly, a precondition for administering a tradable permit system. In developing countries where households collect and burn firewood for heat and cooking, administering a tradable permit system for the resulting GHG emissions seems improbable. For the United States, however, where regulated markets already exist for the fuels consumers use for home energy, administering an upstream tradable permit system seems relatively straightforward.

Another oft-cited example of monitoring difficulties involves automobile tailpipe emissions. For criteria pollutants, such as nitrogen oxides (NO_x) and carbon monoxide (CO), tailpipe emissions depend on the nature of the gasoline, the characteristics of the vehicle, and the behavior of the driver. Regulating or permitting tailpipe emissions directly still seems technologically infeasible. And regulating gasoline, vehicle characteristics, or miles driven in isolation would miss the other components. (A gasoline tax provides no incentive to maintain emission control equipment.) The obvious solution is a combination of policies, such as the gasoline tax and new car subsidy studied by Fullerton and West (2002, 2010) or Walls and Palmer (2001). A key difference, however, between the criteria pollutants (NO_x, CO, etc.) and greenhouse gases is that while criteria pollutant emissions depend on automobile and driver characteristics, GHG emissions depend almost exclusively on the carbon content of the fuel and how much is consumed. So a tradable permit system can be administered quite easily, upstream at the level of the fuel suppliers.[7]

Metcalf and Weisbach (2009) address this point directly. They examine the entire inventory of US GHG emissions, and show that 80 percent of those emissions could be covered by a tax or permit-trading policy governing only about 3,000 taxed or regulated entities. The other 20 percent would have to be regulated with traditional standards. So long as polluters in that remaining 20 percent were not allowed to sell permits to the other 80 percent based on their compliance with those standards, there would be no interaction between the two policy instruments. Metcalf and Weisbach's analysis suggests that the administrative complexity argument used to justify combining tradable permits with traditional regulations for other air pollutants does not apply to GHG emissions in the United States.

In sum, these other market failures and sources of administrative complexity can in theory provide economic rationales for combining cap and trade with more traditional standards, but we must be careful. In many cases the rationales do not apply to the case of US greenhouse gas emissions and climate change, because GHG damages do not depend on the location of emissions, and because GHG emissions are more directly related to the characteristics of fuels and can be effectively administered upstream of final users. The most consistent economic rationale for multiple instruments involves either (a) multiple market failures, as with the R&D externality and the landlord/tenant energy paradox; or (b) administrative difficulty assigning permits to GHG emissions, as with nonpoint sources in developing countries. In other cases the rationale is not so clear, and we should ask whether the multiple-policy legislation enacted in Europe and proposed for the United States has an economic basis.

8.5   Conclusion

Climate policy in the United States is likely to combine tradable permits with more traditional regulatory standards. These standards have the

---

7. See Erin Mansur's chapter 11 in this volume: "Upstream versus Downstream Implementation of Climate Policy."

potential to be harmlessly redundant, to reduce the cost-effectiveness of the tradable permits, or to solve a problem involving multiple-market failures or administrative complexity. In the worst-case scenario, if polluters are allowed to sell permits based on their compliance with nonmarket regulations two things could happen: (a) the nonmarket, traditional regulatory portion of a climate bill could reduce the efficiency gains from the market-based tradable permit portion; and (b), the market-based, tradable permit parts of a climate bill could reduce the environmental gains from the traditional regulatory standards. The root cause of both is the same: polluters forced to meet a costly regulatory standard sell permits, reducing their price, and shrinking the market incentives for abatement from other sources.

To assess in advance whether the traditional regulatory components of new legislation are redundant or interact to reduce the cost-effectiveness of the cap-and-trade components, we need to forecast the compliance costs of both components. But as Harrington, Morgenstern, and Nelson (2000) show, we are likely to overstate the compliance costs of cap and trade, relative to traditional regulations, and therefore to understate the degree to which the traditional regulations erode the cost-effectiveness of cap and trade.

If the nonmarket component of legislation has an economic rationale—a second market failure, or difficulty regulating the externality directly—then in a best-case scenario, polluters would not be allowed to sell emissions permits based on compliance with the nonmarket parts of the law. This is a legislative issue, but the economic rationale is that if polluters can meet their tradable caps by complying with the nonmarket regulation, that regulation is either irrelevant and a waste of administrative resources, or binding and damaging to the cost-effectiveness of the cap-and-trade permit system.

References

Abrell, Jan, and Hannes Weigt. 2008. "The Interaction of Emissions Trading and Renewable Energy Promotion." Dresden University of Technology Working Paper no. WP-EGW-05.

Acemoglu, Daron, Philippe Aghion, Leonardo Bursztyn, and David Hemous. 2009. "The Environment and Directed Technical Change." NBER Working Paper no. 15451. Cambridge, MA: National Bureau of Economic Research, October.

Aldy, Joseph, and William Pizer. 2008. "Issues in Designing U.S. Climate Change Policy." Resources for the Future. RFF Discussion Paper no. DP 08-20.

Atkinson, Scott E., and Donald H. Lewis. 1974. "A Cost-Effectiveness Analysis of Alternative Air Quality Control Strategies." *Journal of Environmental Economics and Management* 1:237–50.

Bennear, Lori Snyder, and Robert N. Stavins. 2007. "Second-Best Theory and the Use of Multiple Policy Instruments." *Environ Resource Econ* 37:111–29.

Bohringer, Christoph, Carolyn Fischer, and Knut Einar Rosendahl. 2010. "The

Global Effects of Subglobal Climate Policies." Resources for the Future. RFF Discussion Paper no. 10-48.

Brunnermeier, Smita, and Arik Levinson. 2004. "Examining the Evidence on Environmental Regulations and Industry Location." *Journal of the Environment and Development* 13 (1): 6–41.

Congressional Budget Office (CBO). 2009a. "Cost Estimate for H.R. 2454 American Clean Energy and Security Act of 2009." June 5. Washington, DC: Author.

———. 2009b. "How Regulatory Standards Can Affect a Cap-and-Trade Program for Greenhouse Gases." September 16. Washington, DC: Author.

Eichner, Thomas, and Rudiger Pethig. 2009. "Efficient CO_2 Emissions Control with Emissions Taxes and International Emissions Trading." *European Economic Review* 53:625–35.

Fischer, Carolyn, and Louis Preonas. 2010. "Combining Policies for Renewable Energy: Is the Whole Less than the Sum of Its Parts?" Resources for the Future. RFF Discussion Paper no. DP 10-19.

Fullerton, Don, Shaun P. McDermott, and Jonathan P. Caulkins. 1997. "Sulfur Dioxide Compliance of a Regulated Utility." *Journal of Environmental Economics and Management* 34:32–53.

Fullerton, Don, and Sarah West. 2002. "Can Taxes on Cars and Gasoline Mimic an Unavailable Tax on Emissions?" *Journal of Environmental Economics and Management* 42:135–57.

———. 2010. "Tax and Subsidy Combinations for the Control of Car Pollution." *B.E. Journal of Economic Analysis & Policy* 10 (1): Iss. 1 (Advances), Article 8.

Gonzalez, Pablo del Río. 2007. "The Interaction between Emissions Trading and Renewable Electricity Support Schemes: An Overview of the Literature." *Mitigation and Adaptation Strategies for Global Change* 12:1363–90.

Goulder, Lawrence. 2002. *Environmental Policy Making in Economies With Prior Tax Distortions.* Northampton, MA: Edward Elgar.

Harrington, Winston, Richard D. Morgenstern, and Peter Nelson. 2000. "On the Accuracy of Regulatory Cost Estimates." *Journal of Policy Analysis and Management* 19 (2): 297–322.

Hassett, Kevin A., and Gilbert E. Metcalf. 1993. "Energy Conservation Investment: Do Consumers Discount the Future Correctly?" *Energy Policy* 21 (6): 710–6.

Hausman, J. 1979. "Individual Discount Rates and the Purchase and Utilization of Energy-Using Durables." *Bell Journal of Economics* 10:33–54.

Jaffe, A. B., R. G. Newell, and R. N. Stavins. 2005. "A Tale of Two Market Failures: Technology and Environmental Policy." *Ecological Economics* 54:164–74.

Johnstone, Nick. 2003. *Efficient and Effective Use of Tradeable Permits in Combination with Other Policy Instruments.* Paris: Organization for Economic Cooperation and Development.

Jones, C., and J. C. Williams. 1998. "Measuring the Social Return to R&D." *Quarterly Journal of Economics* 113 (4): 1119–35.

Krugman, Paul. 2010. "Building a Green Economy." *New York Times.* April 7.

Levinson, Arik, and Scott Niemann. 2004. "Energy Use by Apartment Tenants when Landlords Pay for Utilities." *Resource and Energy Economics* 26 (1): 51–75.

McGartland, Albert M., and Wallace E. Oates. 1985. "Marketable Permits for the Prevention of Environmental Deterioration." *Journal of Environmental Economics and Management* 12 (3): 207–28.

Metcalf, Gilbert. 2008. "Using Tax Expenditures to Achieve Energy Policy Goals." *American Economic Review Papers and Proceedings* 98:90–4.

Metcalf, Gilbert, and David Weisbach. 2009. "The Design of a Carbon Tax." *Harvard Environmental Law Review* 33:499–556.

Newell, Richard G., Adam B. Jaffe, and Robert N. Stavins. 1999. "The Induced Innovation Hypotheis and Energy-Saving Technological Change." *Quarterly Journal of Economics* 114 (3): 941–75.

Oates, Wallace E., Paul R. Portney, and Albert M. McGartland. 1989. "The Net Benefits of Incentive-Based Regulation: A Case Study of Environmental Standard Setting." *American Economic Review* 79 (5): 1233–42.

Oikonomou, V., and C. J. Jepma. 2008. "A Framework on Interactions of Climate and Energy Policy Instruments." *Mitigation and Adaptation Strategies for Global Change* 13:131–56.

Pearlstein, Steven. 2009. "Climate-Change Bill Hits Some of the Right Notes but Botches the Refrain." *Washington Post*, Friday, May 22.

Pizer, William A. 2002. "Combining Price and Quantity Controls to Mitigate Global Climate Change." *Journal of Public Economics* 85 (3): 409–34.

Roberts, Marc J., and Michael Spence. 1976. "Effluent Charges and Licenses under Uncertainty." *Journal of Public Economics* 5 (3–4): 193–208.

Sigman, Hilary. 1996. "Cross-Media Pollution: Responses to Restrictions on Chlorinated Solvent Releases." *Land Economics* 72:298–312.

Sijm, J. 2005. "The Interaction between the EU Emissions Trading Scheme and National Energy Policies: A General Framework." *Climate Policy* 5:79–96.

Sorrell, S., and J. Sijm. 2003. "Carbon Trading in the Policy Mix." *Oxford Review of Economic Policy* 19 (3): 420–37.

Tietenberg, Thomas. 1990. "Economic Instruments for Environmental Regulation." *Oxford Review of Economic Policy* 6 (1): 17–33.

Unger, T., and E. O. Ahlgren. 2005. "Impacts of a Common Green Certificate Market on Electricity and CO_2 Emission Markets in the Nordic Countries." *Energy Policy* 33:2152–63.

Walls, Margaret, and Karen Palmer. 2001. "Upstream Pollution, Downstream Waste Disposal, and the Design of Comprehensive Environmental Policies." *Journal of Environmental Economics and Management* 41 (1): 94–108.

Weitzman, Martin. 1974. "Prices vs. Quantities." *Review of Economic Studies* 41 (4): 477–91.

Comment   Gilbert E. Metcalf

In comparison to the large literature on instrument choice, comparatively little has been written on the rationale for multiple policy approaches for reducing greenhouse gas emissions. Thus Arik Levinson's chapter is a welcome addition. Levinson starts from the simple observation that existing approaches to reducing greenhouse gas emissions rely on a patchwork of overlapping policies of various forms. Is this efficient? Are the policies mutually reinforcing or do they work at cross-purposes? Levinson provides a framework for thinking about these questions.

Gilbert E. Metcalf is deputy assistant secretary for environment and energy, US Department of the Treasury. He is on leave from the Department of Economics at Tufts University. The views expressed are those of the author and do not necessarily reflect those of the US Department of the Treasury.

For acknowledgments, sources of research support, and disclosure of the author's material financial relationships, if any, please see http://www.nber.org/chapters/c12139.ack.

As Levinson notes, the simultaneous reliance on cap and trade and other regulations has been termed a "belts and suspenders" approach. One view is that the policies are mutually reinforcing. Another is that they are redundant. A third—and this is the most troubling—is that they work at cross-purposes and raise the cost of reducing emissions. To put it differently, the suspenders may get tangled in the underwear.

The first part of Levinson's chapter provides a framework for sorting out these different views of multiple policy approaches. Put simply, if the marginal cost of abatement of the binding cap-and-trade policy—in equilibrium equal to the permit price—exceeds the marginal cost of achieving the regulatory standard layered on top of the market-based approach, then the regulatory standard is nonbinding and can be viewed as redundant. Conversely, if the marginal cost of abatement from the regulation exceeds the permit price, then the textbook model tells us that we will achieve no additional emission reductions and the cost of meeting the cap in the cap-and-trade system has just been increased. The explanation is straightforward. Consider a cap-and-trade system that limits emissions to one hundred. Now add a regulation stating that some sector must reduce emissions by fifty and assume that in the absence of the regulation this sector would have reduced emissions by twenty to achieve the cap in the cap-and-trade system. The additional thirty units of emission reductions in this sector free up permits that allow an increase in emissions elsewhere in the economy. The result is emissions are limited to one hundred but we have substituted thirty units of high-cost emission reductions for low-cost emission reductions.

Levinson limits his analysis to regulations such as renewable portfolio standards, low-carbon fuel mandates, appliance standards, and other technology mandates. But cap-and-trade policy interacts as well with tax policy, federal loan guarantees, and other subsidies to clean energy production. His analysis can be easily extended to incorporate these other government initiatives. Generally the result is to raise the cost of reducing emissions. As an example, the most recent US budget analysis of tax expenditures shows a jump of nearly three-quarters of a billion dollars per year for the federal technology tax credits (Office of Management and Budget 2010). Some of this is due to California's implementation of a Renewable Portfolio System with a 20 percent mandate by 2010 and 33 percent mandate by 2020.[1]

Levinson's framework for assessing multiple policies can identify situations in which the additional regulation is redundant or counterproductive but it cannot provide any theoretical support for multiple policies being beneficial. Recognizing this, he next considers possible reasons for why multiple policies could be beneficial focusing on two reasons: logistical complexity and other market failures.

1. The California RPS program is described at: http://www.cpuc.ca.gov/PUC/energy/Renewables/index.htm.

Levinson cites as examples of complexity price uncertainty and the attendant call for cost containment mechanisms in cap-and-trade legislation, spatially differentiated damages, and technological barriers to the use of some market-based instruments (e.g., a tax on automobile tailpipe emissions). But none of these are relevant in the climate change realm. As Levinson notes, cost containment mechanisms like price collars are not really multiple instruments. They are in fact hybrid instruments, as has been previously discussed by Roberts and Spence (1976), among others. Hot spots that call for spatially differentiated permit prices are not relevant in the climate change literature.[2] And the technology example he provides is relevant for road pollutants but not for greenhouse gas emissions. One of the appealing characteristics of coal, for example, from a regulatory standpoint, is that the carbon emissions per ton of coal are unaffected by where you impose the carbon-pricing burden.[3] Thus we do not need to price carbon emissions at the electric socket (taking the downstream approach to its limit) but rather can impose the carbon price at an intermediate level (coal-fired electric generating plants) or upstream at the coal mine.[4]

The existence of other market failures is a compelling reason for multiple-policy instruments, but whether the appropriate additional instruments are being proposed is another matter. While perhaps overly simplistic, the Tinbergen view that one needs at least as many instruments as policy goals (and in many cases an equal number of instruments as goals) is relevant here. Take the example that Levinson discusses from Bennear and Stavins (2007) on fishing catch limits and gear restrictions. There are two goals here: to limit overall catch and to limit the catch of certain species. This could be reframed as goals on the catch of specific fish species—and the attendant need for multiple instruments.

The pure public good nature of research and development is an example of a secondary market failure that merits additional instruments. As Levinson himself notes "nothing about the R&D market failure is particular to the environment, and there is no reason a sensible R&D policy shouldn't be economy-wide." In general this is true but it must be qualified. Acemoglu et al. (2009) show that it can be optimal to combine a carbon price with directed environmental research subsidies. The subsidy provides a stimulus to clean environmental technology while avoiding an overly high tax on the dirty technology that would otherwise be needed to stimulate the new technology with its attendant efficiency costs. While this argument may be rele-

2. But see Borenstein, chapter 6 in this volume, for a challenge to the view that the location of greenhouse gas emissions is irrelevant. As Borenstein notes, our understanding of spatial differences in impacts is highly rudimentary.

3. This abstracts from carbon capture and storage. My point is unaffected by the ability to capture and store carbon emissions from burning fossil fuels.

4. Metcalf and Weisbach (2009) discuss the administrative details of implementing a carbon tax including the advantages and disadvantages of imposing the tax on fossil fuels at different levels.

vant for R&D policies that complement carbon pricing, it does not justify the sorts of technology mandates or renewable portfolio standards that are commonly proposed as complementary policies to carbon pricing.

Another example of a market failure that could justify multiple policies cited by Levinson is the "energy paradox," the unwillingness of households and firms to make investments in energy-saving technologies that have apparently high rates of return. One must be cautious before relying on this observation to justify policy. If the source of the energy paradox is the interaction of volatile returns to efficiency investments and irreversibility, then there is no paradox at all (Hassett and Metcalf 1993). The rates of return measured that do not take these factors into account are upwardly biased. Similarly, the paradox may be due to mismeasured returns to efficiency investments (Metcalf and Hassett 1999). In either case there is no market failure and no need for additional instruments. If, on the other hand, investments are being passed up because of principal-agent problems (e.g., landlord-tenant issues), then there is a role for directed policies at rental structures (Levinson and Niemann 2004). But again this would not suggest the sorts of policies that we in fact see being proposed.

The point to all this is that we do have multiple market failures. At the simplest level, our use of fossil fuels entails many other environmental damages beside greenhouse gas emissions (see the detailed treatment of the social cost of energy in National Research Council [2009]). But the existence of multiple market failures does not justify the array of policies that we now observe. This is not meant as a criticism of Levinson's chapter. If there is a criticism at all it is that he is being too generous to the advocates of the current mix of policies that are being proposed. His theory and framework for thinking about multiple policies is quite helpful. The chapter could push a bit more on the question of whether the policies currently being advocated are sensible given the multiple market failures that we observe.

One could take another approach altogether and argue that the policies that are being proposed are not driven by market failures other than climate change at all but rather by the political expediency of needing to build a coalition to pass energy legislation. This is an entirely different tack that leads to a whole different assessment procedure. If this is the motivation for the constellation of policies we see being proposed, then the right way to assess a policy portfolio is whether this portfolio is the least-cost way to build a successful coalition to pass climate change legislation. But that is a different chapter altogether from the one that Levinson has set out to write. His focus is on efficiency. And even if one believed that political coalition building is the rationale for multiple policies, the sort of analysis that Levinson undertakes is important for measuring the costs of different political coalition proposals.

To sum up, this chapter is a valuable addition to the literature on the design of optimal policy with multiple instruments. It is a clearly written and thoughtful analysis of this critically important issue. It goes beyond the standard textbook treatment of policy interventions to address environmental problems by adding some policy realism. It deserves to be widely read and carefully studied.

References

Acemoglu, Daron, Philippe Aghion, Leonardo Bursztyn, and David Hemous. 2009. "The Environment and Directed Technical Change." NBER Working Paper no. 15451. Cambridge, MA: National Bureau of Economic Research, October.

Bennear, Lori Snyder, and Robert N. Stavins. 2007. "Second-Best Theory and the Use of Multiple Policy Instruments." *Environmental and Resource Economics* 37 (1): 111–29.

Hassett, Kevin A., and Gilbert E. Metcalf. 1993. "Energy Conservation Investment: Do Consumers Discount the Future Correctly?" *Energy Policy* 21 (6): 710–6.

Levinson, Arik, and Scott Niemann. 2004. "Energy Use by Apartment Tenants When Landlords Pay for Utilities." *Resource and Energy Economics* 26 (1): 51–75.

Metcalf, Gilbert E., and Kevin A. Hassett. 1999. "Measuring the Energy Savings from Home Improvement Investments: Evidence from Monthly Billing Data." *Review of Economics and Statistics* 81 (3): 516–28.

Metcalf, Gilbert E., and David Weisbach. 2009. "The Design of a Carbon Tax." *Harvard Environmental Law Review* 33 (3): 499–556.

National Research Council. 2009. *Hidden Costs of Energy: Unpriced Consequences of Energy Production and Use.* Washington, DC: National Academies Press.

Office of Management and Budget. 2010. *Budget of the United States Government, Fiscal Year 2011.* Washington, DC US Government Printing Office.

Roberts, Marc J., and Michael Spence. 1976. "Effluent Charges and Licenses under Uncertainty." *Journal of Public Economics* 5 (3–4): 193–208.

Climate Policy and Voluntary Initiatives
An Evaluation of the Connecticut Clean Energy Communities Program

Matthew J. Kotchen

9.1 Introduction

Concern about climate change is having an increasing influence on how companies pursue corporate strategy and individuals make consumption choices. There exists a large and growing literature that seeks to explain why such voluntary initiatives occur and to evaluate their effectiveness. General areas of inquiry include corporate environmental management (e.g., Lyon and Maxwell 2004), voluntary programs (e.g., Morgenstern and Pizer 2007; Potoski and Prakash 2009), and environmentally friendly consumption (e.g., Kotchen 2005, 2006). Despite great enthusiasm for voluntary initiatives, economic theory casts serious doubt on whether they alone can meaningfully address the challenges of climate change. The incentive for free-riding is simply too strong when it comes to the voluntary provision of public goods—especially ones that are global in scale.

It would be a mistake, however, to ignore voluntary initiatives entirely in the pursuit of climate policy based on more centralized instruments. When individual nations, states, and municipalities seek to reduce greenhouse gas emissions, they are volunteering to incur their own costs in the interest of global, public benefits. But even outside of regulatory frameworks, volun-

Matthew J. Kotchen is associate professor of environmental economics and policy at Yale University and faculty research fellow of the National Bureau of Economic Research.

I am grateful to Bryan Garcia (Yale Center for Business and the Environment), Bob Wall (Connecticut Clean Energy Fund), and especially Greg Clendenning (NMR Group Inc.) for helpful discussions about the programs studied in this chapter and for assisting with data collection. I also thank Lucas Davis and Catherine Wolfram for helpful comments on an earlier version of the chapter. For acknowledgments, sources of research support, and disclosure of the author's material financial relationships, if any, please see http://www.nber.org/chapters/c12132.ack.

tary initiatives warrant attention. Real and substantial expenditures are being made in the interest of climate-related corporate strategy, voluntary programs, and "green" goods and services. We should thus seek to maximize the potential benefit of these activities. What is more, and perhaps less recognized, is that voluntary initiatives are effective at increasing awareness, education, and opportunities for leadership. Apart from their potential to reduce emissions, voluntary initiatives are important because of their influence on public support for climate policy more generally.

This chapter considers a question that has received little attention in the literature: Can simple and relatively low-cost government programs effectively promote voluntary initiatives? In particular, what follows is an evaluation of how the Connecticut Clean Energy Communities (CCEC) program affects whether households voluntarily switch to an electricity provider with generation capacity that comes entirely from renewable sources of energy. The results suggest that within participating communities, offering symbolic rewards—that is, municipal solar panels or some other clean-energy technology—upon reaching green-electricity enrollment targets increases the number of household purchases by 35 percent. In effect, the CCEC program is responsible for 7,020 additional households having purchased green electricity. The reduction in greenhouse gas emissions due to these additional purchases comes from a modest and mostly symbolic subsidy. The CCEC program can thus serve as a model for how basic incentive programs can mobilize voluntary initiatives within communities, promote demand for renewable energy, and reduce greenhouse gas emissions. While a growing body of research investigates how "green nudges" can change individual behavior as it relates to the environment, the present chapter evaluates effectiveness of a green nudge applied at the community level.

9.2   Background

The Connecticut State Legislature established the Connecticut Clean Energy Fund (hereafter CTFund) in 2000 in order to stimulate supply and demand of renewable sources of energy within the state. Three CTFund programs are of interest here. The first is a program targeted at the household level, while the second two are targeted at the municipality level.

CT Clean Energy Options (Options Program): As part of Connecticut's Climate Change Action Plan, the Options Program was established in 2005 as a mechanism for electricity customers to purchase green electricity from their utility company. All customers of Connecticut's two major utility companies are eligible, and they are able to choose from two different green-electricity suppliers, Sterling Planet and Community Energy. The two suppliers offer electricity that comes from a similar mix of wind and small-scale hydro sources, and they charge slightly different prices at 1.19¢ and 1.3¢ per kWh, respectively. Residential customers can purchase the green options at either

50 or 100 percent levels of their actual electricity demand. Participation at the 100 percent level for a household that consumes the average amount of electricity in Connecticut (750 kWh/month) costs either $8.93 or $9.75 per month. As of December 2009, a total of 22,776 Connecticut households were Options Program participants, with 83 percent participating at the 100 percent level.[1]

Connecticut Clean Energy Communities (CCEC): In order to stimulate demand for green electricity, the CTFund simultaneously established the CCEC program to incentivize purchases through the Options Program. Qualifying municipalities receive free photovoltaic panels or some other clean-energy technology in proportion to the number of Options Program purchases. The clean-energy technologies are installed at highly visible, public locations within a municipality, including town halls, schools, fire stations, and police stations. The number of free installations is based on the number of designated points earned. Initially, residential sign-ups at the 50 and 100 percent levels counted as half a point and one point, respectively, but a sign-up at any level began counting as one point beginning in November 2008. In order for a municipality to qualify for the CCEC program, however, it must first meet a threshold of either 100 points or a 10 percent participation rate, in addition to having made a commitment to the 20% by 2010 Clean Energy Campaign.[2]

20% by 2010 Clean Energy Campaign (20by2010): Created by the non-profit marketing organization SmartPower, the 20by2010 campaign began in 2003 and thus predates the other programs. Now administered by the CTFund, 20by2010 challenges communities to purchase 20 percent of their energy from renewable sources by 2010. Participation requires that municipalities pass a resolution or issue an official proclamation committing to the challenge. In return, CTFund and SmartPower provide consultation services that can help municipalities reach the goal. Services include information about technology and cost options, media events, task-force formation, and educational materials for use in schools. The 20by2010 program is entirely voluntary, and there is no consequence for failing to meet its goal.

Figure 9.1 illustrates the status of all 169 Connecticut municipalities with respect to enrollment in the CCEC program and 20by2010 campaign as of December 2009. Recall that all CCEC municipalities must also be participants in the 20by2010 campaign. Three municipalities are ineligible because electrical service is supplied by a municipal provider rather than one of the

---

1. The Options Program is also available to commercial customers, but this chapter restricts attention to residential households. See CTFund (2010) for complete details.

2. Qualification also depends on a few other criteria, including municipal government purchases of clean energy and participation in the US Environmental Protection Agency's Community Energy Challenge. In practice, however, the other criteria are generally not binding constraints and have also changed somewhat through time. While the program description here is simplified to focus on factors influencing residential participation, interested readers should refer to CTFund (2010) for complete details about qualification criteria and point conversions.

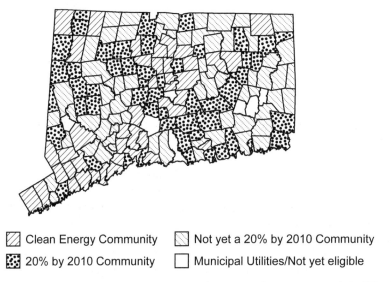

☑ Clean Energy Community   ▨ Not yet a 20% by 2010 Community

▦ 20% by 2010 Community   ☐ Municipal Utilities/Not yet eligible

Fig. 9.1   Most recent program participation among Connecticut municipalities (Recall that CCEC requires 20by2010)
Source: CTFund 2010.

two possible utility companies, United Illuminating and Connecticut Light and Power.

9.3   Data

Data were obtained from the CTFund and were prepared with assistance from NMR Group Inc., the consulting firm that provides ongoing monitoring and evaluation support for administering the CCEC program. The key variables are illustrated graphically in figure 9.2. The upward sloping line indicates the total number of residential households participating in the Options Program by month from June 2005 through December 2009. While only quarterly data is available for 2005, all subsequent observations are monthly counts. The counts sum households participating at the 50 and 100 percent levels through both Sterling Planet and Community Energy.[3] Overall participation increased substantially from 3,383 to 22,776 households. Figure 9.2 also illustrates the percent of eligible municipalities enrolled in the 20by2010 and CCEC programs in each month, and these enrollments have increased substantially over time as well. Participation in the 20by2010 campaign increased from 8 to 57 percent, while participation in the CCEC program increased from 2 to 25 percent. The next section considers how a

3. While future research will investigate differences in participation levels and choices among providers, all types are combined in this chapter to focus simply on overall participation rates.

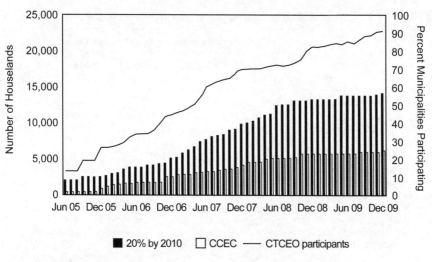

Fig. 9.2   Household and municipality participation in Connecticut green-electricity programs

municipality's enrollment in these programs affects household purchases of green electricity through the Options Program.

Table 9.1 reports descriptive statistics for additional variables at the municipality level that were obtained from the Connecticut Economic Resource Center (CERC) for 2009. Among all municipalities, the mean number of households is 7,699, the mean of median household income is $81,239, and the mean percentage of individuals with at least a bachelor's degree is 47 percent.[4] Moreover, based on municipality data for December 2009, the mean participation rate of households in the Options Program is 2.7 percent. Table 9.1 also reports descriptive statistics separately for municipalities enrolled in the CCEC program, the 20by2010 campaign only, and neither of the two programs. Partitions are based on a municipality's status as of December 2009. Differences in the means reveal that municipalities with more involvement in CTFund programs are larger, have greater incomes, and are more highly educated. Household participation rates in the Options Program are also positively related to involvement in the CTFund's community programs: 4.9 percent for CCEC municipalities, 3.7 percent for 20by2010 only municipalities, and 1.5 percent for municipalities enrolled in neither program.

4. It is worth noting that household income and educational attainment in Connecticut differs substantially from the rest of the nation, where comparable figures are just over $50,000 for median household income and just under 0.30 for the proportion of the population with at least a college degree.

Table 9.1   **Descriptive statistics among different sets of municipalities in 2009**

	Set of municipalities with status as of December 2009			
	All municipalities (Obs. = 166)	Neither program (Obs. = 72)	20by2010 (Obs. = 94)	CCEC (Obs. = 41)
Number of households in	7,699	4,857	9,876	13,482
municipality	(9,302)	(5,924)	(10,756)	(13,346)
Median household income	81,239	79,221	82,784	90,617
($2009s)	(26,254)	(23,514)	(28,200)	(37,131)
Proportion individuals with	0.468	0.421	0.503	0.564
bachelor's degree or more	(0.145)	(0.138)	(0.141)	(0.151)
Options Program participation	0.027	0.015	0.037	0.049
rate in December 2009	(0.036)	(0.012)	(0.044)	(0.058)

Notes: All municipalities are those eligible for the Connecticut Clean Energy Communities (CCEC) program. Statistics reported are means (standard deviations) of the variables. Municipalities listed a CCEC participants are a subset of those reported as 20by2010 participants.

9.4   Analysis

Existing studies use household surveys to investigate variables that explain the decision to participate in price-premium, green-electricity programs (e.g., Clark, Kotchen, and Moore 2003; Kotchen and Moore 2007). The closest thing possible with the Connecticut data is to use municipality characteristics to explain differences in municipality Options Program participation rates. Table 9.2 reports the results of two regression models in which the dependent variable is the natural log of a municipality's Options Program participation rate in December 2009. Model (a) includes number of households, income, and education as explanatory variables. Municipalities that are larger have lower participation rates, more highly educated municipalities have higher participation rates, and municipalities with greater income have lower participation rates.[5] One thousand additional households in a municipality is associated with a 3 percent decrease in the participation rate. A 10 percent increase in the proportion of residents with at least a college degree is associated with a 6 percent increase in the participation rate. Finally, after controlling for education, a $1,000 increase in median income is associated with a 1.6 percent decrease in the participation rate.

Model (b) includes two additional dummy variables for whether in December 2009 the municipality is enrolled in the CCEC program or the 20by2010 campaign. These variables are included in the model because they are expected to affect participation rates, meaning that excluding them

5. Though the model is not reported here in the interest of brevity, it is worth mentioning that income has the opposite sign in a model that does not include education as an explanatory variable. This underscores the importance of controlling for education when estimating the effect of income on purchases of green electricity. Even allowing for nonlinearities, similar results are found on the linear term if income is included as a quadratic.

Table 9.2   **Linear regression models of household Options Program participation rates by municipality in 2009**

	Model	
	(a)	(b)
Number of households in municipality (1,000s)	–0.030*	–0.045*
	(0.005)	(0.004)
Median household income ($10,000s)	–0.163*	–0.124*
	(0.029)	(0.025)
Proportion individuals with bachelor's degree or more	6.040*	4.348*
	(0.503)	(0.477)
Dummy for 20% by 2010 participant	—	0.403*
		(0.088)
Dummy for CCEC participant	—	0.505*
		(0.109)
Constant	–5.278*	–5.038*
	(0.163)	(0.149)
R-Squared (adjusted)	0.590	0.703

Notes: The dependent variable is the natural log of the municipality participation rate. All models include 166 observations. Standard errors are reported in parentheses. An asterisk indicates statistical significance at the 99 percent level.

renders the model susceptible to omitted variable bias. Hence the important thing to note is that the coefficients on number of households, income, and education do not change substantially. While one might be tempted to interpret the new coefficients as estimates of the program effects on household participation rates, this should be done with caution for at least two reasons. First, the variables account for enrollments in December 2009, but municipalities up until that point had been enrolled for different periods of time, as can be seen in figure 9.2. Second, the CCEC variable is susceptible to some degree of endogeneity because threshold participation rates must be met before a municipality is able to qualify. Despite these caveats, the estimates imply that compared to municipalities enrolled in neither program, those in only 20by2010 have 40 percent higher participation rates on average; and those in CCEC have participation rates that are 90 percent higher, where 50 percent of the difference is due to CCEC enrollment over and above the effect of 20by2010. Turning now to an alternative empirical strategy, these differences are shown to be overestimates of the actual program effects.

A more reliable evaluation of the 20by2010 and CCEC programs is possible using the complete panel of data on participation rates in the Options Program for each municipality from June 2005 through December 2009. Consider a model of the form

$$(1) \quad \ln(participation\_rate_{it}) = \alpha 20by2010_{it} + \beta CCEC_{it} + \mu_i + v_t + \varepsilon_{it},$$

where i indexes municipalities, t indexes each month-year, μ_i is a unique intercept for each municipality, v_t is a unique intercept for each month-year, and ε_{it} is an error term. Advantages of specification (1) are that it controls for changes in participation rates through time that are common to all municipalities and for unobserved time-invariant heterogeneity among municipalities (e.g., differences in size, education, and income).[6] Coefficients on the program variables estimate differences in the average household participation rate when a municipality is enrolled in different CTFund programs. Identification comes entirely from changes within a municipality, which are then averaged across municipalities. The estimate of α captures how participation rates differ during periods when municipalities are enrolled in the 20by2010 program compared to no program, and the estimate of β captures the additional effect of periods when municipalities are enrolled in the CCEC program. The sum $\alpha + \beta$ captures the overall CCEC effect on participation rates because CCEC requires enrollment in the 20by2010 campaign.

Model (a) in table 9.3 reports the fixed effects estimates of equation (1) with standard errors clustered at the municipality level to account for serial correlation. The effect of the 20by2010 program is positive but not statistically different from zero. The effect of the CCEC program is positive and has a high degree of statistical significance: within municipalities, CCEC enrollment is associated with a 39 percent higher participation rate compared to 20by2010 enrollment alone. The overall difference in participation rates associated with CCEC enrollment compared to no program enrollment (i.e., the estimate of $\alpha + \beta$) is 41 percent.

A potential concern with the preceding estimate of the CCEC effect on participation rates is still endogeneity due to the participation threshold for enrollment. Municipalities with more participants in the Options Program are more likely to qualify for CCEC enrollment, and this relationship could lead to an overestimate of β. To address this concern, a useful feature of the data is that qualifying municipalities do not always enroll in the CCEC program. In fact, enrollment occurs in only 62 percent of the periods when municipalities satisfy the qualification threshold. As an alternate specification, model (b) in table 9.3 includes an additional dummy variable to control for the average difference in participation rates due to satisfying the qualification threshold, which is distinct from CCEC enrollment. While the coefficients of interest do not change substantially, they do have lower magnitudes. The 20by2010 effect remains statistically indistinguishable from zero, and the additional CCEC effect reduces to 35 percent. Combining the two coefficients in this model, the overall difference in participation rates

6. Models were also estimated for which the dependent variable is simply the participation rate rather than its natural log. Only the results of specification (1) are reported because they fit the data better and are easier to interpret; however, the sign and significance of all coefficients are robust to both specifications.

Table 9.3   **Fixed effects estimates of program evaluation models**

	Model	
	(a)	(b)
Dummy for 20% by 2010 participation (α)	0.026	0.019
	(0.039)	(0.040)
Dummy for CCEC participation (β)	0.387*	0.350*
	(0.088)	(0.089)
Dummy for satisfying CCEC qualification threshold	—	0.054
		(0.042)
Month-year dummies	Yes	Yes
Observations	8,460	8,460
Municipalities	166	166
R-squared (overall)	0.318	0.319
Estimate of $\alpha + \beta$	0.413*	0.369*
	(0.103)	(0.104)

Notes: The dependent variable is the natural log of the municipality participation rate. Reported in parentheses are standard errors clustered at the municipality level. An asterisk indicates statistical significance at the 99 percent level.

associated with CCEC enrollment, compared to no program enrollment, is 37 percent.

9.5   Conclusion

Can symbolic rewards in the form of publically displayed solar panels in municipalities increase the number of households that purchase price-premium, green electricity? Or more generally, can community-level green nudges affect individual behavior? The CCEC program provides evidence that they can: within municipalities that choose to enroll, household participation rates in the Options Program increase 35 percent. Therefore, based on the observed mean participation rate of 4.9 percent among CCEC communities, 1.27 percent is due to the CCEC program. Within these municipalities, the CCEC program is thus responsible for 7,020 additional household participants in the Options Program, which translates into 31 percent of all household participants statewide in December 2009. Counting all residential Options Program participants in the state, assuming average electricity consumption, and using the observed proportion of 50 and 100 percent sign-ups, the estimated reduction in carbon dioxide emissions is 74,528 metric tonnes. Of this total, 31 percent, or 23,104 metric tonnes, is due to the CCEC program having awarded a total solar capacity of 259 kWs in participating municipalities. Assuming Connecticut would have subsidized installation of these solar panels anyway, the CCEC program provides a model for how simple matching grants can promote voluntary initiatives related to climate change. Whether such initiatives will continue to be as effective after passage of more centralized climate policies—that is, whether voluntary and mandatory initiatives are complements or substitutes—is an important question for future research.

References

Clark, C. F., M. J. Kotchen, and M. R. Moore. 2003. "Internal and External Influences on Pro-Environmental Behavior: Participation in a Green Electricity Program." *Journal of Environmental Psychology* 23:237–46.
Connecticut Clean Energy Fund. 2010. Home page. Available at: http://www.ctcleanenergy.com. Accessed April 15.
Kotchen, M. J. 2005. "Impure Public Goods and the Comparative Statics of Environmentally Friendly Consumption." *Journal of Environmental Economics and Management* 49:281–300.
———. 2006. "Green Markets and Private Provision of Public Goods." *Journal of Political Economy* 114:816–34.
Kotchen, M. J., and M. R. Moore. 2007. "Private Provision of Environmental Public Goods: Household Participation in Green Electricity Programs." *Journal of Environmental Economics and Management* 53:1–16.
Lyon, T. P., and J. W. Maxwell. 2004. *Corporate Environmentalism and Public Policy.* Cambridge: Cambridge University Press.
Morgenstern, R. D., and W. A. Pizer, eds. 2007. *Reality Check: The Nature and Performance of Voluntary Environmental Programs in the United States, Europe, and Japan.* Washington, DC: Resources for the Future Press.
Potoski, M., and A. Prakash, eds. 2009. *Voluntary Programs: A Club Theory Perspective.* Cambridge, MA: MIT Press.

Comment   Lucas W. Davis

The chapter by Matthew Kotchen examines voluntary initiatives to reduce greenhouse gas emissions. In particular, Kotchen considers a green electricity program in Connecticut in which households may volunteer to pay a monthly premium of about ten dollars in exchange for receiving electricity from wind and other renewable energy sources. To encourage households to sign up, a state-run program called Connecticut Clean Energy Communities (CCEC) rewards municipalities that reach certain enrollment targets with free photovoltaic panels or other clean energy technologies that are installed in highly visible public locations within the municipality. Kotchen

Lucas W. Davis is assistant professor of economic analysis and policy at the Haas School of Business, University of California, Berkeley; a faculty affiliate at the Energy Institute at Haas; and a faculty research fellow of the National Bureau of Economic Research.

For acknowledgments, sources of research support, and disclosure of the author's material financial relationships, if any, please see http://www.nber.org/chapters/c12133.ack.

documents the success the program has had in signing up households and argues that such programs can serve as a model for how voluntary initiatives can be used to reduce carbon emissions.

This chapter is relevant to a substantial literature in economics that has emerged over the last thirty years on private provision of public goods. See, for example, Bergstrom, Blume, and Varian (1986) and Andreoni (1988). This literature is relevant to carbon policy because voluntary reductions are likely to continue to be an important component of total carbon abatement, particularly if international cooperation cannot be reached on a carbon tax or cap-and-trade program. The starting point in these models is the standard public goods problem; individuals decide how to allocate a fixed amount of resources between a private good and a public good. When deciding how much to contribute to the public good individuals do not take into account the benefits of their contributions to others, so the total level of provision is inefficiently low. This "free rider" problem is particularly severe in this case because carbon abatement is a global public good.

Viewed in this context, green electricity programs are a bit of a puzzle. Kotchen shows that as of December 2009, 23,000 households had signed up with Connecticut's green electricity program. Even though this is not a large fraction of the 1.4 million total households in the state, any private provision of a global public good is difficult to reconcile with the neoclassical model. Perhaps the simplest explanation for this behavior is that households care not only about the overall level of greenhouse gas reductions, but also about their personal contributions. In the "warm glow" model first described in Andreoni (1989) and Andreoni (1990), households derive utility directly from their personal contributions to a public good, so households will contribute even if these contributions have no perceptible change on the total level of contributions.

Even more interesting, Kotchen finds that the participation rate responds sharply to the CCEC program, increasing by 35 percent within municipalities that chose to enroll in the program. These increases in enrollment come in response to modest government investments, raising the possibility that other incentive-based projects could represent similarly "low hanging fruit," yielding sizable decreases in carbon emissions at relatively low cost. An important priority for future work will be to confirm this finding in other real-world settings. Again, the existing theoretical literature provides a roadmap. For example, Andreoni (1998) examines charitable organizations that use leadership grants to encourage private giving. Whereas the neoclassical model would suggest that these grants would crowd out private contributions, the chapter shows that leadership grants can increase private contributions when there are increasing returns to giving such as in the case in which households are trying to reach a municipality-level enrollment target. There is also a fascinating literature in experimental economics that examines giving effects of seed money (List and Lucking-Reiley 2002) and matching grants (Karlan and List 2007). This chapter and related work by Matthew Kotchen is a welcome addition to this rich literature.

References

Andreoni, James. 1988. "Privately Provided Public Goods in a Large Economy: The Limits of Altruism." *Journal of Public Economics* 35:57–73.

———. 1989. "Giving with Impure Altruism: Applications to Charity and Ricardian Equivalence." *Journal of Political Economy* 97:1447–58.

———. 1990. "Impure Altruism and Donations to Public Goods: A Theory of Warm-Glow Giving." *Economic Journal* 100:464–77.

———. 1998. "Toward a Theory of Charitable Fundraising." *Journal of Political Economy* 106:1186–213.

Bergstrom, Theodore C., Lawrence E. Blume, and Hal R. Varian. 1986. "On the Private Provision of Public Goods." *Journal of Public Economics* 29:25–49.

Karlan, Dean, and John A. List. 2007. "Does Price Matter in Charitable Giving? Evidence from a Large-Scale Natural Field Experiment." *American Economic Review* 97 (5): 1774–93.

List, John A., and David Lucking-Reiley. 2002. "The Effects of Seed Money and Refunds on Charitable Giving: Experimental Evidence from a University Capital Campaign." *Journal of Political Economy* 110 (1): 215–33.

Updating the Allocation of Greenhouse Gas Emissions Permits in a Federal Cap-and-Trade Program

Meredith Fowlie

10.1 Introduction

A growing sense of urgency is fueling efforts to pass domestic climate change legislation now, rather than waiting for a coordinated global agreement to emerge. Debates about how and when to implement these policies have been dominated by concerns about potentially adverse impacts on domestic industrial competitiveness, trade flows, and emissions leakage. Policymakers are looking to strike an appropriate balance between curbing domestic greenhouse gas (GHG) emissions and protecting the competitive position of domestic manufacturing in the near-term.

Border tax adjustments offer one approach to "leveling the carbon playing field," as discussed in chapter 3 by Krishna in this volume.[1] This chapter considers an alternative approach. Proposed federal climate change legislation includes provisions that would freely allocate emissions allowances to eligible industries using a continuously updated, output-based formula. These free permit allocations are designed to offset both direct and indirect

Meredith Fowlie is assistant professor of agricultural and resource economics at the University of California, Berkeley, and a faculty research fellow of the National Bureau of Economic Research.

I thank Don Fullerton, Lawrence Goulder, Richard McCann, Renger van Nieuwkoop, and Catherine Wolfram for helpful comments. For acknowledgments, sources of research support, and disclosure of the author's material financial relationships, if any, please see http://www.nber.org/chapters/c12142.ack.

1. An important concern with regard to these countervailing measures is that they may not pass World Trade Organization (WTO) scrutiny. Border tax adjustments included in the House bill were criticized by President Obama who noted that "we have to be very careful about sending any protectionist signals" ("Rust Belt Democrats say Obama was 'wrong' to criticize trade provisions," *E&ENews PM,* 07/07/2009). Available at: http://www.eenews.net/public/eenewspm/2009/07/07/1.

compliance costs in eligible sectors, while preserving some incentive for individual firms to reduce their emissions intensity.

The potential benefits of these proposed allocation provisions, including the mitigation of emissions leakage and the moderation of adverse competitiveness impacts, have been well documented (US EPA, EIA, and Treasury 2009). This chapter draws attention to the fact that these benefits come at a cost. When output-based rebates are offered to a subset of the sources in an emissions trading program, a greater share of the mandated emissions reductions must then be achieved by sources excluded from rebating provision. This can significantly undermine the economic efficiency of permit market outcomes.

The chapter makes two important contributions. First, it extends the previous literature on output-based allocation updating in order to characterize cost-benefit trade-offs inherent in proposed output-based allocation updating provisions.[2] A simple analytical model is used to investigate the welfare consequences of allocating permits via output-based updating in one or more industries in a GHG emissions trading program. In a first-best policy setting, output-based permit allocation updating reduces welfare vis-à-vis auctioning or lump-sum permit allocations.[3] If emissions regulation is incomplete (meaning that a subset of the emitting sources are exempt from the regulation for some reason), the benefits of output-based rebating can exceed the costs. The net welfare implications of output-based rebating depend on a variety of factors, including the elasticity of domestic demand and supply, the emissions intensity of domestic and foreign production, and the price responsiveness of imports.

Second, the chapter illustrates how cost-benefit trade-offs can inform decisions about the appropriate scale and scope of these allocation-based incentives. Among the most fundamental questions in the design of cost mitigation measures is: Who should be eligible for this assistance? From an economic efficiency perspective, output-based rebates should only be offered in cases where the benefits to the industry receiving the rebate exceed the costs imposed on other sectors and stakeholders. The analytical model is used to derive eligibility criteria that are consistent with a standard, albeit stylized, welfare maximization concept. This exercise helps to highlight qualitative differences between the eligibility criteria defined in proposed legislation and those derived from a theoretical welfare maximization exercise.

2. A growing literature investigates the efficiency implications of output-based allocation updating. Previous work has demonstrated how output-based allocation updating will generally undermine the efficiency of permit market outcomes in first-best policy settings (Bohringer and Lange 2005; Fischer 2001; Sterner and Muller 2008) and that allocation updating has the potential to be advantageous when there are preexisting distortions to contend with (Bernard, Fischer, and Fox 2007; Fischer 2003; Fischer and Fox 2007).

3. A first-best setting, in this context, is one that is free of market distortions or failures, other than the environmental externality that the emissions regulation is designed to address.

Although this chapter is germane to ongoing policy debates, it is important to put this analysis in context. The underlying model assumes a fairly stylized objective function for the policymaker; political constraints are ignored entirely. In practice, the political viability of any federal climate change policy is going to depend significantly on the distribution of costs and benefits across politically powerful constituencies. Permit allocation is the most important lever that policymakers have to use to alter the distributional implications of an emissions cap-and-trade program, so it seems inevitable that concessions will be made in order to design an emissions trading program that is supported by key stakeholders. An important objective of this chapter is to draw attention to the welfare costs incurred when these concessions come in the form of output-based rebates.

The chapter proceeds as follows. The next section provides an overview of permit allocation design in cap-and-trade programs, with an emphasis on the political economy of these design decisions. Section 10.3 briefly summarizes the output-based rebating provisions in the proposed federal climate change legislation currently being considered by Congress. Section 10.4 presents a simple analytical framework that can be used to characterize the advantages and disadvantages of output-based updating provisions. Section 10.5 brings the analysis to bear upon the eligibility issue. Section 10.6 concludes.

10.2   Permit Allocation as Industry Compensation

Historically, policymakers have chosen between two types of permit allocation approaches: auctioning and grandfathering. Under an auction regime, emissions permits are sold to the highest bidder. In contrast, grandfathered permits are freely distributed in lump-sum to regulated sources based on predetermined, firm-specific characteristics.

In theory, provided standard assumptions are met, the efficiency properties of the permit market equilibrium are achieved regardless of whether permits are auctioned or grandfathered.[4] This so-called "independence property" has important policy implications (Hahn and Stavins 2010). If the initial distribution of permits plays no role in the determination of emissions and abatement outcomes in equilibrium, emissions permits can be freely allocated to pursue political objectives (such as establishing a constituency for the market-based regulation).

Economists have generally argued in favor of auctioning permits when

4. Assumptions include: perfectly competitive input and output markets, no preexisting regulatory distortions (such as factor taxes), zero transaction costs, complete information, lump-sum free allocations, and compliance-cost-minimizing firms. This result is closely related to a seminal paper by Coase (1960) and has been formally demonstrated in an emissions permit market context by Montgomery (1972).

auction revenues can be used to offset factor taxes or other preexisting distortions.[5] However, policymakers have routinely chosen to forego auction revenues in favor of handing permits out for free to regulated entities.[6] The ability to make concessions to adversely impacted and politically powerful stakeholders via grandfathering has played an essential role in securing widespread support for the adoption of emissions trading programs.

A pure grandfathering approach is unlikely to be a politically feasible option in the context of a federal GHG trading program, primarily due to the unprecedented value of the permits to be allocated.[7] A lump-sum allocation of all GHG permits to regulated sources would likely result in significant overcompensation (Bovenberg and Goulder 2001). Pure auctioning is also unlikely because politically powerful industry stakeholders are united in their opposition to this approach (at least in the near term).[8]

In this politically charged climate, output-based updating of permit allocations has emerged as something of a Goldilocks solution. Proposed output-based updating provisions are designed to offset the average effect that emissions regulation would otherwise have on producers' variable operating costs. Industry is compensated—but not overcompensated—for the compliance costs incurred. Because the number of permits a firm is freely allocated is increasing with its output, equilibrium levels of domestic manufacturing activity will exceed those associated with auctioning or grandfathering. This in turn implies larger domestic market shares in trade-exposed markets, fewer manufacturing jobs lost, and less emissions leakage.

The economic benefits and political advantages of output-based updating come with strings attached. An important drawback is that the independence property no longer holds. Making future permit allocations conditional on current production choices undermines the efficiency of the permit market outcome by dampening (or eliminating) incentives for consumers to reduce their consumption of goods produced by industries receiving output-based rebates. Increased production (and emissions) in these industries shifts more of the compliance burden to sources outside the provision. Contingent allocation updating therefore introduces important trade-offs between

5. A summary of the literature that considers the permit allocation design choice in the presence of distorted factor markets is provided by Goulder and Parry (2008).

6. A majority of permits are distributed freely to regulated entities in Southern California's RECLAIM program, the European Union's Emissions Trading Program (EU ETS), the National Acid Rain Program (ARP), and the regional NO_x Budget Trading Program.

7. The Congressional Budget Office estimates that emissions permits allocated annually under the federal cap-and-trade system proposed by the Senate in 2009 could be worth up to $300 billion a year by 2020 (CBO 2009).

8. The US Climate Action Partnership, a nonpartisan coalition comprised of twenty-five major corporations and five leading environmental groups, has urged Congress to use some portion of allowances to buffer the impacts of increased costs to energy consumers, and to provide transitional assistance to trade-exposed and emissions-intensive industry (United States Climate Action Partnership [USCAP], "A Blueprint for Legislative Action," January 2009). Available at: http://www.us-cap.org.

reducing the compliance cost burden for a specific sector and minimizing the overall economic cost of achieving mandated emissions reductions.

10.3   Proposed Measures to Address Near-Term Competitiveness Impacts

Climate change legislation passed in the House in 2009, but ultimately dismissed by the Senate, would have establishd a multisector cap-and-trade system in which a subset of industries are eligible for rebates (in the form of a free permit allocation) for direct and indirect compliance costs.[9] Figure 10.1 illustrates the proposed eligibility criteria. Eligibility is determined at the six-digit NAICS industry-classification level. The size of each industry-specific circle reflects annual greenhouse gas emissions in 2006. The horizontal axis measures energy expenditures as a share of the value of domestic production. The vertical axis measures the combined value of exports and imports as a share of the value of domestic production plus imports. This measure is intended to capture the extent to which an industry is exposed to foreign competition.

An industry is defined to be "presumptively eligible" for output-based rebates if energy intensity or greenhouse gas emissions intensity is at least 5 percent and import penetration is at least 15 percent. Industries with energy or emissions intensities exceeding 20 percent are also eligible regardless of trade intensity. The broken line in figure 10.1 traces out this eligibility threshold. Industries lying to the right of this line are presumptively eligible to receive rebates under this provision.

Recent analysis suggests that forty-four manufacturing industries are presumptively eligible based on these criteria. Taken together, these industries account for 6 percent of all manufacturing employment and 12 percent of the total value of annual manufacturing shipments (US EPA, EIA, and Treasury 2009). Approximately 15 percent of the total allocation is set aside for output-based rebating. This annual set-aside exceeds the total emissions of presumptively eligible industries in 2006.

The potential benefits of this output-based rebating provision have been analyzed in detail. Multiple recent studies of H.R. 2454 predict that output-based rebating will significantly mitigate, if not eliminate, negative impacts on energy-intensive manufacturing outputs and emissions leakage (EIA 2009; US EPA 2009; US EPA, EIA, and Treasury 2009). Although much work has been done to document the benefits of this compensating provision, there have been few, if any, attempts to estimate the costs.

---

9. Direct compliance costs are calculated as the product of the eligible entity's output two years prior and the greenhouse gas emissions intensity for all entities in the sector. Rebates for indirect emissions costs are based on the eligible entity's electricity use, the average electricity intensity in the sector, and an estimate of the emissions intensity of the electricity consumed by the eligible entity.

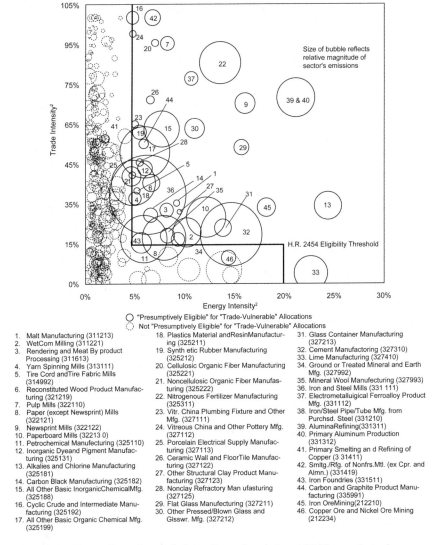

O "Presumptively Eligible" for "Trade-Vulnerable" Allocations
⊙ Not "Presumptively Eligible" for "Trade-Vulnerable" Allocations

1. Malt Manufacturing (311213)
2. WetCom Milling (311221)
3. Rendering and Meat By product Processing (311613)
4. Yarn Spinning Mills (313111)
5. Tire Cord and Tire Fabric Mills (314992)
6. Reconstituted Wood Product Manufacturing (321219)
7. Pulp Mills (322110)
8. Paper (except Newsprint) Mills (322121)
9. Newsprint Mills (322122)
10. Paperboard Mills (32213 0)
11. Petrochemical Manufacturing (325110)
12. Inorganic Dye and Pigment Manufacturing (325131)
13. Alkalies and Chlorine Manufacturing (325181)
14. Carbon Black Manufacturing (325182)
15. All Other Basic Inorganic Chemical Mfg. (325188)
16. Cyclic Crude and Intermediate Manufacturing (325192)
17. All Other Basic Organic Chemical Mfg. (325199)
18. Plastics Material and Resin Manufacturing (325211)
19. Synth etic Rubber Manufacturing (325212)
20. Cellulosic Organic Fiber Manufacturing (325221)
21. Noncellulosic Organic Fiber Manufasturing (325222)
22. Nitrogenous Fertilizer Manufacturing (325311)
23. Vitr. China Plumbing Fixture and Other Mfg. (327111)
24. Vitreous China and Other Pottery Mfg. (327112)
25. Porcelain Electrical Supply Manufacturing (327113)
26. Ceramic Wall and Floor Tile Manufacturing (327122)
27. Other Structural Clay Product Manufacturing (327123)
28. Nonclay Refractory Man ufasturing (327125)
29. Flat Glass Manufacturing (327211)
30. Other Pressed/Blown Glass and Glsswr. Mfg. (327212)
31. Glass Container Manufacturing (327213)
32. Cement Manufactoring (327310)
33. Lime Manufacturing (327410)
34. Ground or Treated Mineral and Earth Mfg. (327992)
35. Mineral Wool Manufacturing (327993)
36. Iron and Steel Mills (331 111)
37. Electrometalluigical Ferroalloy Product Mfg. (331112)
38. Iron/Steel Pipe/Tube Mfg. from Purchsd. Steel (331210)
39. Alumina Refining (331311)
40. Primary Aluminum Production (331312)
41. Primary Smelting an d Refining of Copper (3 31411)
42. Smltg./Rfg. of Nonfrs.Mtl. (ex Cpr. and Almn.) (331419)
43. Iron Foundries (331511)
44. Carbon and Graphite Product Manufacturing (335991)
45. Iron Ore Mining(212210)
46. Copper Ore and Nickel Ore Mining (212234)

Fig. 10.1   Energy intensity, trade intensity, and emissions of US manufacturing sectors at the six-digit NAICS code-level

Source: "The effects if H.R. 2454 on international competitiveness and emissions leakage in energy-intensive and trade-exposed industries: An interagency report responding to a request from Senators Bayh, Specter, Stabenow, McCaskill, and Brown." December 2, 2009; and EPA analysis.

Notes: 1. Petroleum refining is not depicted because it is explicitly excluded from H.R. 2454's allocations to "trade-vulnerable" industries. Also, 91 other sectors, with 126MMTCO2e of emissions, are not depicted due to lack of trade-intensity data. One of these, iron and steel pipe and tube manufacturing (331210; 2.5 MMTCO2e) is expected to be eligible based on language in the bill. Four others meet the energy-intensity threshold, each with two to three MMTCO2e of emissions: beet sugar manufacturing, broadwoven fabric finishing mills, steel foundries (except investment), and metal heat treating. Twelve sectors with a calculated trade intensity greater than 100 percent are depicted here with an intensity of 100 percent (the maximum possible intensity). The two copper sectors (212234 and 331411) do not meet the energy or trade intensity thresholds specified in H.R. 2454, but are expected to be eligible based on other language in the bill.

2. Energy-intensity and trade-intensity measures are as defined in H.R. 2454 and elsewhere in this report.

10.4   The Costs and Benefits of Output-Based Rebating

This section provides a framework for analyzing the cost-benefit trade-offs inherent in output-based allocation updating. To keep the analysis tractable and intuitive, I make several simplifying assumptions:

1. General equilibrium effects, including interactions with preexisting factor taxes, are not considered.

2. Throughout the analysis, the permit price τ is an exogenous parameter, equivalent to assuming that the aggregate marginal abatement cost curve is flat in the neighborhood of the constraint imposed by the emissions cap. This assumption is likely to be approximately true in a federal GHG trading program.[10]

3. I focus exclusively on the short-run implications of output-based rebates. Because output-based rebating is intended as a temporary stop-gap measure, an analysis that conditions on initial technological characteristics is important.[11]

4. Operating costs and emissions rates are assumed to be immutable technology characteristics in the short run. In fact, many industries have some ability to reduce their emissions intensity in the short run through fuel switching or input substitution. Short-run abatement opportunities will lower the costs of output-based updating, all else equal.

5. The model does not capture heterogeneity in cost structure and emissions intensity across producers within an industry. This rules out any reallocation of production to relatively clean firms (which would reduce the costs of output-based rebating).

6. Social welfare is defined to be the value of consumption less the costs of industrial production less costs associated with greenhouse gases emitted as a consequence of this production and consumption.

10.4.1   Rebating Compliance Costs in an Autarkic Industry

I first consider a perfectly competitive industry in which there is no trade with unregulated jurisdictions (i.e., the "autarkic" case). This exercise helps to lay the foundation for the more complicated, trade-exposed industry case. It is relevant to any permit regime that would make industries with no trade exposure, but exceptionally high emissions intensities, eligible for output-based allocations.

---

10. Keohane (2009) estimates the slope of the marginal abatement cost curve in the United States to be 8.0×10^7 \$/GT CO_2 for the period 2010 to 2050 (expressed in present-value terms and in 2005 dollars). If this value is used to crudely approximate the slope of the permit supply function, a 10 percent reduction in the emissions of "presumptively eligible" industries over this forty-year period is associated with only a \$0.25/ton decrese in permit price.

11. Output-based allowance allocations for emissions-intensive US industry are portrayed as a "stop-gap measure." "The Carbon Leakage Prevention Act (H.R. 7146) Output-Based Allowance Allocation for Emissions-Intensive U.S. Industry Rep. Jay Inslee (D-WA) and Rep. Mike Doyle (D-PA)." Available at: http://otrans.3cdn.net/5c61e8367815ece533_7om6bhijz.pdf.

The industry is comprised of N identical sellers producing a homogeneous good q and generating greenhouse gases. These producers have convex cost functions $C(q_i)$, linear marginal costs cNq_i, and a constant emissions rate e per unit of output. Market output is denoted $Q = \sum_{i=1}^{N} q_i$. The inverse demand function is $p(Q) = a - bQ$.

Firms in this industry are required to participate in a greenhouse gas emissions trading program. To remain in compliance, producers must hold sufficient permits to offset their emissions eq. I assume that all firms comply with the program and that the aggregate cap binds such that $\tau > 0$. A firm's short-run profit function is:

$$\pi_i = p(Q)q_i - C(q_i) - \tau(1 - s)eq_i + \tau L_i,$$

where $C(q_i)$ captures firm-level operating costs and s is the rate at which compliance costs are rebated to firms, $s \in (0,1)$.

This simple model nests the three classes of permit allocation regimes under consideration. The firm's lump-sum permit allocation is L_i. Let $\bar{E}$ represent the total number of permits to be allocated for free to this industry. Under complete auctioning, $L_i = 0 \ \forall \ i$ and $s = 0$. Under complete grandfathering, $\sum_i L_i = \bar{E}$ and $s = 0$. Under complete output-based rebating, $L_i = 0 \ \forall \ i$ and $s = \bar{E}/Q$.

The assumption of identical firms implies that $Q = nq_i$. Profit maximization implies that the equilibrium level of output in this perfectly competitive industry is:

(1)
$$Q_A^* = \frac{a - \tau e + s\tau e}{b + c},$$

where the subscript A denotes the autarkic case and c denotes the slope of the aggregate (i.e., industry) marginal cost curve.

Conditioning on the model parameters τ, a, b, and c, we can express the welfare implications of production and pollution activities in this industry as a function of s:

(2)
$$W_A(s) = \int_0^{Q(s)} p(x)dx - \int_0^{Q(s)} C(x)dx - \tau e Q(s).$$

This welfare measure captures the benefits from consumption less the costs of production less damages from industry emissions.

The net welfare impact of offering an output-based rebate (relative to the welfare obtained under a more standard auctioning or grandfathering permit allocation regime) can thus be expressed as:

(3)
$$W_A(s) - W_A(0) = -\frac{e^2 \tau^2 s^2}{2(b+c)} < 0.$$

Figure 10.2 provides a graphical illustration of the partial-equilibrium welfare consequences of output-based allocation updating at a rate of $s = 1$

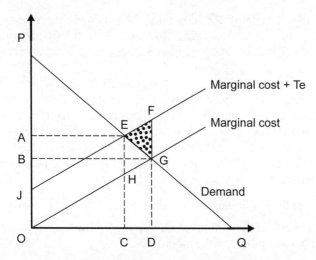

Fig. 10.2   Welfare impacts of an output-based rebate of environmental compliance costs in an emissions-intensive industry with no trade exposure

relative to a baseline policy regime in which permits are grandfathered or auctioned (such that $s = 0$). In the baseline case, quantity C is sold at price A. When compliance costs are rebated in full, a quantity D is sold at a price of B. The net increase in producer and consumer surplus induced by the output-based permit allocation updating is area EGH. Note that rebating also induces an increase in industry emissions of $(D - C)e$.

System-wide emissions are subject to the same binding cap across all allocation regimes, so any rebate-induced increase in emissions from this industry must be offset elsewhere. Put differently, when output-based rebates are offered to this industry, abatement in other industries under the cap or purchases of permits from other countries must rise relative to grandfathering or auctioning levels. By assumption two, there is a sufficient supply of abatement from sources outside the industry to offset this increase in emissions at a per unit cost of τ. The costs of permit allocation updating manifest as an increase in the abatement costs incurred at sources outside this industry. In figure 10.2, this cost is represented by area $EFGH$. Subtracting this rebate-induced cost from the benefits yields a welfare cost equal to the shaded area EFG.[12]

Two insights from this autarkic case are worth highlighting. First, auctioning or grandfathering welfare dominates output-based allocation up-

12. Figure 10.2 also helps to illustrate some of the distributional consequences of output-based rebating. Producers in this industry will prefer the output-based rebating to an auctioning regime; profits increase from AEJ under auctioning to BGO with a full output-based rebate. However, producers will most prefer grandfathering if producer surplus $AEJ + \tau L > BGO$.

dating.[13] This is because the rebate-induced decrease in abatement costs incurred by the industry receiving the rebate is smaller (in absolute value) than the rebate-induced increase in abatement costs incurred in other sectors under the cap.

Second, the net welfare cost of output-based rebating (vis-à-vis grandfathering or auctioning) is increasing with emissions intensity.[14] The costs of output-based updating manifest as increases in the overall costs of achieving the mandated emissions cap. Intuitively, the more emissions-intensive the industry, the larger the effect of a given output-based rebate s on total industry emissions in equilibrium, the greater the required increase in emissions abatement among other sectors and sources.

10.4.2   Rebating Compliance Costs in a Trade-Exposed Industry

In order to extend the analysis to a trade-exposed industry, a linear import supply schedule is added to the model:

$$(4) \qquad p(Q^M) = d + gQ^M,$$

where Q^M represent the quantity of imports supplied at price p. At any price below d, import supply is zero. As the slope of the import supply schedule g approaches infinity, this model reduces to the autarkic case.

Subtracting import supply from aggregate demand yields the residual demand curve faced by domestic producers:

$$(5) \qquad p(Q^D) = \frac{ag+bd}{b+g} - \frac{gb}{b+g}Q^D.$$

Profit maximization by price-taking firms implies that domestic production in equilibrium is:

$$(6) \qquad Q^{D^*} = \frac{bd - b\tau e + bs\tau e + g(s\tau e + a - \tau e)}{bc + g(b+c)}.$$

Note that as the slope of the import supply curve approaches infinity (and import pressure approaches zero) this quantity approaches Q^*_A. Solving for the equilibrium price and substituting into equation (4), imports in equilibrium are:

$$(7) \qquad Q^{M^*} = \frac{ac - bd - cd + b\tau e - bs\tau e}{bc + bg + cg}.$$

13. The analysis in the text omits the following two examples of second-best considerations. First, in an imperfectly competitive industry, the implicit production subsidy can mitigate the preexisting distortion associated with the exercise of market power, and output-based allocation updating can welfare-dominate auctioning or grandfathering, even in the autarkic case. Second, output-based allocations can be used to reduce the distortionary effects of factor tax distortions (Fischer and Fox 2007).

14. To see this, note that the derivative of equation (3) with respect to e is negative. In figure 10.2, the height of the area that defines the net welfare cost of updating is τe. The area of this parallelogram is increasing with e.

Note that equations (6) and (7) together imply that import market share in the absence of emissions regulation, $Q^M/(Q^M + Q^D)$, is $c/(c + g)$.[15]

With imports added to the model, two additional arguments are added to the welfare function:

$$(8)\ W_{TE} = \int_0^{Q(p,s)} p(x)dx - \int_0^{Q^D(p,s)} C(x)dx - pQ^M(p) - \tau e^D Q^D(p,s) - \tau e^M Q^M(p).$$

The third argument in equation (8) captures expenditures on imports. The last argument measures damages from import-related emissions. The emissions intensity of imports is e^M. Emissions in foreign jurisdictions are penalized at the same rate as domestic emissions (τ per unit of emissions). This assumes that the domestic permit price serves as an adequate measure of marginal emissions damages and that the damages caused by an incremental change in emissions are independent of the source. This will be true for greenhouse gases provided there are no co-emissions of local pollutants. The welfare measure in equation (8) ignores any surplus accruing to foreign firms; only costs and benefits affecting domestic stakeholders are accounted for.

Substituting equations (5), (6), and (7) into (8) yields a measure of welfare in terms of the model parameters $a, b, c, d, e, g, \tau, s$. Subtracting $W_{TE}(0)$ from $W_{TE}(s)$ captures the welfare effect of allocation updating vis-à-vis grandfathering or updating. A comprehensive analysis of how this effect varies systematically with different model parameters is beyond the scope of this chapter. Instead, a more general and conceptual discussion provides the essential intuition.

In a trade-exposed and emissions intensive industry, the relative welfare effect $W_{TE}(s) - W_{TE}(0)$ can be decomposed into three parts:

1. The effect on domestic economic surplus (measured by the first three arguments in equation [8]). This effect will be positive for two reasons. Similar to the autarkic case, an increase in the level of production and consumption generates more producer and consumer surplus. Add to this the transfer of surplus from foreign to domestic producers as the share of the domestic market served by foreign imports decreases under updating.

2. The effect on domestic emissions (and associated costs). As in the autarkic case, the rebate-induced increase in production leads to an increase in domestic emissions. All else equal, this increases abatement costs incurred in other industries subject to the cap.

3. The effect on foreign emissions. Foreign imports are reduced under output-based updating, as are the emissions associated with those imports.

---

15. This measure of trade exposure is intended to be analogous to the measure used to determine eligibility for output-based rebates (see figure 10.1). In this simple modeling framework, domestic production and imports are valued at the equilibrium output price and exports are assumed to be zero.

This mitigation of emissions leakage is an important benefit of output-based updating in a trade-exposed industry.

In sum, allocation updating in a trade-exposed industry increases the direct costs of achieving the mandated emissions reductions. However, unlike the autarkic case, it confers additional welfare benefits in the form of leakage mitigation and a transfer of surplus from foreign to domestic producers. These additional benefits will, in some trade-exposed industry contexts, justify the costs of allocation updating. For any given set of model parameters a, b, c, d, τ, s, and g, there is a corresponding threshold emissions intensity below which the benefits of updating exceed the costs.

10.5   Welfare Implications of Output-Based Rebates

The foregoing analysis has implications for determining which industries should receive output-based rebates. In this section, I derive the eligibility criteria used by a policymaker seeking to maximize social welfare as defined by equation (8). In keeping with the provisions in proposed federal legislation, I assume that the output-based rebates will refund compliance costs in full (i.e., $s = 1$) and that eligibility determinations will be based on two observable industry characteristics: a measure of import penetration ($c/[c + g]$), and emissions intensity e.

The derivation proceeds as follows. First, in order to define eligibility criteria in terms of emissions intensity and import penetration parameters exclusively, I must assume values for the other model parameters τ, a, b, c, d, and e^M. Let θ represent a given set of these parameter values. Conditional on θ, I identify all of the e and $c/(c + g)$ combinations that are associated with a welfare level under updating $W_{TE}(1)$ that is greater than or equal to the corresponding welfare level under auctioning or grandfathering $W_{TE}(0)$.

Figure 10.3 illustrates results for two different θ values (θ_1 and θ_2).[16] The solid line represents the welfare-maximizing eligibility threshold associated with θ_1. This line connects all of the combinations of e and $c/(c + g)$ which, given θ_1, yield equivalent welfare outcomes $W_{TE}(1) = W_{TE}(0)$. All points to the left (right) of this line are associated with industry contexts in which output-based updating welfare dominates (is welfare dominated by) auctioning or grandfathering regimes. The broken line is the eligibility threshold associated with a different set of assumed parameter values θ_2.

The most striking difference between the derived thresholds in figure 10.3 and the proposed threshold in figure 10.1 is that the relationship between emissions intensity and eligibility status is reversed. Under proposed allocation designs, the most emissions-intensive industries are presumptively eligible for output-based compensation, presumably because these industries stand to benefit the most from the provision. In figure 10.3, industries with

---

16. The model is not parameterized to represent any industry in particular. Simple parameter values are chosen to maximize expositional clarity (values are reported in figure notes).

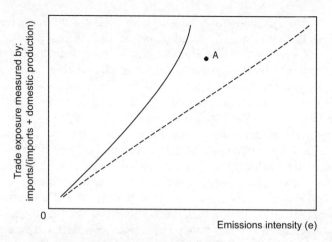

Trade exposure measured by: imports/(imports + domestic production)

0

Emissions intensity (e)

Fig. 10.3   Welfare-maximizing eligibility thresholds

Notes: These eligibility thresholds are derived from the unconstrained welfare-maximization exercise described in the text. Lines connect all points that correspond with a net welfare impact of zero given parameters in θ. Points to the left of the curve are associated with positive welfare changes (i.e., output-based rebating is welfare improving). Points to the right are associated with negative welfare changes. Assumed parameter values associated with the solid line: $\theta_1 = \{a = 50; b = 1; c = 1; d = 0; e^m = 1; \tau = 5; s = 1\}$. The broken line is associated with a set of parameters θ_2 that is identical to θ_1 except that $e^m = 3$. An industry at point A is ineligible given θ_1 because costs exceed the benefits accruing from output-based rebates. This industry is eligible given θ_2 because the benefits—including increased benefits associated with leakage mitigation—outweigh the costs.

high emissions intensities are not eligible for output-based rebates because the benefits accruing to the industry receiving the rebate are smaller than the costs to the economy as a whole.

Figure 10.3 also helps to illustrate how the sign of the net-welfare effect of allocation updating cannot be completely determined based on emissions intensity and import share alone. Put differently, when eligibility rules are determined based on emissions intensity and trade exposure measures exclusively, there is no one eligibility threshold that fits all industries. Parameter values in θ_1 and θ_2 are identical except that the import emissions intensity parameter e^M is higher in θ_2. An industry located at point A is eligible if it can be described using the parameter values in θ_1, but ineligible if it is described by the values in θ_2. Intuitively, the benefits from allocation updating will be greater when imports are more emissions intensive and the emissions leakage potential is greater.

10.6   Conclusion

This chapter presents a framework for thinking about the cost-benefit trade-offs inherent in output-based allocation updating. A simple analytical model is used to examine the welfare impacts of providing output-based rebates to an industry regulated under market-based environmental regulation. In a perfectly competitive industry with no exposure to competition from unregulated imports, these welfare impacts are unambiguously negative. However, when domestic producers compete with firms in less stringently regulated jurisdictions, the benefits of output-based updating may exceed the costs. In this context, the net welfare impacts of introducing output-based rebates will depend on a number of factors, including the emissions intensity of domestic production and the price elasticity of supply and demand.

The chapter concludes with an analysis of one of the most fundamental issues in allocation-based cost mitigation: eligibility. The model is used to demonstrate the stark contrast between the eligibility criteria contained in proposed legislation and those implied by economic welfare maximization.

Although the eligibility requirements in figure 10.1 differ qualitatively from those derived in this chapter, they are consistent with interest group theories of regulation. When policy impacts are concentrated among few and costs are diffusely distributed among many, these few have an incentive to advocate for surplus redistribution (or compensation) at the expense of the larger, but relatively disinterested, many (Olson 1965; Stigler 1971). Output-based rebates offer a politically palatable means of redistributing surplus from foreign firms and the majority of industries where compliance costs are expected to be relatively insignificant (industries to the left of the eligibility threshold in figure 10.1) to a minority of industries that expect to experience significant adverse impacts under federal GHG emissions regulation (industries to the right of the threshold in figure 10.1). A politically viable climate policy regime will need to shelter these politically powerful industries from significant adverse impacts. This chapter draws attention to the costs incurred when output-based rebates are chosen as the vehicle for transferring surplus to these important industries.

References

Bernard, A. L., C. Fischer, and A. K. Fox. 2007. "Is There a Rationale for Output-Based Rebating of Environmental Levies?" *Resource and Energy Economics* 29 (2): 83–101.

Bohringer, C., and A. Lange. 2005. "Economic Implications of Alternative Allocation Schemes for Emission Allowances." *Scandinavian Journal of Economics* 107 (3): 563–81.

Bovenberg, A. Lans, and Lawrence H. Goulder. 2001. "Neutralizing the Adverse Industry Impacts of CO_2 Abatement Policies: What Does It Cost?" In *Behavioral and Distributional Effects of Environmental Policies*, edited by C. Carraro and G. Metcalf. Chicago: University of Chicago Press.

Coase, R. 1960. "The Problem of Social Cost." *Journal of Law and Economics* 3:1–44.

Congressional Budget Office. 2009. *Congressional Budget Office Cost Estimate, December 16, 2009 S. 1733 Clean Energy Jobs and American Power Act As Ordered Reported by the Senate Committee on Environment and Public Works on November 5, 2009.* Washington, DC: CBO.

Energy Information Administration. 2009. *Energy Market and Economic Impacts of H.R. 2454, the American Clean Energy and Security Act of 2009.* Office of Integrated Analysis and Forecasting. US Department of Energy, Washington, DC: EIA.

Fischer, Carolyn. 2001. "Rebating Environmental Policy Revenues: Output-Based Allocations and Tradable Performance Standards." Resources for the Future Discussion Paper no. 01-22.

———. 2003. "Output-Based Allocation of Environmental Policy Revenues and Imperfect Competition." Washington, DC: Resources for the Future. Unpublished Manuscript.

Fischer, C., and A. K. Fox. 2007. "Output-Based Allocation of Emissions Permits for Mitigating Tax and Trade Interactions." *Land Economics* 83 (4): 575–99.

Goulder, Lawrence H., and Ian W. Parry. 2008. "Instrument Choice in Environmental Policy." *Review of Environmental Economics and Policy* 2 (2): 152–74.

Hahn, Robert W., and Robert N. Stavins. 2010. "The Effect of Allowance Allocations on Cap-and-Trade System Performance." Harvard Kennedy School, Mossavar-Rahmani Center for Business and Government. Faculty Research Working Paper no. RWP10-010.

Keohane, Nathaniel O. 2009. "Cap and Trade, Rehabilitated: Using Tradable Permits to Control U.S. Greehouse Gases." *Review of Environmental Economics and Policy* 3 (1): 42–62.

Montgomery, D. W. 1972. "Markets in Licenses and Efficient Pollution Control Programs." *Journal of Economic Theory* 5:395.

Olson, Mancur. 1965. *The Logic of Collective Action.* Cambridge, MA: Harvard University Press.

Sterner, T., and A. Muller. 2008. "Output and Abatement Effects of Allocation Readjustment in Permit Trade." *Climatic Change* 86 (1–2): 33–49.

Stigler, George J. 1971. "The Theory of Economic Regulation." *Bell Journal of Economics and Management Science* Spring:3–21.

US EPA. 2009. *Analysis of the American Clean Energy and Security Act of 2009: H.R. 2454 in the 111th Congress (June 23, 2009).* Available at: http://www.epa.gov/climatechange/economics/economicanalyses.html.

US EPA, EIA, and Treasury. 2009. *The Effects of H.R. 2454 on International Competitiveness and Emission Leakage in Energy-Intensive Trade-Exposed Industries: An Interagency Report Responding to a Request from Senators Bayh, Specter, Stabenow, McCaskill, and Brown.* Available at: http://www.epa.gov/climatechange/economics/economicanalyses.html#interagency.

Comment   Lawrence H. Goulder

Output-based emissions allowance allocation (OBA) has been proposed in climate policy discussions as a way of avoiding international emissions leakage and of preserving the competitive position of energy-intensive, trade-exposed firms. The OBA differs from allocation based on auction-

Lawrence H. Goulder is the Shuzo Nishihara Professor in Environmental and Resource Economics and chair of the Economics Department at Stanford University, a university fellow of Resources for the Future, and a research associate of the National Bureau of Economic Research.

For acknowledgments, sources of research support, and disclosure of the author's material financial relationships, if any, please see http://www.nber.org/chapters/c12143.ack.

ing or grandfathering in that it awards allowances to firms in proportion to the level of their regular output. As a result, OBA effectively subsidizes the output of the qualifying firms.

As Meredith Fowlie's chapter indicates, there have been several prior studies of the potential impacts of OBA on firms' competitive position and on international emissions leakage, but virtually no studies of the implications for efficiency. Her chapter stands out in assessing the efficiency implications of OBA relative to the alternatives of auctioning or grandfathering. The analysis is very clear and carefully done. Fowlie has a knack for taking the elements of a complex system, distilling them, and arriving at very clear analytical expressions showing the equilibrium aspects of the system. The analysis is deceptive: it is so cleanly performed that the analytical problem seems easier to solve than it is!

The efficiency results derived in her chapter are correct (subject to a few qualifications mentioned later). Beyond showing the efficiency impacts, the chapter examines the eligibility criteria for OBA under recent legislation, and compares these criteria with those that would be consistent with economic efficiency. Interestingly, the chapter reveals that the eligibility criteria used by recent legislation are essentially the reverse of what efficiency would call for.

In these comments I will first lay out and comment on Fowlie's results, taking as given the assumptions of her model. Later I will discuss how alternative assumptions might influence the results and offer some broader perspectives.

Efficiency Implications of Output-Based Allocation

The Autarky Case

Fowlie examines the efficiency impacts in an autarkic setting, that is, a setting where there is no international trade and thus no potential for emissions leakage outside of the domestic economy. The model assumes pure competition, no preexisting tax distortions, and fixed marginal abatement costs for firms not qualifying for the output-based allocation. Fowlie finds that, under these circumstances, social welfare is lower than under auctioning or grandfathering. Her results are correct, given the assumptions of the model.

I feel the analysis would be a bit clearer if it decomposed the welfare impacts more. I would recommend decomposing the difference between OBA and the other forms of allocation as follows:

1. Gain in producer and consumer surplus to qualifying firms
2. Loss of taxpayer surplus (under auctioning) or in value of free allowances received by qualifying firms (under grandfathering)
3. Increase in abatement costs to non-qualifying firms

Fowlie's chapter does not separate factors (1) and (2). It considers only the combination, which is described as the "net change in producer and consumer surplus." It took me a bit of work to figure out what is in this net change. It seems clearer to keep factor (2) separate from factor (1)—especially in the case of auctioning, where the loss of taxpayer surplus applies to individuals other than the consumers or producers associated with production by the OBA-eligible firms.

The fundamental reason for Fowlie's (correct) bottom-line result—that moving from auctioning or grandfathering to OBA involves a loss of efficiency in the autarkic case—is that OBA leads to too much output and emissions by the qualifying firms (from an efficiency point of view) and eliminates the equality of marginal abatement costs between qualifying and nonqualifying firms. The abatement costs of nonqualifying firms are higher than the costs of avoiding emissions by forgoing the extra output and emissions that qualifying firms generate under OBA.

The Case with International Trade

Fowlie shows that the efficiency impacts differ in the presence of international trade. In this setting OBA has the potential to increase efficiency relative to other forms of emissions allowance allocation. Her analysis indicates that in this situation there are two other influences on efficiency:

4. Avoidance of environmental damage stemming from international production

5. Transfer of surplus from foreign to domestic producers (as share of domestic market served by foreign imports declines)

The chapter claims that both of these factors contribute positively toward the relative efficiency of OBA.

Factor (4) represents the fact that, by leading to lower foreign-generated emissions (relative to the amount under the other forms of allocation), the domestic policy promotes a larger worldwide reduction in emissions and the environmental damage associated with global emissions.

Factor (5) represents the fact that OBA transfers demand (and surplus) from foreign to domestic firms. This item seems to deserve qualification. Over the longer term, it might not affect efficiency. Balance of payments considerations imply that changes in the value of imports are accompanied by equal-value changes in exports or in international capital inflows. In this case, the overall efficiency impact of item (5) disappears.

The overall efficiency impact—the combined impact of factors (1) through (5)—reflects two fundamental and opposing economic considerations. On the one hand, the basic finding from factors (1) through (3) is that OBA hurts efficiency by removing the equality of marginal abatement costs across domestic firms and leading to too much production by OBA-eligible firms relative to other firms. On the other hand, it supports efficiency by lowering international emissions leakage and the associated environmental damages (factor [4]). Which of the two effects is stronger depends on parameters. The chapter makes an important contribution by showing that, other things equal, as the relative efficiency impact of OBA increases, the *lower* is the emissions intensity of the qualifying firms. This suggests that, if efficiency is the goal, it is the low-emissions-intensity firms that should qualify for OBA. Yet, as nicely demonstrated in the chapter, the climate bill passed by the US House of Representatives in June 2009 provides OBA to the firms with the *highest* emissions intensity. As indicated in the chapter, this seems to be a case where distributional considerations (and associated political factors) are much more powerful than efficiency concerns.

Qualifications and Extensions

How robust are the results from Fowlie's analysis? Here are some additional factors that could influence the results.

Differences between the allowance price and marginal environmental damages. Fowlie's analysis assumes that the market price of allowances (τ) is equal to the marginal environmental damage. If in fact the cap-and-trade system is less (more) stringent, so that the allowance price is lower (higher) than the environmental damage, then the relative efficiency of OBA would be greater (lower) than that in Fowlie's analysis. The reason is that the environmental benefit from reducing international leakage and foreign production would be higher (lower) than that captured in the model.

Preexisting distortionary taxes. Fowlie acknowledges that preexisting distortionary taxes could affect the results. The OBA leads to lower output prices than the other allocation methods. Fischer and Fox (2007) have shown that this mitigates the costly "tax interaction effect" of prior taxes, and thus is an advantage of OBA.

Terms of trade effects. If the domestic economy has monopsony power on international markets, then the OBA-induced reduction in demand for foreign goods could lead to a reduction in the relative price of imports. (See, for example, Neuhoff, Martinez, and Sato 2006.) This would buttress the efficiency impact of OBA.

Rising marginal abatement costs for ineligible domestic firms. This would attenuate the potential efficiency gains from OBA.

Broader Perspectives

This chapter makes an important contribution by laying out the efficiency implications of OBA—and showing that the overall impact depends on competing forces. Additional considerations not captured by the model also work in opposite directions and the overall efficiency impact cannot be determined analytically. It becomes an empirical matter.

As noted in the chapter, much of the support for OBA reflects concerns about international competiveness and leakage rather than efficiency. Fowlie's analysis might offer comfort to the more efficiency-minded, since it indicates that OBA need not be inferior to the alternatives on these grounds.

The chapter also provides hints as to what sort of policy might be more efficient than OBA in addressing leakage. Border taxes are another way to address the leakage problem, and such taxes in principle can avoid distorting domestic abatement efforts (thus avoiding the net cost of factors [1] through [3]) while yielding the benefit from factor (4). Thus border taxes might have an efficiency advantage over OBA. However, it is not clear whether border taxes would violate World Trade Organization rules; furthermore, such taxes might invite retaliatory actions by foreign governments more than OBA would.

References

Fischer, C., and A. K. Fox. 2007. "Output-Based Allocation of Emissions Permits for Mitigating Tax and Trade Interactions." *Land Economics* 83 (4): 575–99.

Neuhoff, K., K. K. Martinez, and M. Sato. 2006. "Allocation, Incentives, and Distortions: The Impact of EU ETS Emissions Allowance Allocations to the Electricity Sector." *Climate Policy* 6 (1): 73–91.

III

Design Features of Climate Policy

11

Upstream versus Downstream Implementation of Climate Policy

Erin T. Mansur

11.1 Introduction

This chapter examines the trade-offs of regulating greenhouse gases (GHG) upstream versus downstream. Upstream regulation focuses on firms producing or importing raw materials that contain GHG like coal, natural gas, and refined petroleum products. In contrast, downstream regulation typically refers to regulating the direct sources of GHG, including motor vehicles, farms, power plants, and other stationary sources. The implications of which sectors to target will depend on four issues discussed in the following: cost-effectiveness, transactions costs, leakage, and offsets.

Before examining these issues, this chapter explores the terms "upstream" and "downstream." Regulation may occur at many different segments of a vertical chain. For this reason, I will refer to the choice of upstream versus downstream regulation as one of regulatory vertical segment selection, or *vertical targeting.* Some industries have short chains, while others have many links.

For example, consider the regulation of carbon dioxide (CO_2) emissions from personal vehicles. The chain begins with worldwide exploration and extraction of crude oil. Firms extract most of the oil used for US transportation internationally. The United States only produces a third of the oil that it consumes (United States Energy Information Administration [EIA] 2008).

Erin T. Mansur is associate professor of economics at Dartmouth College and a research associate of the National Bureau of Economic Research.

I thank Severin Borenstein, Jim Bushnell, Meredith Fowlie, Don Fullerton, Matt Kotchen, Kerry Smith, Chris Snyder, Rob Williams, and participants at the NBER Design and Implementation of US Climate Policy Conference for useful comments on this chapter. For acknowledgments, sources of research support, and disclosure of the author's material financial relationships, if any, please see http://www.nber.org/chapters/c12146.ack.

In the second vertical segment, firms transport crude by pipeline or tanker. Third, the oil reaches a refinery, most likely one of the 150 refineries in the United States. Imports account for approximately 12 percent of US motor gasoline consumption (EIA 2008). Fourth, after refining the crude oil into motor gasoline, the product moves, typically by pipeline, to about 390 major wholesale racks.[1] Fifth, trucks bring it to approximately 105,000 US gasoline stations (United States Census Bureau 2010). Sixth, consumers purchase and pump the gasoline into over 244 million private and commercial registered motor vehicles in the United States (Department of Transportation 2009). While firms and consumers release CO_2 emissions in all six links, in this case, the vast majority occurs during consumption of the final product.

This example illustrates two points regarding vertical targeting. First, the number of firms or consumers involved in each step may differ dramatically. As discussed in the following, optimal regulation occurs at the pollution source (assuming an otherwise functioning market). However, the number of refineries pales in comparison to the number of registered vehicles. If few opportunities exist to abate CO_2 downstream of refining—namely, if wholesale racks, gasoline stations, and motor vehicles cannot sequester some of the carbon content in the gasoline at marginal costs equal to or below carbon prices—then regulating at the refinery level will result in small losses in cost-effectiveness from potential trades but great savings in transactions costs.

Second, the terms "upstream" and "downstream" do not define a specific vertical segment. The upstream industry could mean any one of several industries. In this example, upstream typically refers to refineries, while downstream refers to vehicles. However, in other contexts, "upstream" might mean the polluters and "downstream" might mean consumers. For example, in electricity markets, upstream regulation targets power plants, while downstream refers to regulating retailers, the load serving entities (LSEs). Downstream regulation would require estimating the source of electricity for each LSE and using a carbon price at that level of the vertical chain. The terminology of upstream and downstream must be understood in context. This chapter aims to address: (a) why, in a general setting, regulating polluters directly maximizes social welfare and (b) why this might not apply for carbon policy.[2] In particular, if policies do not target polluters, would a regulation upstream of the pollution source be more cost-effective, or would a downstream one be preferred?

In the following sections, I develop a theoretical model that explains why regulating the source of pollution lowers abatement costs. In particular, if firms can reduce emissions at the end of the pipe, upstream regulations

---

1. Oil Price Information Service (OPIS) collects wholesale gasoline and diesel prices for over 390 racks (http://www.opisnet.com/rack.asp, accessed April 15, 2010).

2. For simplicity, I will refer to all GHG emissions and regulations as carbon emissions and carbon policy, respectively. See the Intergovernmental Panel on Climate Change Fourth Assessment Report (Solomon et al. 2007) for an explanation of the science of converting various GHG emissions into carbon dioxide equivalent emissions.

may miss these options. Next, I discuss three mechanisms that may affect regulators choice of vertical targeting and how one could account for them in determining a least-cost policy. First, transactions costs from monitoring and enforcing regulations differ dramatically along the vertical chain given the number of consumers or producers involved at each segment. Second, while policy discussions include concerns of leakage, I note how the choice of vertical targeting will affect the degree of leakage. Namely, the supply elasticity of unregulated firms varies by segment. Last, if the point of regulation lies upstream of the pollution source, offsets can reward firms for choosing to abate downstream. I discuss how these offset programs may affect the total costs of a regulation for a given vertical chain. Many consider offsets to provide a trade-off: lower abatement costs but increase total emissions. I show that offsets may even increase both costs and emissions. Taking account for all four aspects of vertical targeting—cost-effectiveness, transactions costs, leakage, and offsets—this chapter provides a model of how costs vary along a vertical chain. The chapter concludes with a brief discussion of other potential issues with vertical targeting and a summary of the main findings.

11.2   Theory of Cost-Effectiveness

This section examines the relative cost-effectiveness of upstream versus downstream regulation.[3] Suppose that firm i produces a single good that results in carbon emissions. The firm maximizes profits π with respect to its output q, the carbon content of its fuel F (measured in carbon/q), and its end-of-pipe emissions rate r (measured as the fraction of a fuel's carbon emitted):

$$(1) \qquad \max_{q,F,r} \pi = P(Q)q - c(q) - a(q, F, r),$$

where the price of the good sold (P) depends on the total industry output Q, and firm costs are denoted $c(q)$ for production (given no carbon regulation) and $a(q, F, r)$ for abatement. Note that Fr equals the typical emissions rate definition. For a given competitive quantity-choosing environment, an unregulated firm will set marginal revenue (MR $\equiv \partial P(Q)q/\partial q$) equal to marginal cost (MC $\equiv c'(q)$) and not abate: $r = 1$, $a = 0$.

Next I write $a(q, F, r)$ as two additive components: $a_{in}(q, F)$ depending only on inputs, and $a_{out}(q, F, r)$ for "end-of-pipe" technologies. Switching to a lower carbon fuel (for example, a vehicle switching from oil-based diesel to biodiesel, or a power plant switching from coal to natural gas) would be

---

3. This chapter relates to several literatures. Schmalensee (1976) compares upstream versus downstream welfare measurements of input-based taxes. The environmental costing literature notes the practical importance of making both inputs and outputs reflect social costs (Smith 1992). Burrows (1977) modeled the input substitution implications of pollution taxes relative to standards. Carlton and Loury (1980) consider the entry and exit implications of taxation policy.

in a_{in}. a_{out} includes other technologies, like installing carbon capture and sequestration (CCS) technology on a power plant, but also any other type of abatement decision that would not be covered by changing inputs. For example, if a refinery changed the product mix to produce more asphalt (which would sequester carbon), then this would also be part of a_{out}.

Consider two possible regulations: a carbon price as an input-based regulation t_{in}, and a carbon price as an end-of-pipe regulation τ_{out}. We can rewrite the firm's objective function in equation (1) as:

$$(2) \quad \max_{q,F,r} \pi = P(Q)q - c(q) - t_{in}F\bar{r}q - \tau_{out}Frq - a_{in}(q,F) - a_{out}(q,F,r),$$

where $\bar{r}$ corresponds to the emissions rate of the firm's unregulated fuel choice. As mentioned in the preceding, an unregulated firm would not abate, $\bar{r} = 1$. In this setting, I write the first-order conditions as:

$$(3) \qquad q: t_{in}F + \tau_{out} = MR - c'(q) - \frac{\partial a_{in}}{\partial q} - \frac{\partial a_{out}}{\partial q},$$

$$(4) \qquad F: t_{in}q + \tau_{out}rq = \frac{-\partial a_{in}}{\partial F} - \frac{\partial a_{out}}{\partial F},$$

$$(5) \qquad r: \tau_{out}Fq = \frac{-\partial a_{out}}{\partial r}.$$

A cost-effective regulation would allow firms to use any means of abating pollution, whether it be end of pipe, input based, or just producing less output. In this case, the regulator would need to be able to monitor the actual emissions rate, r. When feasible, like in the case of power plants that use a continuous emissions monitoring system (CEMS), firms will choose among all possible ways of reducing carbon. To enact this, regulators would set $t_{in} = 0$ and, if socially optimal, $\tau_{out} = MD$, the marginal damages from carbon emissions.[4] From equations (3), (4), and (5), we see that firms have an incentive to reduce pollution on *all* margins and to continue to abate until the carbon price τ_{out} equals the marginal abatement cost (MAC):

$$(6) \qquad \tau_{out} = MAC_{out} = \frac{MR - c'(q) - (\partial a_{in}/\partial q) - (\partial a_{out}/\partial q)}{Fr}$$

$$= -\frac{(\partial a_{in}/\partial F) + (\partial a_{out}/\partial F)}{rq} = -\frac{\partial a_{out}/\partial r}{Fq}.$$

All regulated firms would have similar incentives. Hence, the marginal cost of abatement will be equal across all techniques and all firms: the result being cost-effective.

In contrast, an input-based regulation would set $\tau_{out} = 0$ and, in order

4. Under a tax, regulators would levy a tax τ_{out}, while under a cap-and-trade regulation, permits would be auctioned or grandfathered such that the expected permit price equals τ_{out}.

to be allocatively efficient, t_{in} = MD.[5] In this case, from equation (5), we see that firms have no incentive to abate using end-of-pipe technologies. Furthermore, only under an end-of-pipe regulation, the marginal abatement cost from reducing output or changing inputs depends on the choice of r. While firms will still have incentives to reduce output and improve the carbon content of fuels, some opportunities to abate will be forgone. In equilibrium, all firms would set:

$$(7) \qquad t_{in} = \mathrm{MAC}_{in} = \frac{MR - c'(q) - (\partial a_{in}/\partial q) - (\partial a_{out}/\partial q)}{F}$$

$$= -\frac{(\partial a_{in}/\partial F) + (\partial a_{out}/\partial F)}{q}.$$

If such an approach had been used for sulfur dioxide regulation twenty years ago, firms would only have incentive to switch to low-sulfur coal and not to install scrubbers. Given the number of scrubbers that have been installed because of Title IV of the 1990 Clean Air Act Amendments, an input-based regulation may have been quite costly in that case. In the context of CO_2, CCS's high capital costs may make end-of-pipe opportunities less relevant.

In order to measure the additional costs of using an input-based regulation, one would need to be able to estimate the marginal abatement cost for all techniques. Figure 11.1 depicts how these costs might be determined. As Metcalf and Weisbach (2009) note, a narrow policy will miss out on some opportunities and will result in a steeper marginal abatement cost curve. Figure 11.1 shows this in a slightly different way. The horizontal axis shows the overall amount of abatement required, aggregating over all polluters, by the policy $\hat{A}$. The left vertical axis maps input-based marginal abatement costs, MAC_{in}, as in equation (7). The right vertical axis represents the marginal costs only for end-of-pipe abatement, MAC_{end}. This includes those incentives outlined in equation (6) but not in equation (7):

$$(8) \qquad \mathrm{MAC}^{-1}_{end}(A) \equiv \mathrm{MAC}^{-1}_{out}(A) - \mathrm{MAC}^{-1}_{in}(A).$$

In other words, MAC_{end} accounts for the abatement options resulting from changing r. Where the marginal costs equate ($\mathrm{MAC}_{in} = \mathrm{MAC}_{end}$) at A^*, firms achieve the least-cost option. The shaded area shows the additional costs (AddCost) that firms incur by only being rewarded for changing q and F:

$$(9) \qquad \mathrm{AddCost} = \int_{A^*}^{\hat{A}} \mathrm{MAC}_{in}(x)dx - \int_{A^*}^{\hat{A}} \mathrm{MAC}_{end}(x)dx.$$

---

5. This section looks at extremes of regulating only one vertical segment. However, some combination of upstream and downstream policies could provide incentives for lowering abatement costs but also keep transactions costs low (for example, see Fullerton and Wolverton 2000). The discussion of offsets revisits this issue.

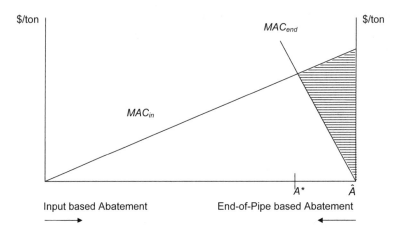

Fig. 11.1   Depiction of marginal abatement costs broken into input-based and other, end-of-pipe abatement

Notes: The horizontal axis is the total amount of abatement required under the cap. The shaded area is the additional costs incurred by only allowing input-based abatement methods to be used.

Under the theoretical assumptions in the preceding, flexibility achieves the lowest overall costs. As a starting point, downstream regulation appears to be the cost-effective policy. Furthermore, dynamic incentives may exacerbate this finding. Firms would have incentive to develop, and invest in, new end-of-pipe abatement technologies if the carbon price were on emissions but not if they face an input-based policy.

11.3   Three Main Concerns of Vertical Targeting

However, regulating at the source of pollution may fail to realize these gains from trade for several reasons. This section highlights three: transactions costs, leakages, and offsets. Transaction costs recognize that monitoring and enforcement become more complex when a vertical segment includes many polluters. Leakage occurs when unregulated firms emit more because of the policy. Vertical targeting will affect leakage: unregulated firms in some vertical links will be more price elastic than others. Upstream policies coupled with offsets may allow for cost-effectiveness. However, asymmetric information could result in greater emissions and greater costs with offsets than without them. The following section discusses some other issues that have been raised on this issue.

11.3.1   Transactions Costs

Transactions costs pose a major hurtle for establishing an end-of-pipe regulation: the cost of monitoring and enforcing regulation for millions of pollution sources could dwarf the benefits from some downstream regulations. In contrast, a regulation upstream of pollution sources could substantially reduce these costs. Metcalf and Weisbach (2009) note that regulating a few thousand fossil-fuel producing companies would account for 80 percent of GHG emissions in the United States. By including some select nonfossil polluters, an additional 10 percent of total emissions would be regulated. Metcalf and Weisbach (2009) argue that the transactions costs of adding these polluters would be modest.

I modify the theory from the previous section to account for these costs. Suppose that regulators incur a cost κ in determining emissions from *each* source. In addition, monitoring the usage and carbon content of each fuel also results in costs. For simplicity, assume the same constant cost κ that society incurs on each input supplier. Furthermore, assume that the decision to regulate upstream or downstream—that is, input-based or end of pipe—is jointly determined for all n pollution sources and m fuel suppliers. A regulator trying to minimize costs now faces a trade-off: regulate end of pipe and incur costs $n\kappa$, or regulate inputs and incur higher abatement costs and some transactions costs AddCost $+ m\kappa$. Note that if $m > n$, then end-of-pipe regulation will always be lower cost (assuming similar transactions costs per firm).

As discussed in the motor vehicle example at the start of this chapter, many segments in the vertical chain could be regulated. In order to minimize overall costs, regulators may consider all V options, where V equals the number of vertical links associated with carbon emissions from one particular sector or industry. Let v^* solve the cost minimization problem:

$$(10) \qquad v^* = \arg \min_{v \in \{1,...V\}} \{\text{AddCost}_v + l_v \kappa\},$$

where l_v equals the number of agents in segment v (e.g., n or m). Note that for the polluting segment, AddCost $= 0$.

In general, moving further upstream (or downstream) from the source of pollution results in forgoing some abatement opportunities. Hence, I expect AddCost to increase monotonically with vertical distance from the pollution source. However, the number of regulated firms may increase or decrease along the vertical chain. In the vehicle example, while the number of vehicles vastly exceeds refineries, more firms extract oil worldwide than own US refineries.

Finally, note that transactions costs depend on technology. In the future, technology will likely improve such that collecting and using information for enforcement becomes even easier. As a result, the cost of regulating more complex vertical levels will likely fall; regulating 250 million vehicles may become feasible. In other words, the optimal vertical targeting of regulation may change over time.

11.3.2   Leakage

Leakage poses a second major concern of upstream versus downstream regulation. If all nations do not harmonize carbon prices, then incomplete regulation will affect the types of goods produced and consumed. Leakage occurs when partial regulation results in an *increase* in emissions in unregulated parts of the economy.[6] The vertical targeting of the policy will affect the magnitude of leakage. Here, leakage could be an issue with either upstream or downstream regulation.

Define the market demand for a good as $Q^D(p)$. We can write the residual demand for regulated firms' output as $Q^{DR}(p) = Q^D(p) - Q^{SU}(p)$, where Q^{SU} represents the supply of firms not regulated. In particular, Q^{SU} will include output from foreign firms. Note that not all foreign production need be unregulated, as firms in some countries already face a carbon price. In addition, many policy proposals include a discussion of border adjustments (for example, see Metcalf and Weisbach 2009). Fischer and Fox (2009) compare the effects on leakage of border taxes versus rebates.

Decomposing market demand into its two components—$Q^{DR}(p)$ and $Q^{SU}(p)$—is useful in understanding the relationship between leakage and vertical targeting. In particular, if market prices increase in equilibrium, residual demand for domestic firms will fall for two reasons. Consumers buy less, which reduces emissions, but also foreign firms produce more, which will increase emissions. These unregulated emissions cause damage. If marginal damages are (locally) constant and equal the carbon price τ, then regulating segment v will result in additional damages (AddDmg):

(11) $$\text{AddDmg}_v = \tau \bar{F} \bar{r} [Q^{SU}(p_1) - Q^{SU}(p_0)],$$

where $\bar{F}$ and $\bar{r}$ represent unregulated firms' fuel carbon content and end-of-pipe emissions rate, and p_1 and p_0 denote the price of good v with and without regulation, respectively. All else equal, a policy that aims at the part of the vertical chain with the least elastic foreign supply will result in the greatest welfare.

This also applies to a multiproduct setting. When close substitutes, more leakage occurs in markets for unregulated goods. In general, more precisely defined markets will have greater substitutes, so fine-tuned regulations may cause greater leakage. Note that this perspective has focused narrowly on the prices of the regulated good. In a general equilibrium setting, prices throughout the vertical chain, and in the rest of the economy, will also be affected. As such, leakage could occur in many industries.

---

6. Many recent papers examine leakage. For example, Fowlie (2009) develops a theory of incomplete regulation. She shows how leakage can, in some cases, increase total emissions relative to no regulation, and in other cases, decrease emissions relative to full regulation. Bushnell and Chen (2009) simulate the Western US electricity grid to examine how various proposals on how permits are allocated would affect the degree of leakage.

One particular type of leakage deserves further examination. Reshuffling occurs when firms do not change production (firms' location, output, and methods stay fixed) but do change where they sell the goods. In electricity markets, reshuffling may occur if regulation requires LSEs to document the sources of purchased power (Bushnell, Peterman, and Wolfram 2008). Unlike leakage, where the location and amount of production of carbon-intensive goods physically changes, reshuffling looks more like an accounting exercise. Producers sell the relatively clean power to the regulated LSEs and the relatively dirty power to others. For goods where transportation is inherently difficult to track, like electricity, regulators may find reshuffling particularly problematic.

Regulators face the issue of reshuffling for other goods with heterogeneous carbon intensities. Within biofuels, for example, some fuels have carbon rates well below that of oil, while others may exceed crude's carbon content. Even with consumer goods, heterogeneity arises due to production technology differences. Suppose that an import tariff were enacted, and regulators could accurately measure the carbon content of the imported goods. We would expect that some reshuffling would take place with only the clean goods coming to the United States and the dirty goods staying in the other country. Unlike with leakage, emissions may not increase with reshuffling.[7] However, import tariffs will only apply to the cleanest goods in equilibrium, limiting their effectiveness in *reducing* emissions.

11.3.3 Offsets

If regulators decide to use upstream regulation, they may consider giving firms credit for choosing options that reduce GHGs downstream. Regulators offer offset programs to lower overall abatement costs while still reducing emissions to a set level (i.e., the cap). However, asymmetric information may cause unintended consequences.

Suppose that regulators have imperfect information regarding how much firms would emit without regulation (i.e., the baseline). Define $\bar{e} \equiv \overline{qFr}$ as regulators' expected baseline. Firms have private information; they know the actual unregulated emissions e^0. After opting in, regulators and firms observe actual emissions $e \equiv qFr$. Finally, I denote actual abatement as $\alpha \equiv e^0 - e$ and regulators' expected abatement as $\bar{\alpha} \equiv \bar{e} - e$.

The objective function for firms facing input-based regulation with offsets is:

$$(12) \quad \max_{q,F,r} \pi = P(Q)q - c(q) - t_{in}Fq - a_{in}(q,F) - a_{out}(q,F,r) + \sigma(r,\bar{e}).$$

---

7. If firms reshuffle through electronic transfers, then emissions will not increase. On the other hand, if reshuffling requires that goods be physically moved to different locations, this would (presumably) increase emissions due to additional transportation.

The subsidy σ commonly takes the form of pollution credits for perceived abatement $\bar{\alpha}$. Regulated firms can use offset credits in lieu of using pollution permits and, thus, equal the carbon price in equilibrium: $\sigma(r,\bar{e}) = t_{in}\bar{\alpha}$.

Asymmetric information over α can result in adverse selection (Montero 1999). Unlike with an end-of-pipe regulation, firms have a choice to opt into an offset program. For a continuous, differentiable abatement technology, a firm will opt in if the marginal subsidy exceeds the marginal abatement costs, $\partial\sigma/\partial r > \partial a_{out}/\partial r$. If marginal abatement costs lie below the carbon price t_{in}, then such adoption could lower total abatement costs across all firms.

Regulators will likely either understate or overstate baseline emissions e^0, and *both* cases may lead to adverse effects. First, if $\bar{e}$ falls substantially below e^0, then a firm with low marginal abatement costs may lack the incentive to reduce r. Even though the firm could reduce emissions at low social costs, the subsidy would be insufficient to provide it with incentives to do so. This type of error will result in forgone cost savings to society. However, these opportunities would also be missed in an input-based regulation without offsets.

The second type of error could actually increase social costs relative to a no-offset regime. In this case, a particularly lucrative subsidy may entice even a firm with high marginal abatement costs to opt in. This will occur if the regulator substantially overstates the baseline emissions, $\bar{e} > e^0$. Given continuous and differentiable abatement costs, a firm could abate just a small amount, $|\Delta r| < \varepsilon$, and receive a large subsidy. The number of credits awarded equal the *perceived* abatement, $\bar{\alpha} > 0$, even though actual abatement α is near zero. In this case, when virtually no actual abatement occurs, society incurs no costs (even those firms receive transfers).

However, for "lumpy" investments, this type of error can result in costs to society. Lumpiness may result from a technological characteristic (CCS may have large capital costs and low marginal costs) or a policy (if regulators can only monitor large changes in r). In either case, firms must now either make a large investment or none at all.

Offsets provide net benefits to society equal to the actual value created (i.e., the carbon price times actual abatement) less the firms' abatement costs: $t_{in}\alpha - a_{out}$. Under a cost-effective policy, firms abate only if the social benefits exceed social costs. If $e^0 = \bar{e}$, offsets would be cost-effective. However, firms with larger predicted baselines, $\bar{e} > e^0$, may have incentive to abate even if doing so reduces social welfare. Even with unbiased estimates, measurement error in the regulators' perceived baseline results in higher costs due to adverse selection. To see this, note that a firm will opt in only if it receives payments greater than cost, $t_{in}\bar{\alpha} > a_{out}$. Thus, offsets increase abatement costs when firms have incentive to opt in ($t_{in}\bar{\alpha} > a_{out}$) even though doing so results in a net loss to society ($t_{in}\alpha < a_{out}$), or:

(13) $$t_{in}\bar{\alpha} > a_{out} > t_{in}\alpha.$$

Some high-cost firms will opt in, and some low-cost firms will opt out.[8]

Furthermore, offsets can result in a form of leakage.[9] If firms abate α but earn credits for $\bar{\alpha}$, then overall emissions increase by $\bar{\alpha} - \alpha$. These additional emissions increase the damages associated with climate change. If damages are locally linear and, if marginal damages equal the carbon price, then these additional emissions cost society $t_{in} \cdot (\bar{\alpha} - \alpha)$.

Combining the net benefits from offsets with the damages from additional emissions, one can measure the overall net losses from offsets (OffLoss) across all firms in link v as:

$$(14) \qquad \text{OffLoss} = \sum_{i=1}^{1} \{[-(t_{in}\alpha - a_{out}) + t_{in} \cdot (\bar{\alpha} - \alpha)] \cdot \mathbf{1}[t_{in}\bar{\alpha} > a_{out}]\},$$

where $\mathbf{1}[\cdot]$ indicates opting in. Note that OffLoss may be positive or negative.

While regulators cannot observe e^0 for each firm, they may know its distribution. In this case, the expected net losses from offsets, $E[\text{OffLoss}_v]$, can help determine the least costly policy. Combining all four components— cost-effectiveness, transactions costs, leakage, and offsets—the link v^{**} minimizes total social costs:

$$(15) \quad v^{**} = \arg \min_{v \in \{1...,V\}} \{\text{AddCost}_v + l_v\kappa + \text{AddDmg}_v + E[\text{OffLoss}_v]\}.$$

11.4   Other Issues of Vertical Targeting

Next, I briefly discuss several other issues that have been raised in the context of upstream versus downstream regulation. These include imperfect competition, regulatory treatment, tax salience, and integrating markets.

11.4.1   Imperfect Competition

With regard to upstream regulation, some raise a concern that imperfect competition amplifies carbon price pass-through. In particular, some argue that input-based carbon prices will be marked up repeatedly in a chain of industries with market power. In contrast, they posit, a downstream carbon price will only affect the last segment of the chain.

Consider three issues regarding imperfect competition and carbon price pass-through. First, while firms with market power have incentives to increase prices above marginal costs, this does not imply that an additional carbon cost will increase market prices by more than the additional cost. Firms optimize by setting marginal revenue equal to marginal costs, and the slope of marginal revenue may be either greater or less than the slope of

---

8. Note that these distortions can persist in the long run as the subsidy reduces the permit price below the cost-effective price τ_{out}.

9. This occurs only if regulators tie the offset program to the cap-and-trade regulation. However, if separate government subsidies or voluntary markets fund offsets and regulated firms cannot use these offsets for compliance, then the additional supply of offsets will not reduce abatement in the regulated market.

inverse demand. Second, when firms exert market power, the theory of the second best applies, and the optimal tax need not equal marginal damages (see, for example, Buchanan 1969). Third, with fixed proportions (whereby firms cannot substitute other inputs to change emissions, that is, $\bar{r} = r$), upstream and downstream regulation will result in the same equilibrium. Chiu, Mansley, and Morgan (1998) refer to this as the irrelevance result.

To see this last point, I use an example of a chain of imperfectly competitive industries. In particular, suppose that a monopolist in one market sells to another downstream monopolist, who then sells to customers. The upstream firm maximizes profits (π_u) by producing q_u at an input price w. The upstream firm incurs costs $c(q_u)$. The downstream firm maximizes profits (π_d) by producing q_d, for which consumers pay p. The downstream firm pays $wq_u + k(q_d)$. Using notation from the previous sections, the regulator will impose either an input-based or an end-of-pipe carbon price. The resulting profit functions equal:

$$\pi_u = w(q_u)q_u - c(q_u) - t_{in}\bar{r}Fq_d$$

$$\pi_d = p(q_d)q_d - wq_u - k(q_d) - \tau_{out} rFq_d.$$

For simplicity, let $q_d = q_u$ and $F = 1$. I write the firms' first order conditions as:

$$w + w'q = c' + t_{in}\bar{r}$$

$$p + p'q = w + k' + \tau_{out}r,$$

or, rearranging terms, the downstream firm's response function as $w = p + p'q - k' - \tau_{out} r$. Thus, solving backward, the upstream firm's first-order condition becomes:

$$p + 3 p'q + p''q^2 - c' - k' - k''q = t_{in}\bar{r} + \tau_{out}r.$$

Note that if $r = \bar{r}$, then an upstream carbon price equates to downstream policy.[10]

11.4.2   Regulation

Metcalf and Weisbach (2009) discuss how regulated industries may treat upstream and downstream policies differentially. For example, if electric utilities face direct, end-of-pipe regulation and receive grandfathered permits, then regulators may limit their ability to pass on marginal cost increases: the opportunity cost of permits in hand may not be treated the same as a purchased permits. In contrast, the same utility may easily pass on higher input prices under upstream regulation. Note that from a social welfare perspective, fully incorporating increases in marginal costs in deter-

---

10. For perfectly competitive downstream markets, firms' first-order condition imply $w = p - k' - \tau_{out}r$. The upstream monopolist maximizes profits by solving $p + p'q - c' - k' - k''q = t_{in}\bar{r} + \tau_{out}r$. Again, the policies are equivalent. Chiu, Mansley, and Morgan (1998) reach the same conclusion for an upstream monopolist selling to downstream Cournot oligopolists.

mining the market equilibrium price will be efficient. Namely, the optimal price would be where marginal social costs equal marginal social benefits, not where price equals average costs.

11.4.3   Tax Salience

Some promote downstream regulation by arguing that a carbon price near the point of emissions (e.g., power plants or gasoline stations) will make the policy more salient for the polluter and, therefore, result in greater response. This argument stems from findings of behavioral economists, who posit that consumers respond more to easily computed taxes. Chetty, Looney, and Kroft (2009) look at state-level alcohol consumption from 1970 to 2003. They find a greater change in consumption with taxes already included in the shelf price (excise taxes) than with taxes applied at the point of sale (sales taxes). Consumers find those taxes already imbedded in the price of the good to be the most salient. Note that these findings suggest that *any* policy in which firms account for carbon costs in the "shelf" price (whether it be because of an increase in fuel prices from input-based regulation or because of an increase in marginal costs directly from an end-of-pipe regulation) would be more effective at changing end users' behavior than a carbon price placed on consumers afterward.

11.4.4   Integrating Markets

The optimal vertical segment of regulation for one emissions source's vertical chain may differ across sources. For example, regulating refineries may minimize costs in the case of vehicles' carbon, while emission source regulation may minimize costs for stationary facilities.

In integrating these different regulations, it will be important, from a cost-effective perspective, that chains do not "cross." Namely, cost-effectiveness will fail if firms pay the carbon price more than once: for example, if a refinery faces a carbon price and then sells its fuel oil to a power plant already paying for emissions, then the outcome will not be least cost. On the other hand, in integrating regulations across markets, establishing trading ratios so that refineries and power plants can trade permits (in dollars per ton of carbon dioxide, for example) will enable greater gains and lower overall costs. If power plants can reduce emissions at a lower marginal cost than can a refinery, then allowing firms to trade across sectors will lower overall costs.

11.5   Conclusions

This chapter sets out some key issues in deciding what level of a vertical chain of industries to target in designing regulation. After developing a model of cost-effectiveness, the chapter examines several reasons why potential gains from trade may not be realized. First, upstream regulation could substantially reduce transactions costs. Regulating a few thousand fossil-fuel producing companies would account for 80 percent of GHG emissions (Metcalf and Weisbach 2009). Second, if all nations do not harmonize carbon prices, then incomplete regulation will affect the types of goods produced, traded, and consumed. The magnitude of regulatory leakage depends on whether policy regulates firms upstream or downstream. Third, offsets have been considered in order to give firms facing upstream regulation with the incentive to choose some downstream options to reduce emissions. While these offsets may result in lower overall abatement costs, they may also have unintended consequences that result in less overall abatement (Montero 1999). This chapter discusses how cost-effectiveness, transactions costs, leakage, and offsets relate to the issue of regulatory vertical segment selection.

References

Buchanan, James. 1969. "External Diseconomies, Corrective Taxes, and Market Structure." *American Economic Review* 59 (1): 174–77.

Burrows, Paul. 1977. "Pollution Control with Variable Production Processes." *Journal of Public Economics* 8:357–67.

Bushnell, James, and Yihsu Chen. 2009. "Regulation, Allocation, and Leakage in Cap-and-Trade Markets for CO_2." NBER Working Paper no. 15495. Cambridge, MA: National Bureau of Economic Research.

Bushnell, James, Carla Peterman, and Catherine Wolfram. 2008. "Local Solutions to Global Problems: Climate Change Policies and Regulatory Jurisdiction." *Review of Environmental Economics and Policy* 2 (2): 175–93.

Carlton, Dennis W., and Glenn C. Loury. 1980. "The Limitations of Pigouvian Taxes as a Long-Run Remedy for Externalities." *Quarterly Journal of Economics* 95 (3): 559–66.

Chetty, Raj, Adam Looney, and Kory Kroft. 2009. "Salience and Taxation: Theory and Evidence." *American Economic Review* 99 (4): 1145–77.

Chiu, Stephen, Edward C. Mansley, and John Morgan. 1998. "Choosing the Right Battlefield for the War on Drugs: An Irrelevance Result." *Economics Letters* 59 (1): 107–11.

Fischer, Carolyn, and Alan K. Fox. 2009. "Comparing Policies to Combat Emissions Leakage: Border Tax Adjustments versus Rebates." RFF Discussion Paper no. 2009-02. Washington, DC: Resources for the Future.

Fowlie, Meredith. 2009. "Incomplete Environmental Regulation, Imperfect Competition, and Emissions Leakage." *American Economic Journal: Economic Policy* 1 (2): 72–112.

Fullerton, Don, and Ann Wolverton. 2000. "Two Generalizations of a Deposit-Refund System." *American Economic Review: Papers and Proceedings* 90 (2): 238–42.

Metcalf, Gilbert, and David Weisbach. 2009. "The Design of a Carbon Tax." *Harvard Environmental Law Review* 33 (2): 499–556.

Montero, Juan-Pablo. 1999. "Voluntary Compliance with Market-Based Environmental Policy: Evidence from the US Acid Rain Program." *Journal of Political Economy* 107:998–1033.

Schmalensee, Richard. 1976. "Another Look at the Social Valuation of Input Price Changes." *American Economic Review* 66 (1): 239–43.

Smith, V. Kerry. 1992. "Environmental Costing for Agriculture: Will It Be Standard Fare in the Farm Bill of 2000?" *American Journal of Agricultural Economics* 74 (5): 1076–88.

Solomon, S., D. Qin, M. Manning, Z. Chen, M. Marquis, K. B. Averyt, M. Tignor, and H. L. Miller. 2007. *Contribution of Working Group I to the Fourth Assessment Report of the Intergovernmental Panel on Climate Change, 2007.* Accessed April 12, 2010. http://www.ipcc.ch/publications_and_data/ar4/wg1/en/contents.html.

United States Census Bureau. 2010. *Quarterly Census of Employment and Wages.* Accessed April 10, 2010. http://www.bls.gov/iag/tgs/iag447.htm.

United States Department of Transportation. 2009. "State Motor-Vehicle Registrations—2008." http://www.fhwa.dot.gov/policyinformation/statistics/2008/pdf/mv1.pdf.

United States Energy Information Administration. 2008. *Annual Energy Review.* DOE/EIA-0384. Accessed April 10, 2010. http://www.eia.doe.gov/emeu/aer/pdf/aer.pdf.

Comment   Roberton C. Williams III

Erin Mansur's chapter provides a concise, clear, and thorough description of the trade-offs between upstream and downstream regulation of an environmental externality, with a particular focus on regulation of greenhouse gases (GHGs). In my comments, I will begin with a brief summary of the chapter's main points and then will go on to describe one additional potentially important factor to consider and to provide further discussion of the immediate policy implications of these points for climate policy.

The comparison of upstream and downstream regulation is often presented as a dichotomous choice, but the chapter points out that there are many different stages in the production process that could be regulated. Nonetheless, the terms are still useful: "upstream" refers to regulation closer to the beginning of the value chain (the stage where polluting inputs first enter the economy) and "downstream" refers to regulation closer to the end of the chain (where consumers use polluting goods).

Regulation provides the most efficient incentives to reduce emissions when it is targeted at the stage where those emissions occur. Regulating upstream of this point provides less efficient incentives. There may be ways to reduce use of a polluting input without actually reducing emissions at all (perhaps by switching from a regulated polluting input to an unregulated but equally

Roberton C. Williams III is associate professor of agricultural and resource economics at the University of Maryland, a senior fellow of Resources for the Future, and a research associate of the National Bureau of Economic Research.

For acknowledgments, sources of research support, and disclosure of the author's material financial relationships, if any, please see http://www.nber.org/chapters/c12147.ack.

polluting input), and upstream regulation would provide an incentive to do so. There may also be ways to reduce emissions without reducing use of polluting inputs (by installing end-of-pipe abatement, for example), and upstream regulation would not provide an incentive for this. Thus, in either case, regulation at the stage where emissions occur would provide the correct incentives, and regulation further upstream would not.

Similarly, regulation downstream of the stage where emissions occur would also be inefficient. It might be possible to reduce use of a polluting final good without reducing emissions or to reduce emissions without reducing use of a polluting final good. In either case, regulating at the stage where emissions occur would provide the correct incentives, and regulation further downstream would not.

The chapter also outlines several other potentially important factors. By targeting regulation at the stage where it is easiest to enforce, regulators can minimize transaction costs. For example, the United States has almost 250 million cars and trucks but only 150 refineries that produce motor fuels, so targeting GHG regulation at the refinery level will likely lead to far lower transaction costs than targeting regulation at individual cars and trucks. Leakage—substitution away from regulated parts of the economy into unregulated but still polluting alternatives (such as industries exempt from regulation or foreign countries that do not regulate GHGs)—is also a concern. Targeting regulation at the stage of production where it is hardest to substitute to unregulated alternatives will minimize leakage. And targeting regulation at the stage at which a given regulated entity's contribution to emissions is easiest to measure will lower costs by improving the accuracy of regulation.[1]

Mansur's chapter does an excellent job of describing all of those issues and providing a simple and clear theoretical framework that incorporates all of them. However, I see one additional issue that could be quite important. In practice, environmental regulation often exempts some industries, firms, or consumers from regulation for primarily political reasons. For example, a politically powerful industry may be left unregulated (or regulated less stringently), as has often been the case for the coal industry. Another example is that older cars are often exempt from emissions rules. These exemptions tend to be quite inefficient. The likelihood of such exemptions depends greatly on which stage of production is regulated. A GHG regulation enforced at the refinery level would make it much harder to exempt older cars than would a regulation enforced at the vehicle level, for example. Thus, this represents

---

1. The chapter describes this last point as being about offsets, but it in fact applies more broadly. Offsets are commonly used in cases where measuring emissions is hard, so it is natural to think of this as being a point about offsets. But if it is difficult to measure a firm's emissions accurately, then this will lead to inefficiency in regulation, regardless of whether the regulation uses offsets or some other regulatory instrument such as tradable permits or an emissions tax.

another potentially important trade-off between upstream and downstream regulation.

My one substantive criticism of Mansur's chapter is that it does not put enough emphasis on the policy implications of all these issues for GHG regulation. For a generic pollutant, all of these issues are potentially important, and so it is not clear whether upstream or downstream regulation will be more efficient. But for the GHG case, some of these issues are likely to be very important, while others are likely to be insignificant. Thus, we can draw clearer conclusions about the relative cost-effectiveness of upstream versus downstream regulation of GHGs.

Carbon capture and storage technology is not yet economically viable and seems unlikely to become economically viable anytime soon. Therefore, carbon emissions are directly proportional to the use of fossil fuel inputs. As a result, there is no cost advantage at present to regulating GHGs at the stage where they are actually emitted versus regulating them further upstream. Conversely, the transaction cost issue is highly important. Upstream regulation entails several orders of magnitude fewer regulated entities than downstream regulation, and, thus, transaction costs will be far lower under upstream regulation.

These points might change in the future (if carbon capture and storage becomes economically viable, for example, or if new technology greatly reduces the transaction costs of downstream regulation). But at present, they imply that upstream regulation will enjoy a substantial cost advantage over downstream regulation.

The Economics of Carbon Offsets

James B. Bushnell

12.1 The Motivation for Offset Markets

The evolution and growth of offset markets is recounted in Lecoq and Ambrosi (2007) and Grubb et al. (2010). The most significant current global offset program, the Clean Development Mechanism, emerged from the Kyoto Treaty. It combined the desires for flexible market-based mechanisms with the goal of financing a low-carbon development trajectory in emerging economies. Offset mechanisms comprise a prominent part of the proposed US CO_2 market articulated in H.R. 2454 (the Waxman-Markey Bill). There are also important roles for offsets in regional US carbon markets such as in California and the northeast United States as well as for voluntary carbon offset markets.

The primary distinction between offset programs and other forms of regulation are that offsets pay firms to *reduce* their emissions rather than raise the costs of continuing to emit. The entire concept of offset programs is, therefore, closely related to the question of the "reach" of traditional regulations. If all sources of emissions would fall under traditional regulations, there would be no need to extend those regulations through offsets. There are many reasons why traditional regulatory measures may be constrained. In the case of greenhouse gas (GHG) emissions, the most obvious is that emissions from any given jurisdiction hold consequences for the entire world. The fact that environmental damages span boundaries far greater than the reach

James B. Bushnell is associate professor of economics at the University of California, Davis, and a research associate of the National Bureau of Economic Research.

I am grateful for helpful discussion and comments from Severin Borenstein, Tristan Brown, Kala Krishna, Erin Mansur, and Stephen Holland. For acknowledgments, sources of research support, and disclosure of the author's material financial relationships, if any, please see http://www.nber.org/chapters/c12156.ack.

of even international organizations makes the consistent application of traditional regulation almost impossible. A second reason is practicality. The effective implementation of cap-and-trade mechanisms requires reasonable monitoring and transactions costs for the sources falling under the cap. Some nonpoint sources of GHG emissions, such as those associated with land use, would be difficult to integrate into a cap-and-trade program under any circumstances. A third reason is political; some sectors may simply be more successful at convincing governments that they should be exempted from mandatory emissions limits.

If we accept the fact that some countries and economic sectors are unlikely to fall under a mandatory limit on their GHG emissions, the question then becomes how best to motivate those sectors to reduce emissions. Ideally, those actions would be coordinated in some fashion with the sectors that are directly regulated. This is where offset markets come into play.

Although the fundamental need for offsets is rooted in the limits of regulatory jurisdiction, today's programs are, in fact, motivated by a host of goals. A primary goal for many regulated industries is cost control. The prospect of a deep pool of offset projects providing a potentially low-cost supply of reductions creates an effective cap on allowance prices in a cap-and-trade system.[1] Among developing nations and many nongovernmental organizations (NGOs), offset mechanisms have been seen as an important new source of capital to aid in development and the alleviation of poverty. For firms and individuals outside of sectors that might fall directly under a cap, such as the US agricultural sector, an offset mechanism offers a potentially lucrative new source of revenue.[2]

From the perspective of economic efficiency, the great promise of an offset market is the potential for reducing GHG emissions at a much lower cost. To the extent that low-cost options for reducing emissions exist in sectors that are not directly regulated under a cap, an offset market allows for these "low-hanging fruit" to be harvested in place of more expensive reductions from the capped sector. For example, if the marginal source of abatement under the European emission trading system (ETS) costs twenty euro per ton, and the opportunity cost of preventing a similar ton of CO_2e through deforestation in Africa is two euro per ton, the same level of CO_2e emissions could be achieved at a cost of eighteen less euros if a European firm were allowed to offset its ETS emissions by financing the African project.

This relates closely to the notion of extending the jurisdiction of a cap. If there were no jurisdictional or measurement issues, these same efficiencies could be reaped by simply placing all relevant sectors under the same cap-and-trade regime. As I discuss in the following, this maximum-cap approach

1. The economic analysis of proposed GHG regulations by agencies such as the US Environmental Protection Agency and the California Air Resources Board highlight the sensitivity of future allowance prices to the cost and availability of offsets.
2. See United States Department of Agriculture (2009).

would also avoid many of the information and incentive problems that are of such concern in offset markets. However, the reality given both domestic and international political and legal constraints is that important sectors and countries will be outside of a binding cap-and-trade regime. The question is, therefore, whether the informational and incentive problems with offset markets can be sufficiently overcome to capture these potential savings.

12.2   Criticisms of Offset Markets

Despite the alluring potential of offset mechanisms for reducing mitigation costs and overcoming jurisdictional boundaries, the programs remain quite controversial. At the heart of most criticisms of offset programs is the concern that the programs are not, in fact, yielding the emissions reductions implied by their transacted quantities. In this section, I discuss the various types of enforcement concerns in the context of the more general economic and regulatory issues to which they are related. In the following section, I explore the various methodologies that have been applied to mitigating these problems.

One can attribute most potential verification and enforcement problems to three key institutional factors that dominate offset programs. An important observation to which I will return, however, is that two of these three factors would apply to *any* regulations directed at mitigating GHG emissions although the interaction of these factors does make the problem worse in the context of offsets. The first factor is jurisdictional. Offset programs test the limits of international regulatory cooperation in that differing regulatory agencies in different countries need to at least agree on consistent measurement and reporting metrics, and officials in the "host" countries of projects need to provide or allow access to data for verification purposes. Another complication from jurisdictional limits are the many types of indirect impacts that climate policies can have on land use, energy consumption, and industrial activity in other jurisdictions. These effects include the leakage of emissions to other jurisdictions as well as the types of indirect land-use questions that have come to play a large role in biofuels and forestry policy. All these indirect impacts have the consequence of reducing the actual net reductions of GHG emissions from the level one might measure by focusing only on "local" reductions.

The second institutional issue relates to the strength of regulatory and governance institutions within many of the countries that might seem to be prime candidates for selling offsets. This is perhaps most pronounced in the context of land-use related offsets.[3] Unfortunately, the development of

3. Murray, Lubowski, and Shongen (2009) highlight the fact that about half the potential GHG savings from the forestry sector comes from Africa and that governance and infrastructure improvements are likely necessary before much of that potential can be reliably tapped.

strict environmental measurement, let alone enforcement, practices is likely beyond the resources of the regulatory institutions in many of these countries. This problem is greatly complicated by the fact that the incentives of officials in differing jurisdictions are often not aligned. Developing countries would like to get access to the capital provided by offset programs and may be less directly concerned about the true mitigation associated with any given project. At least in the context of an offset regime, the enforcement powers in effect reside outside of local jurisdictions. Final accreditation decisions are made by an international governing body in the case of the clean development mechanism (CDM), and by the US Environmental Protection Agency (EPA) in the case of H.R. 2454.

The third issue, to which I will devote the bulk of our attention in this chapter, relates to the fundamental aspect of offset programs. This is the fact that offset programs require a determination of an *emissions baseline* from which the attributable reductions can be measured. Assuming the institutional issues described in the preceding could be overcome, regulators should be able to reliably verify the actual emissions of a facility, or at least a sector. However, baselines (e.g., the emissions in the absence of an offset) by definition cannot be *observed* because they are the product of a "what-if" exercise. The regulator can hope to accurately measure the emissions of a facility after it registers for an offset but can only estimate what those emissions would have been if the facility had not sold any offsets. By contrast, under a cap-and-trade program, the baseline is essentially zero, and firms must provide emissions allowances to offset any emissions observed above zero.

By structuring a program around the concept of paying firms to *reduce* emissions, offset regimes become vulnerable to two classic regulatory problems; moral hazard and adverse selection. The latter involves paying too much to firms with already low emissions, while the former involves firms actively taking steps to inflate their baselines. I discuss each of these issues in the following subsections.

12.2.1   Moral Hazard

The moral hazard, or perverse incentive, problem stems from the fact that emissions baselines are not only the private information of firms, but can also in some cases be readily influenced by those firms. In the offset context, this can take two forms. Firms (or countries) could actively pursue investments in high-carbon sources, with the intent of earning offset payments to drop those investments. Alternatively, firms or countries could delay investments that would lower emissions from existing sources with the same intention.

One of the most controversial offset initiatives has been the funding of hydrofluorocarbon-23 (HFC-23) mitigation under the CDM. This is an extremely potent GHG that is a by-product of industrial coolant manufacturing. Because of its potency, investments to capture HFC-23 emissions qualified for large CDM credits whose value arguably far exceeded the value of the product for which this pollutant was a by-product. In the face of these obviously perverse incentives, it has been argued that firms expanded or maintained operations solely to qualify for CDM payments to capture their by-product.[4] New projects for the capture of HFC-23 may no longer qualify for CDM credits, and activities to capture industrial gases claim an increasingly modest share of newly qualified projects.

12.2.2 Adverse Selection

The primary concern in offset markets is the phenomenon that offset sales will be particularly attractive to firms whose true baselines are lower than the regulators' estimates. These firms can essentially be paid for "reductions" that would have happened anyway. In the jargon of offset policy, this problem is known as *additionality*. In H.R. 2454, additional is defined as:

> The term additional, when used with respect to reductions or avoidance, or to sequestration of greenhouse gases, means reductions, avoidance, or sequestration that result in a lower level of net greenhouse gas emissions or atmospheric concentration than would occur in the absence of an offset project.

The additionality problem has come to dominate the debates over offset markets, and there is a large amount of enforcement language and effort put into trying to mitigate it. There is also a rich literature on environmental regulation under imperfect information that has also focused on this problem. In this literature, the main culprit is adverse selection. Particularly relevant for this discussion is the work of Montero (1999, 2000), which examines the consequences of voluntary "opt-in" to a cap-and-trade program. These opt-in provisions, such in the US SO_2 program, bear many similarities to offset mechanisms. In Montero's derivation, allowing opt-in produces a trade-off between the efficiency gains of lower-cost abatement and the "excess emissions" resulting from adverse selection.

However, some of this focus on additionality and the mechanisms deployed to combat it may be misguided as not all additionality problems may stem from adverse selection. A key issue is the extent to which an overestimate of baselines is a firm-specific or aggregate phenomenon. The regulators information about aggregate emissions is also a factor. If the additivity problem stems from the fact that the regulator overestimated the baselines from the entire sector, then the implications of an offset program can be very different. The result is still less "abatement" than expected, but this does not necessarily translate into more emissions than expected.

Consider the case of the Chinese power sector. As Wara (2008) documents, an impressive percentage of new Chinese power plants received

4. See Wara (2008) Grubb et al. (2010) argue that, despite the incentive problems, the program did result in meaningful early reductions in a very potent GHG.

CDM credits by virtue of their *not* being coal plants. Almost certainly, as Wara argues, some of these plants would have utilized noncoal technology in the absence of an offset payment.[5] However, consider the possibility that future projections of Chinese business as usual (BAU) emissions and, consequently, emissions caps in the developed world assumed that new power plants *would* utilize coal. If this were true, then the BAU projection for the entire Chinese power sector and, therefore, of future global emissions was overstated. Viewed in this light, the CDM provided new information about aggregate emissions and could, in theory, allow for reductions from the capped sector to adjust to this new information.

In the sale of offsets, the key information asymmetry lies in the estimates of BAU emissions, in particular for the uncapped sector. It is common in the mechanism design literature to assume that the regulator knows the distribution of information (here expected emissions, or "baselines") but does not know where any specific firm falls in that distribution. This is the asymmetry framework utilized by Montero (2000). In related work (Bushnell 2010), I represent this as a special case, but it is also important to consider the very real prospect that the regulator may not have perfect information about even the aggregate distribution of baselines. In particular, the regulator may be wrong about the expected mean baseline.

Independently Distributed Baselines

First consider the case where the regulator does know the distribution of baselines but not the baseline of an individual firm. For any given firm, the actual marginal costs of providing offsets might be lower or higher than that of the average firm in the uncapped population. This is because their true baseline emissions from which they must abate may be higher or lower than the regulator's estimate. This "true" cost of offsets reflects the *actual* costs of reducing emissions from a baseline level that differs from the regulator's estimate. Thus, the firms with the lowest actual baselines have the lowest "costs," and, in a competitive market, these will be firms selling offsets. Conversely, it is the high baseline firms for whom offset sales are most expensive who stay out of the offset market.

Because the low-baseline firms participate and the high baseline firms do not, the actual reductions from the uncapped sector will be less than the offsets traded, and total emissions from the uncapped sector will be greater than the official estimate of reductions. Although the regulator's estimate of total baseline emissions from the uncapped sector are correct, the self-selection of low-baseline firms into the offset program leaves only high-baseline firms without abatement. The result, after offsets are transacted, is

5. Haya (2009) provides many examples of energy projects in India that funded under the CDM were not considered additional even by their developers. Lewis (2010), by contrast, emphasizes what she considers a critical role offsets have played in providing financing for Chinese power projects.

more emissions than anticipated from the uncapped sector and, therefore, more emissions overall.

This is essentially the framework examined by Montero (2000). If I assume that the cap is set with optimal desired emissions levels in mind, this excess of pollution becomes a potentially serious problem. There are also savings as the capped sector spends less to abate. Montero demonstrates these trade-offs.[6]

Correlated Baselines among Uncapped Firms

An alternative implication emerges as the baseline levels become more highly correlated. Consider the possibility that regulators overestimate the BAU emissions from the entire uncapped sector. The offset costs of most firms are now lower than the actual costs of abatement because most have to do less abatement than expected. Prices for offsets and allowances are, therefore, lower, and participation in the offset market increases. Although there are more offsets being sold, there is now much less abatement going on, and the share of emissions from the capped sector increases quite a bit relative to the case with no offsets. However, total *emissions* are actually below the aggregate expected level. This is because of the large negative shock to emissions in the uncapped sector. I define excess emissions as additional emissions *above the cap* that are created by introducing offsets. In the case of highly correlated baselines, total emissions from the uncapped sector (vertical striped area) can be much lower than expected, even though there is a considerable amount of emissions reductions that are not "additional." This is because the low baselines of firms who are selling offsets also imply low baselines even from firms who are *not* selling offsets.

Note that introducing offsets does increase emissions relative to the no-offset case. In the absence of offsets, aggregate emissions are well below the cap because the low-emissions shock fell outside the cap.[7] The low realization of baseline emissions make compliance with the cap easier, and allowance prices adjust accordingly.[8]

In this example, the baselines of most uncapped firms are overestimated.

---

6. If unlimited transfers are allowed, optimal emissions levels can still be obtained by anticipating the adverse selection and reducing the cap in the capped sector by the amount of excess emissions produced by the offsets.

7. This discussion assumes that the cap is set in terms of emissions, rather than an outcome-based measure such as atmospheric concentration of GHG.

8. This result is similar but not necessarily identical to what would happen if both sectors were capped. If both were capped, then the lower baselines could lower the aggregate abatement necessary without requiring active abatement from the uncapped sector. This can be more efficient as active abatement (the portion of offset sales require action) could still cost more than the equilibrium permit price. If the abatement quantity required from the capped sector yields a marginal abatement cost, after accounting for the lower baselines, that is less than the cost of abatement from the uncapped sector, it would be more efficient for all active abatement to come from the capped sector—even though less-active abatement would be required due to the lower baselines. In this case, the maximal cap would be more efficient.

The excess emissions of offset markets are not symmetric to the baseline realization, however. If the baseline emissions are underestimated, this simply reduces the amount of offsets sold. In the extreme, if the baselines of all firms are underestimated, then there is no adverse selection problem, in the sense that no firm is being paid to do what it would have done anyway absent a payment. In fact, uncapped firms would have to do more abatement than they would receive credit for. While underestimating the BAU emissions of uncapped firms can lead to problems stemming from overall regulations that are, ex post, too lax, these problems are not exacerbated by the existence of an offset market.

In summary, the implications of the adverse selection problem is tied strongly to the assumptions about the distribution of "errors" in the forecast of business as usual emissions. If this error is independently distributed across firms, offsets can produce underabatement. If the errors are highly correlated, however, the offset market can reveal information about the aggregate baseline and allow the abatement decisions of firms in the capped sector to adjust accordingly.

12.2.3   Discussion

As the previous section demonstrates, the question of additionality can be viewed in two lights; the adverse selection view, in which offsets pay the "wrong" firms to reduce, while other firms more than make up the difference, and one in which uncapped firms benefit from a coincidental, surprisingly clean development path. In some circumstances, there can be an important distinction between the two types of additionality. If the offset market were dominated by the latter "pleasant surprise" phenomenon, offsets can play a useful role despite the additionality problem.

Of course, the degree to which this distinction matters is closely linked to the level of the cap in the capped sector. In the context of Kyoto Treaty, the reductions required of the signatories are extremely modest. Any prospect of a pleasant surprise among nonsignatories would not come close to constituting the overall reductions called for by the Intergovernmental Panel on Climate Change (IPCC) and other groups. In short, most view the Kyoto Treaty as so lax that the world needs every ton of reductions it can produce. This is reflected in the fact that there has been relatively little market for excess reduction credits from Annex 1 Kyoto nations, such as Russia and the United Kingdom, because those excess credits are viewed as coincidental. These credits, known as "hot air," have largely been shunned, although this picture could change as Kyoto deadlines approach.[9] The distinction also has less meaning in the context of voluntary offset markets, where there is no mandatory cap to be adjusted.

Looking forward to a post-Kyoto world, however, the implications change

9. See Grubb et al. (2010).

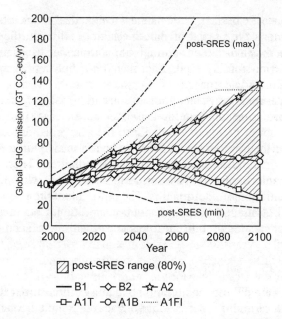

Fig. 12.1   Emissions trajectories of Intergovernmental Panel on Climate Change (IPCC) scenarios
Source: IPCC (2000).

somewhat. If a significant share of developed nations commit to proposed targets of 50 percent to 80 percent reductions, a pleasant surprise scenario could influence thinking about the needed stringency of those caps.[10] The potential stringency of future caps is largely dependent upon a political process, and the potential role of offsets plays a part in those negotiations. Those close to this process acknowledge that a tighter cap in the United States would be much more likely to gain acceptance if offsets are a part of the picture. If caps in the developed world are set ambitiously enough, this may not be the kind of Faustian bargain that critics of offset markets make it out to be.

On the other hand, if the worst-case IPCC scenarios materialize, even 80 percent reductions from developed nations would be insufficient to achieve a stabilization of GHG at levels deemed acceptable by the IPCC. Active abatement would have to be pursued in developing countries. Even under these circumstances, offsets can play an important role for some sectors of developing countries.

An examination of the IPCC scenarios (figure 12.1) for future BAU GHG emissions reveals just how much scope there is for an impact of a coinci-

---

10. The Annex I nations under the Kyoto Protocol account for roughly half of global GHG emissions today, but under the IPCC A2 scenario, this share would decline to under one-third.

dentally clean development path. There is a great deal of uncertainty about future emissions, with much of that uncertainty falling in the developing world. While fossil-fuel intensive, high population scenarios imply roughly a tripling of emissions by 2100, other scenarios imply a peak around 2050 followed by a steady decline.

Another key question is, therefore, whether additionality is likely to reflect adverse selection or common low baselines. In the case studied by Montero (2000), power plants that opted into the SO_2 program had low baselines because their output was reduced to be replaced by other plants. The case studies of the CDM appear to be different matters. There is evidence that many projects earned emissions reduction credits while not meeting the broad definition of additionality. The power plant projects identified in India and China may very well have not been additional, but their construction did not imply higher output from some other power facilities.

12.3   Implications for Offset Market Design

The preceding discussion attempts to highlight three implications. First, not all forms of additionality should be viewed as equally onerous to the effectiveness and efficiency of emissions caps. Second, the perverse incentives to manipulate baselines are an equally serious concern with no redeeming qualities. Third, offset markets can produce several other types of unintended consequences such as leakage, but those risks apply to almost any measures directed at reducing GHG emissions at less than a global scale. The current regulatory focus on additionality tends to paint all these problems with a broad brush without consideration of the context or their implications.

With these observations in mind, it is useful to consider the various policy tools that have been adopted or considered in order to address the perceived difficulties with offset mechanisms. Importantly, two frequently mentioned solutions, capping the number of offsets and discounting their effectiveness, do not address these problems very well. A cap on the number of offsets allowed into a market can limit the overall severity of the adverse selection problem, but by less than commonly thought. If adverse selection is a serious problem, the projects that are allowed would be the ones with the lowest baseline draws. If the baselines in the uncapped sector are instead highly correlated and much lower than expected, then limits on offsets restrict the ability of the mechanism to adjust to the "pleasant surprise" and allow for fewer reductions in the capped sector.

A devaluation of offsets treats all projects as equally nonadditional. As I have argued in the preceding, if this truly were the case and caps were strict enough in the capped sector, this is precisely when additionality does not reduce efficiency. In fact, it produces the exact same outcome as if the uncapped sector were under a mandatory cap and had been allocated allowances equal to its expected baseline. In either case, emissions are reduced and the uncapped sector reaps a windfall. However, both sectors benefit from the added participation of the uncapped sector relative to a case where that participation is limited. If instead baselines are uncorrelated and additionality is a serious problem, only the most extreme nonadditional projects are likely to be financially viable at the reduced returns provided by a genetic devaluation.

The solution identified by Montero (2000) is very different. A first-best reduction can be achieved if the cap were further tightened in anticipation of the excess emissions yielded from adverse selection in the uncapped sector. This allows full participation by the uncapped sector but still reaches the same overall emissions aggregated over both sectors. Unfortunately, this approach is both politically difficult and depends upon accurately predicting, on a sectoral level, the severity of the adverse selection problem.

To date, the primary bulwark against additionality concerns has been a review process that has been simultaneously criticized as too onerous to allow for substantial investment and also inadequate in weeding out nonadditional projects.[11] While some are concerned this may fatally delay investments, others feel that the incentive problems can only be adequately managed within a small program.

Those concerned with streamlining the review process are attracted to a shift away from project-specific review to a more programmatic approach. This offers several potential benefits. First, a programmatic approach can greatly lower the transactions costs of review and certification relative to the value of the offsets produced. Second, such an approach can help access a broader array of activities including energy efficiency and prevention of deforestation that have been largely absent from markets such as the CDM. Last, a program-level review can focus on risks, at an industry level, of the "bad" form of adverse selection while being less concerned with correlated, coincidental reductions. For example, investments in building efficiency may very well prove to be economic in the absence of offset programs and, therefore, not truly additional. But even if that is the case, increased efficiency in one building is unlikely to imply *worse* efficiency in others. A programmatic approach can also mitigate the moral hazard problem at the facility level by reducing the importance of the actions at a specific facility. However, there are still concerns about government level incentives.

Last, one tool that has not been applied to offset markets is the application of randomized trials. For example, a population of applications could be chosen to supply offsets, while another set is retained as a control group against which to judge the actions of the accepted population. This may be usefully combined with a shift in focus to evaluation at the program or sector level. Such approaches have been usefully applied to address similar adverse

---

11. See Grubb et al. (2010) and Wara and Victor (2008).

selection and moral hazard problems in programs that pay for reductions in energy use.[12] Atypical increases in emissions from countries or firms that become eligible for offsets relative to those that are not would indicate an inflation of baselines. Measuring the reductions from offset eligible projects *relative* to others can detect adverse selection relative to a common baseline, but it would also discount gains from commonly shared (e.g., coincidental) reductions. Because, returning to the earlier discussion, there are circumstances in which it is beneficial to allow credits for those coincidental reductions, the treatment of these shared effects would depend upon the stringency of overall caps.

References

Bushnell, James B. 2011. "Adverse Selection and Emissions Offsets." Energy Institute at Haas Working Paper no. 222.

Grubb, M., T. Laing, T. Counsell, and C. Willan. 2010. "Global Carbon Mechanisms: Lessons and Implications." *Climatic Change* 104 (3–4): 539–73.

Haya, B. 2009. "Measuring Emissions against an Alternative Future: Fundamental Flaws in the Structure of the Kyoto Protocols Clean Development Mechanism." Energy and Resources Group Working Paper no. ERG09-001.

Intergovernmental Panel on Climate Change (IPCC). 2000. *IPCC Special Report: Emissions Scenarios.* Geneva, Switzerland: IPCC.

Lecoq, F., and P. Ambrosi. 2007. "The Clean Development Mechanism: History, Status, and Prospects." *Review of Environmental Economics and Policy* 1 (1): 134–51.

Lewis, J. 2010. "The Evolving Role of Carbon Finance in Promoting Renewable Energy Development in China." *Energy Policy* 38 (6): 2875–86. doi:10.1016/j.enpol.2010.01.020.

Montero, J. P. 1999. "Voluntary Compliance with Market-Based Environmental Policy: Evidence from the U. S. Acid Rain Program." *Journal of Political Economy* 107 (5): 998–1033.

———. 2000. "Optimal Design of a Phase-In Emissions Trading Program." *Journal of Public Economics* 75:273–91.

Murray, B., R. Lubowski, and B. Shongen. 2009. "Including International Forest Carbon Incentives in Climate Policy: Understanding the Economics." Nicholas Institute Report no. NI R 09-03.

United States Department of Agriculture. 2009. "A Preliminary Analysis of the Effects of HR 2454 on U.S. Agriculture." Office of the Chief Economist. Economic Research Service. Washington, DC: U.S. Department of Agriculture, July 22.

Wara, M. 2008. "Measuring the Clean Development Mechanism's Performance and Potential. *UCLA Law Review* 55:1759–803.

Wara, M., and D. Victor. 2008. "A Realistic Policy for Carbon Offsets." PESD Working Paper no. 74. Stanford, CA: Program on Energy and Sustainable Development.

12. See Wolak (2010).

Wolak, F. 2010. "An Experimental Comparison of Critical Peak and Hourly Pricing: The PowerCentsDC Program." Stanford University, Unpublished Manuscript.

Comment   Kala Krishna

The goal in this very interesting chapter by Bushnell is to analyze some issues related to the implementation of carbon offsets in an overall plan to reduce emissions through the use of tradeable emissions permits.

There has been considerable discussion in the literature on such offsets. The main problem with offsets is in their implementation. While emissions levels are possible for a government to keep track of and penalize, changes in emissions levels require more effort to keep track of and are subject to potentially more manipulation by agents. Not only do past emissions need to be verified, but strategic manipulation by agents also needs to be policed. For example, agents will find it worthwhile to raise or misreport their emissions at the baseline to gain more from "reductions" in the future. Ted Gayner, in an article in the *American,* June 23, 2009, entitled "Offsets Chipping away at the Cap" illustrates this difficulty using the following example:

> In 2007, the House of Representatives launched its "Green the Capitol" initiative, which took on the goal of making House offices carbon neutral. After purchasing compact fluorescent light bulbs and shifting its electricity production from coal towards natural gas, the House still found itself far short of reaching its goal. To make up the difference, it bought 24,000 metric tons of carbon offsets [and] spent $14,500 to pay farmers for carbon-reducing "no-till" farming, even though the practice was started prior to the purchase of the offsets.

This example is related to the issues raised in Bushnell's chapter, which points out that offsets are more likely to be taken up by fake emissions reducers, as in the preceding, than by real emissions reducers because fake ones find it less costly to take up offsets than real ones. Why might this be so? This comes out most clearly if we model the technology behind emissions and work out a simple example, which is what I do in the following. I will first explain intuitively where the demand for emissions comes from and then try and embed what Bushnell does in a very simple example that might help the reader come to grips with what lies behind the slightly more abstract setting that is dealt with in the chapter.

Kala Krishna is the Liberal Arts Research Professor and professor of economics at the Pennsylvania State University, and a research associate of the National Bureau of Economic Research.

For acknowledgments, sources of research support, and disclosure of the author's material financial relationships, if any, please see http://www.nber.org/chapters/c12157.ack.

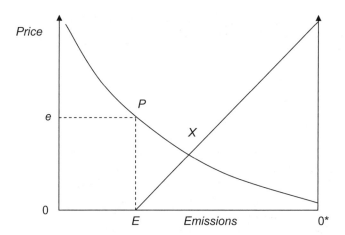

Fig. 12C.1   Offsets: Room to trade

Think of emissions as an input into production. Then the demand for emissions will be a derived demand; that is, it will be derived from the demand for the final good produced by the firm just like the demand for any other factor of production. Suppose there are two countries, H and F. H has binding emissions permits, and F does not. Figure 12C.1 depicts the demand for emissions from firms in H from the left-hand-side origin, while the demand for emissions from firms in F are measured from the right-hand-side origin. Emissions from H are depicted as a downward sloping curve and are limited at OE so that the price of emissions is e. Note that the demand for emissions will also represent the firm's willingness to pay for a permit or the marginal cost of abatement. Emissions are not regulated in F so that their price is zero, and, as a result, F emits a total of O^*E. The basic idea is that the marginal cost of abatement is lower in F than in H so that a Pareto improvement is possible if F could be made to reduce its emissions, while H raises its own. This is what offsets could do. A firm in H could pay a firm in F to reduce its emissions by a unit allowing the firm in H to raise its own by the same as total emissions would not be affected by this action. If unlimited offsets were allowed, the marginal cost of abatement would be equalized by trade in offsets at the intersection of the two curves at X, and the gain in efficiency of EPX would ensue.

Bushnell focuses on an adverse selection issue that may arise in this area. To get more particular about where this might come from, consider an example. Suppose the final good uses emissions, E, which has a price e, and an aggregate input, which we can call L, with price w. Assume each firm can produce a single unit of output that can be sold at a given price p after investing I to begin with, in one of two techniques with which to make the good. Moreover, assume both are fixed coefficient techniques. Technique

C, for clean, uses five units of L and one unit of E, while technique D, for dirty, uses five units of E and one unit of L. Changing technologies costs I.

We could think of firms as *actually* being of two types, C and D. However, they may be *perceived* by the authorities as being of a type other than their true one. For example, the authorities may know of the emissions and output and, hence, emissions per unit of output, for a firm in 2000, but, in 2010, unbeknownst to the authorities, this firm may have chosen a different technology. Thus, firms may be one of four "types": CC, CD, DC, DD. A type DC firm is one that had a dirty technology in 2000, but has a clean one in 2010 and who is thus wrongly classified by the authorities based on their 2000 information as a D firm in 2010. If such a DC firm chose to, it could reclassify itself as a DD firm and sell offsets for four emissions units without investing I as it has already done so. Note that there would be no reduction in emissions. If a DD firm chose to, it could reclassify itself as a DC firm and sell offsets for four emissions units, but it would have to invest I to change its technology. Note that there would be a reduction in emissions in this case.

Suppose that the price of an offset is e. Let us see what the *private gain* is from taking up this offer for each type of firm. A DC firm is already using the clean technology so that it will gain $4e$ from taking up the offer as it has already invested I. A DD firm will have costs of $5e + w$ versus costs of $e + 5w + I$ if it changes its technology, and its profits will change by:

$$[p - (e + 5w + I)] - [p - (5e + w)].$$

It profits will rise if:

$$e > w + \frac{I}{4}.$$

DC firms will always gain more from taking up this offer than DD firms and will always take up the offer. But only if $e > w + I/4$ will both DD and DC firms will take up the offer. A CC firm or a CD firm cannot gain from taking up this offer on offset sales.

What is the actual reduction in their own emissions from each type of firm taking up the offset? Type DC firms do not actually reduce their emissions but are paid as if they did. Type DD firms do actually reduce their emissions as the authorities expect. Thus, if we allowed offsets that were taken up by type DD firms only, total world emissions would be unchanged, but if they were taken up by type DC firms, world emissions would rise!

The government could be wrong about the distribution of types of firms. For example, it might think all firms are type DD, while firms are really all type DC, in which case it would overestimate baseline emissions and, thus, overestimate the extent of emissions reductions due to allowing offsets. This case corresponds to the expected baseline emissions exceeding the actual levels in Bushnell's terminology so that actual emissions reductions fall short of expected ones. On the other hand, the government might think all firms are type DC or CC, while firms are actually type DD. In this case, no firms would be willing to take up the offsets, but this would not lead to any increase in global emissions from allowing offsets.

Even if the government knows the true distribution of firms, it need not know the type of each firm. As argued in the preceding, if e is below $w + I/4$, only type DC firms will take up offsets, and world emissions will rise by the full amount of the offsets. If e is above $w + I/4$, then both DD and DC firms will take up the offsets, and world emissions will rise, but by less than the full amount of the offsets.

The issues raised in the chapter may thus be very real. But the question to ask is how to deal with them! Because the problem arises from an information distortion, the principle of targeting would suggest improving the information of the goverment. In the preceding example, this would involve sending inspectors out to verify the technology used in 2009 and not rely on information from 2001! Another way to deal with them is to implement a policy that does not requre government to have such information. Here I think that it is worth noting that allocating the rest of the world tradeable emissions permits of O^*E in figure 12C.1, on condition that emissions require a permit abroad as well as at home, would have the same outcome as perfectly implemented offset trade. They may also be easier to implement as government would not need to have information on the technology used by firms.

13

Monitoring and Enforcement of Climate Policy

Hilary Sigman

Without effective enforcement, public climate policies may not cause changes in private actions. However, the economics literature has not devoted much attention to enforcement of climate policies, with a few notable exceptions (Kruger and Pizer 2004; Johnstone 2005; Kruger and Egenhofer 2006; Silva and Zhu 2008). In part, the inattention to these issues may stem from the view that climate policies will be easy to enforce, at least relative to other air pollution controls. The empirical evidence reviewed in this chapter does support the view that restrictions on carbon dioxide emissions from point sources can be enforced with moderate cost.

However, enforcement of other aspects of climate policy can be daunting. Enforcement is sometimes a dominant consideration in the design of responses to greenhouse gases other than carbon dioxide and to carbon dioxide from sources other than fossil fuels. As a consequence, climate policy poses the novel challenge of integrating easy-to-enforce and difficult-to-enforce components in one policy.

This chapter investigates monitoring and enforcement of climate policy in practice and suggests several lessons from this experience. First, under the European Union (EU) Emission Trading System (ETS), incentives for compliance may derive at least as much from informal costs as official penalties. Second, prices in the EU ETS do not suggest much concern about differential validity of allowances from within the capped sector and offsets

Hilary Sigman is professor of economics at Rutgers University, a nonresident fellow of Resources for the Future, and a research associate of the National Bureau of Economic Research.

I am grateful to Severin Borenstein, Howard Chang, Don Fullerton, and participants at the NBER conference for helpful comments. For acknowledgments, sources of research support, and disclosure of the author's material financial relationships, if any, please see http://www.nber.org/chapters/c12140.ack.

from sources outside this sector. Finally, the empirical evidence points to substantial variation in monitoring and enforcement costs across these different compliance methods.

Given the differences in monitoring and enforcement costs across compliance methods, the government may be tempted to restrict climate policies to areas where enforcement is relatively easy. However, a simple model of enforcement illustrates that expanding markets may not lower compliance. Although deterrence may fall when the government must monitor more complicated activities, a broader market may also lower allowance prices and decrease the incentive to violate the policy. Thus, the effect of expanding the market will depend on the relative strength of these two opposing effects.

13.1   Incentives for Compliance with a Climate Policy

In this section, I present a basic model of enforcement of incentive-based environmental policy that has been used extensively in the prior literature (e.g., Harford 1978; Stranlund and Dhanda 1999; Stranlund, Chavez, and Field 2002). The model yields one simple insight that I use to analyze practical enforcement issues in the rest of the chapter.

13.1.1   The Compliance Decision

The standard environmental enforcement model considers a risk-neutral emitter who seeks to minimize the sum of compliance cost plus the expected punishment.[1]

Compliance costs depend on the form of the public policy. With a performance standard, compliance costs are just the costs of reducing emissions, $c(e_i, \gamma_i)$, where e_i is the emission level of emitter i, and γ_i reflects cost heterogeneity across the emitters. An incentive-based policy adds to the compliance cost a term that reflects net outlays (purchases or sales) of allowances or tax paid on emissions. Under a cap-and-trade program, an emitter with initial allowance allocation of Q_i thus has a compliance cost of $c(e_i, \gamma_i) + p \times (q_i - Q_i)$, where p is the equilibrium permit price and q_i the quantity of permits the emitter applies to its own emissions. A carbon tax is similar, but q_i is the level of emissions the emitter reports as its tax base, p is the tax, and $Q_i = 0$. The important implication is that an incentive-based policy gives the emitter choices on two margins, e_i and q_i.

The expected penalty depends on the chance a violation is detected, $D(v_i)$, and the fine, $F(v_i)$, each of which is, in general, a function of the magnitude of the violation v_i. For either emissions trading or a carbon tax, the violation is $v_i = e_i - q_i$, the difference between actual emissions and q_i.

In addition to a fine, most environmental policies require the violator to

---

1. Polluters may be risk averse, which would tend to strengthen the incentives for compliance, but not fundamentally change the problem. Malik (1990) models emission-market enforcement with risk averse polluters.

"fix" the violation. This requirement attempts to reduce the probability that violating the law is the least-cost option. Emission trading systems often implement this requirement by having violators surrender enough allowances to cover their emissions, perhaps withholding them from the violator's next-year allocation. Thus, the penalty is the fine plus the value of permits surrendered: $F(e_i - q_i) + p \times (e_i - q_i)$ (ignoring discounting if permits are surrendered next year). This penalty is multiplied by the chance the violation is detected, $D(e_i - q_i)$, to form the expected penalty.

Thus, the emitter's problem is to minimize total expected cost subject to the constraint that the violation is nonnegative:

$$(1) \quad \min_{e_i, q_i} \quad c(e_i, \gamma_i) + p \times (q_i - Q_i) + D(e_i - q_i)[F(e_i - q_i) + p \times (e_i - q_i)]$$
$$\text{s.t.} \quad e_i - q_i \geq 0$$

The first-order condition with respect to q_i is important to the following analysis. If λ_i is the shadow value of the constraint that the violation is nonnegative for source i, this condition is

$$(2) \quad p - [D'(v_i)(F(v_i) + p \times v_i) + D(v_i)(F'(v_i) + p)] + \lambda_i = 0.$$

If $e_i - q_i$ is strictly positive (i.e., the emission source does not fully comply), then $\lambda_i = 0$. The term in brackets is the marginal expected penalty. Thus, equation (2) implies that a partially compliant emitter sets its marginal expected penalty equal to the price. For an emitter to consider full compliance, the price must be less than the marginal expected penalty (because $\lambda_i > 0$).

To simplify this equation for the following applications, assume that the probability of detection is constant and equal to d (so $D'(v_i) = 0$). In addition, assume the fine is just a fixed amount per unit of violation, so $F'(v_i) = f$. This sort of fine is used in several emissions trading systems (see table 13.1). Thus, the first-order condition (2) becomes:

$$(3) \quad p + \lambda_i = d \times (f + p).$$

13.1.2   The Government's Choices

The government influences the private compliance decision through both the probability of detection, d, and the cost of a violation, f. The probability of detection, d, depends on the level and distribution of public monitoring resources. However, nongovernmental actors may also affect d. Whistleblowers, often employees of noncompliant firms, account for a high share of substantive environmental violations detected (Heyes and Kapur 2009). In addition, nonprofit environmental organizations play a substantial role in detecting violations of current environmental laws (Thompson 2000).[2]

2. Both of these forms of private enforcement are likely to result in a probability of detection that rises with the violation and, thus, a higher marginal expected penalty than assumed in the simplified condition in equation (3).

The government also has some control over the penalty, f. As Becker (1968) famously argued, high fines can substitute for costly monitoring in raising the expected penalty. However, high fines are rarely used in practice. The reasons may include horizontal equity concerns and judgment-proof problems (firms cannot be fined more than the depth of their pockets). The government may face political obstacles to imposing Draconian fines. Finally, high fines may trigger costly litigation, as violators have incentives to spend more to fight them.

In an emission trading system, non-Draconian fines can play the role of a "safety valve," allowing polluters to avoid buying permits during price spikes and, thus, effectively setting a ceiling on the marginal cost of carbon reductions (Montero 2002; Kruger and Pizer 2004). However, a requirement that facilities forfeit missing allowances discourages the use of fines as a safety valve. To use fines as a safety valve, the government might eliminate this requirement or allow the emitter to delay forfeiting allowances until allowance prices fall.

A Beckerian high-fine regime could also produce a low expected marginal penalty that could act as a safety valve if the government chooses a low enough d. In such a regime, polluters would not disclose their violations and would face a small risk of high fines. Although it would lower the government's enforcement costs, such a regime would be less transparent than a fine set as an explicit safety valve.

13.1.3   Penalties and Compliance in Practice

Fines in emission trading programs have mostly been modest in practice. Table 13.1 presents a summary of fines in the EU ETS and the US SO_2 allowance program, with price information for scale.

Compliance with emission trading systems seems to have been high.[3] The United Kingdom reports no detected violations of the EU ETS from 2006 through 2008 and 99.7 percent compliance in 2005 (UK Department of Energy 2009). Landgrebe (2009) suggests the following numbers of German facilities with some sort of violation, relative to a total of 1,665 facilities issued allowances: 2005, 174 installations; 2006, 28 installations; 2007, 20 installations. Kruger and Egenhofer (2006) report only twenty-one excess emissions penalties under the US SO_2 allowance program in its first ten years.[4]

High compliance rates are something of a puzzle because of the low level

3. Two frauds recently perpetrated on the EU ETS are exceptions. One scam exploited cross-border collection of the EU value added tax (VAT); the perpetrators purchased allowances without paying the VAT and then resold them, claiming to collect tax they actually pocketed (Europol 2010). A "phishing" scam also targeted the EU ETS (Kanter 2010). However, neither fraud seems to reveal an enforcement problem specific to climate policy.

4. RECLAIM is an exception to the high compliance rates with 85 to 95 percent compliance in early years. Stranlund, Chavez, and Field (2002) attribute the lower compliance to penalties that are less automatic and to higher prices relative to penalties.

Table 13.1   **Penalties with comparison to allowances prices**

Program	Fine	Forfeit next period?	Allowance price Average	Allowance price Maximum
EU ETS, 2005–2007	€40	Yes	€18	€30
EU ETS, 2008–2012	€100	Yes	€17	€29
US SO$_2$ allowance program (in 2008)	$3,337	Yes	$380	$550

Notes: European Union (EU) Emission Trading System (ETS) prices calculated from Blue-Next are for 2006 (first trading period) and for 2008–2009 (second trading period). The SO$_2$ fine is adjusted for inflation, from a base of $2,000 in 1990 dollars. SO$_2$ prices in the table are approximate.

of fines. To assure complete compliance, the first-order condition (2) implies that the marginal expected penalty must exceed the price. With the simplifying assumptions behind equation (3), full compliance requires $p < d \times (f + p)$. For the first trading period of the EU ETS, the penalty for a violation was €40. Therefore, if we believe compliance was, in fact, virtually complete, detection rates had to be greater than $d = 18/(40 + 18)$, or 31 percent, at the average price of €18. At the peak price of €30, they had to exceed 43 percent. The necessary probabilities would have declined with the higher penalties in the second period, but would still have been high.[5]

The perceived chance of detection seems unlikely to be so high, particularly for small violations.[6] Perhaps widespread violations do occur but are not detected. More likely, firms expect costs from noncompliance in addition to the official fines, so the preceding calculations understate the private costs of noncompliance. Noncompliance may tarnish the firm's image with its consumers, host community, potential employees, and regulators. These concerns may loom especially large in a carbon market with ongoing government allocation of valuable allowances: the participants may worry that current noncompliance will lower their future allowance allocations.

If firms perceive a large informal penalty, full compliance requires a lower risk of detection, d, than it would have required with official fines only. The possibility of substantial informal penalties has two policy implications. First, if the government faces constraints on the magnitude of official fines, it might try to raise informal penalties. For example, press releases with the names of violators might draw attention, lowering the required d and, thus, the government's enforcement costs. Second, high informal penalties make it difficult for the government to use fines as a safety valve. Even if the official

5. Stranlund, Chavez, and Field (2002) conduct similar calculations of required detection rates for the SO$_2$ allowance program.

6. Perceived chances of detection may dramatically overstate the reality. Research on income tax compliance shows households consistently overestimate their risk of an audit (Andreoni, Erard, and Feinstein 1998). However, the large firms involved in carbon emissions are likely to be more savvy about actual monitoring systems and detection risks.

fines are low enough to provide a safety valve at a relevant price level, firms may still have strong incentives to comply because of these other costs of violation.

13.2   Heterogeneous Monitoring Costs

Relative to the enforcement problems that have been studied previously, carbon markets add the complication of especially heterogeneous monitoring costs. Because such heterogeneous costs may raise novel issues for policy design, this section presents information on the cost differential for market participants. Information on costs for public enforcement agencies is not available but likely shows the same sort of variation as private monitoring costs.

13.2.1   Direct Costs of Monitoring

Large facilities that emit carbon dioxide probably do not face high costs when trying to demonstrate compliance. As with air pollution generally, the government may allow facilities to demonstrate compliance either through mass-balance approaches or continuous emissions monitoring (CEM). With a mass-balance approach, the facility uses the characteristics of its inputs and production technology to infer pollution without directly monitoring it. For carbon dioxide emissions from most large sources, this inexpensive approach yields an accurate accounting of emissions. The EU ETS allows many types of sources to use this approach to establish compliance (European Commission 2007).

The alternative, CEM, involves equipment that measures facilities' releases directly. The US Clean Air Act already requires CEM of CO_2 for the large coal-fired power plants that account for vast majority of CO_2 from power plants (Ackerman and Sundquist 2008). Thus, even the choice to rely extensively on CEM for enforcement of a US climate policy would probably not generate large new costs.[7]

The EU ETS requires third-party verification of emissions from facilities subject to its controls. This approach partially privatizes enforcement and creates a system analogous to the verification system for offsets. A verification market participant reports that "verification costs ranged from €5,000–€7,500 . . . for a simple site to €10,000–€20,000 . . . or more for a more complex site" (Kruger and Pizer 2004, 19) in the voluntary UK Emissions Trading Scheme, which ran from 2002 to 2006. Third-party verifica-

---

7. The US SO_2 allowance program requires CEM for large sources, although facilities could probably have calculated emissions with fairly high precision. Ellerman et al. (2000) find that CEM has been costly, contributing to private monitoring costs equal to 7 percent of total compliance costs. However, they argue that this approach has the advantage of separating true compliance activities from monitoring and helped convince skeptics of the environmental effectiveness of tradable permit programs.

tion probably raises social costs by less than this amount, however, because verification substitutes for public monitoring and for activities the source might have conducted internally.

A survey by Jaraite, Convery, and Di Maria (2009) of Irish firms in the EU ETS first trading period provides data on overall private monitoring costs. It finds that "monitoring, reporting, and verification" (MRV) costs averaged €0.04 per ton of CO_2 or about €25,000 per year per respondent. Thus, monitoring costs averaged only about 0.1 percent of the total compliance costs, if we assume average compliance costs are a quarter of marginal costs (the allowance price).[8] Jaraite, Convery, and Di Maria also report that 40 percent of MRV costs are for external consultants, which confirms the market participant report from Kruger and Pizer (2004).

Private monitoring and verification costs for other sources, such as those proposed as the basis for offsets, are probably much higher for several reasons. The emissions may not be from point sources, raising challenges for any direct measurement of releases. Verifying all the values necessary to calculate greenhouse gas reductions from nonpoint sources may be complex. For example, the net effect of land use changes on greenhouse gas concentrations may vary greatly with specific agricultural or forestry practices and characteristics of the land. Originators of offsets may bear the burden of establishing "additionality," that is, that pollution is reduced relative to some meaningful baseline (Montero 1999; Bushnell, chapter 12 in this volume; Hahn and Richards, Forthcoming). Finally, the relevant activities may take place abroad and possibly in countries with more corruption, adding to the complexity of assuring compliance.

Antinori and Sathaye (2007) provide an estimate of monitoring and verification costs for offsets. The twenty-eight greenhouse gas reduction projects that they study report average monitoring and verifications costs of $0.30 per metric ton of CO_2, although the variance is high and a few large projects report much lower costs.[9] When compared to the average monitoring costs of €0.04 (about $0.06) per ton of CO_2 for covered facilities in Jaraite, Convery, and Di Maria (2009), these estimates suggest monitoring costs may be several times higher for offsets than for covered facilities.

8. Ellerman, Convery, and De Perthuis (2010) report a lack of ex post estimates of the total costs of the EU ETS first trading period and assume the average costs are half the marginal costs (a linear marginal cost curve).

9. Alternatively, one can estimate the costs of monitoring offsets by comparing prices for emissions reductions with more rigorous and less rigorous certification. Conte and Kotchen (2010) analyze carbon offset prices from an online listing in 2007, 13 percent of which have the more rigorous certification that would make them eligible for use in compliance with Kyoto obligations and the EU ETS. They estimate that certified permits cost 30 percent more than other projects with similar observable characteristics. Although many demand and supply factors may underlie this price differential, the costs of the certification probably contribute part of it. If even 10 percent of Conte and Kotchen's low-end estimate of a 30 percent price difference is monitoring costs, these costs are $0.54 per ton of CO_2.

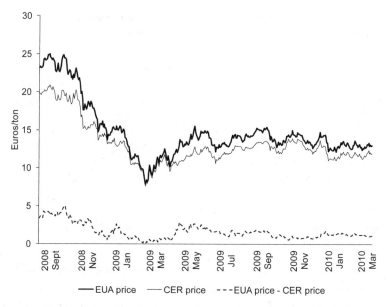

Fig. 13.1 Spot prices of European Union Allowances (EUAs) and secondary Certified Emissions Reductions (CERs) and their difference
Source: BlueNext (www.bluenext.fr).

13.2.2   Differential Enforcement Risks

Higher monitoring costs probably reduce private monitoring. With less thorough monitoring, allowances may be subject to greater risk that the government will find them invalid and conclude that the emitter is out of compliance. The variation in private monitoring costs may lead to variation in what I will call the "validity" of the allowance: the chance that the emitter is deemed to be in compliance when using that allowance. Market prices may reflect any differences in validity across different sorts of allowances and, thus, provide indirect evidence of differential monitoring costs.

Figure 13.1 presents the history of the differential between two types of allowances in Europe. Facilities subject to EU ETS restrictions may cover their emissions either with the European Union Allowances (EUAs), which the EU issues to point sources of CO_2, or with Certified Emissions Reductions (CERs). CERs result from greenhouse gas emission reduction projects undertaken through the Kyoto Protocol's Joint Implementation (JI) or Clean Development Mechanism (CDM).[10]

The figure compares spot market prices of EUAs and "secondary" CERs

10. The vast majority of CERs originate in China and derive from hydroelectric and wind projects (Capoor and Ambrosi 2009).

on one of the major exchanges, BlueNext.[11] "Secondary" CERs are being resold, as opposed to "primary" CERs sold by the originating project. The average price differential between EUAs and CERs from August 2008 through February 2010 is €1.64; the maximum of €5.03 occurred early in the period when allowance prices were highest.

We would expect EUAs and CERs to be perfect substitutes for complying EU facilities; thus, the existence of a price difference requires explanation.[12] One possibility is that the public relations consequences of using EUA and CERs differ, even if the two types of allowances are equally valid from an enforcement perspective. The public may view CERs less favorably than EUAs because CERs relax the constraint that Europe has put upon its own carbon dioxide emissions. However, the public also might prefer CERs to EUAs as "charismatic carbon"; CERs may promise nonclimate benefits, such as reducing local air pollution or protecting natural ecosystems. A second possibility is that market participants perceive a greater risk of being found out of compliance with CERs than with EUAs. The price differential then measures the disparity in the expected validity of the two types of allowances.

Suppose the risk that the government finds a violation is d_{EUA} for EUAs and d_{CER} for CERs. The penalty is the same with either type of permit because it consists of a fine and forfeit of EUAs from next year's allocation. Using the simplified first-order condition in equation (3), the difference in the marginal expected penalties and, thus, the price premium is $p_{EUA} - p_{CER} = (d_{EUA} - d_{CER}) \times (f + p_{EUA})$. With the official fine of $f = €100$, an average EUA price of €25 over the period of price premium data, and an average premium of €1.64, the detection probabilities would differ by 1.3 percentage points, a modest amount.

However, a major objection to this calculation is that the EU ETS places liability for compliance on sellers. Thus, the buyer of CERs might not believe it faced any higher expected penalty than if it had purchased EUAs. On the other hand, public opinion may not respect the legal allocation of compliance obligations, so a violation may still have public relations costs for the buyer. Depending on the comparison between the marginal public-relations cost and the official fine, the 1.3 percentage point disparity may be either too high or too low.

13.2.3   Policy Design with Heterogeneous Enforcement Costs

The variation in monitoring costs across different sources of allowances (e.g., the EUA-CER differential) and the resulting differences in validity give

---

11. Mizrach (2010) discusses the exchanges and analyzes various spot and futures prices in international carbon markets, including the EUA-CER spread.

12. The EU ETS does place caps on the number of CERs each country may use cumulatively over the second trading period. However, this country-level constraint does not affect an individual source's current ability to substitute freely between the two types of allowance and, thus, does not imply different current spot prices.

rise to a number of questions about policy design. One question is whether sources of allowances with high monitoring and enforcement costs ought to be excluded from the market. For example, a US climate policy might allow only domestic offsets or no offsets at all.

The simple enforcement model suggests that broadening the program might not reduce the compliance rate, despite spreading enforcement resources more thinly. Expanding the possible sources of allowances brings additional low cost sources of greenhouse gas abatement into the market, lowering the price of allowances. This reduction in price means that the marginal expected penalty required for full compliance falls and, thus, a lower detection rate can sustain full compliance.[13] Sigman and Chang (2011) present an expression for the conditions under which allowing offsets increases compliance: the effect on compliance depends on the relative costs of abatement and of auditing compliance in the capped and offset sectors and the number of offsets claimed.[14]

To illustrate the possible magnitude of the effect, consider a broadening of the market that causes the allowance price to fall from p to δp. Using equation (3), the probability of detection required for full compliance falls from $d_0 = p/(f+p)$ to $d_1 = \delta p/(f+\delta p)$. For example, the US Environmental Protection Agency (2009) estimates that elimination of international offsets would nearly double the allowance price (from \$13–\$17 to \$25–\$33 in 2015; from \$17–\$22 to \$33–\$44 in 2020) for the Waxman-Markey climate policy. If the fine were set at five times the initial allowance price (along the lines of the EU ETS), including the international allowances would allow the d required for full compliance to fall to 55 percent of the d in the narrower market.[15]

The net effect on compliance depends upon the relationship between government outlays and d in the narrower and broader markets. Obviously, if detection is too difficult and fraud rampant in the broader market, overall compliance will decline with the expansion. Nonetheless, the reduction in the required detection rate does suggest at least the possibility that market expansion improves compliance, despite apparent enforcement difficulties.[16]

13. An analogous effect might arise with nonenvironmental taxes. Lower marginal tax rates decrease the incentive to evade taxes (Allingham and Sandmo 1972; Clotfelter 1983). Thus, a revenue-neutral tax reform that reduces marginal tax rates by broadening the tax base might improve compliance, even if it increased the difficulty of monitoring all taxed activities.

14. Sigman and Chang (2011) also point out that costly enforcement may be a reason to use offsets, rather than expanding the cap to include the second sector: the government need only audit claimed offsets, not the entire sector.

15. This example illustrates possible magnitudes only. The actual Waxman-Markey legislation set the excess emission fine at twice the allowance price (H.R. 2454, 111th Congress, Section 723). This rule would reduce compliance incentives along with compliance costs and not give rise to the effect in the text.

16. This analysis takes a narrow view of "compliance" for offsets, considering only whether actions promised are undertaken, not whether they contribute to an overall reduction in atmospheric greenhouse gases. Elsewhere in this volume, Bushnell (chapter 12) and Borenstein (chapter 6) consider broader issues in expanding the sources of greenhouse gas abatement.

13.3   Conclusions

A climate policy that controls domestic CO_2 emissions from fossil fuels may not present too great an enforcement challenge. Experiences with the EU ETS and the US SO_2 trading program suggest a high degree of compliance with emission trading, despite modest penalties. High compliance may partly result from public relations costs for violators.

Previous experience suggests that monitoring costs vary substantially across different types of allowances in current markets. This variation raises some interesting questions for future analysis. For example, it would be useful to study whether enforcement agencies could improve the overall efficiency of the program by narrowing the difference in the validity of allowances from different sources.

A policy response to the variation in enforcement costs could be to restrict the market to areas of low enforcement cost. However, the simple model presented here suggests that expanding the market to include activities that require more costly enforcement may not lower compliance if inclusion of these abatement opportunities reduces allowance prices sufficiently. This analysis shows the importance of recognizing that enforcement strategies can respond to market conditions and that market conditions may be sensitive to these strategies. Both aspects of this relationship deserve additional consideration in climate policy design.

References

Ackerman, Katherine V., and Eric T. Sundquist. 2008. "Comparison of Two US Power-Plant Carbon Dioxide Emissions Data Sets." *Environmental Science and Technology* 42 (15): 5688–93.

Allingham, Michael G., and Agnar Sandmo. 1972. "Income Tax Evasion: A Theoretical Analysis." *Journal of Public Economics* 1:323–38.

Andreoni, James, Brian Erard, and Jonathan Feinstein. 1998. "Tax Compliance." *Journal of Economic Literature* 36 (2): 818–60.

Antinori, Camille, and Jayant Sathaye. 2007. "Assessing Transaction Costs of Project-Based Greenhouse Gas Emissions Trading." Report no. 57315. Berkeley, CA: Lawrence Berkeley National Laboratory.

Becker, Gary S. 1968. "Crime and Punishment: An Economic Approach." *Journal of Political Economy* 76 (2): 169–217.

Capoor, Karan, and Philippe Ambrosi. 2009. *State and Trends of the Carbon Market, 2009.* Washington, DC: The World Bank.

Clotfelter, Charles T. 1983. "Tax Evasion and Tax Rates: An Analysis of Individual Returns." *Review of Economics and Statistics* 65 (3): 363–73.

Conte, Marc N., and Matthew J. Kotchen. 2010. "Explaining the Price of Voluntary Carbon Offsets." *Climate Change Economics* 1 (2): 93–111.

Ellerman, A. Denny, Frank J. Convery, and Christian De Perthuis. 2010. *Pricing Carbon: The European Union Emissions Trading Scheme.* Cambridge, UK: Cambridge University Press.

Ellerman, A. Denny, Paul Joskow, Richard Schmalensee, Juan-Pablo Montero, and Elizabeth Bailey. 2000. *Markets for Clean Air: The U.S. Acid Rain Program.* Cambridge, UK: Cambridge University Press.

European Commission. 2007. *2007/589/EC: Commission Decision of 18 July 2007 Establishing Guidelines for the Monitoring and Reporting of Greenhouse Gas Emissions.* http://eur-lex.europa.eu/LexUriServ/LexUriServ.do?uri=OJ:L:2007:229 :0001:0085:EN:PDF.

Europol. 2010. "Further Investigations into VAT Fraud Linked to the Carbon Emissions Trading System." https://www.europol.europa.eu/content/press/further -investigations-vat-fraud-linked-carbon-emissions-trading-system-641.

Hahn, Robert, and Kenneth Richards. Forthcoming. "Environmental Offset Programs: Survey and Synthesis." *Review of Environmental Economics and Policy.*

Harford, Jon D. 1978. "Firm Behavior under Imperfectly Enforceable Pollution Standards and Taxes." *Journal of Environmental Economics and Management* 5 (1): 26–43.

Heyes, Anthony, and Sandeep Kapur. 2009. "An Economic Model of Whistle-Blower Policy." *Journal of Law, Economics, and Organization* 25 (1): 157–82.

Jaraite, Jurate, Frank Convery, and Corrado Di Maria. 2009. "Assessing the Transaction Costs of Firms in the EU ETS: Lessons from Ireland." SSRN eLibrary. http://ssrn.com/abstract=1435808.

Johnstone, Nick. 2005. "Tradable Permits for Climate Change: Implications for Compliance, Monitoring, and Enforcement." In *Climate-Change Policy,* edited by Dieter Helm, 238–52. Oxford, UK: Oxford University Press.

Kanter, James. 2010. "Hackers Hit Europe's Carbon Market." *The New York Times,* February 4.

Kruger, Joe, and Christian Egenhofer. 2006. "Confidence through Compliance in Emissions Trading Markets." *Sustainable Development Law and Policy* 6 (2): 2–14.

Kruger, Joseph, and William A. Pizer. 2004. "The EU Emissions Trading Directive: Opportunities and Potential Pitfalls." Resources for the Future Discussion Paper.

Landgrebe, Jurgen. 2009. "Verification in the EU ETS: German Perspective." Paper presented at the ICAP China Conference, October 12–13. http://www .icapcarbonaction.com/phocadownload/china_conference/Presentations191009/ icap_china_conf_plenary4_landgrebe.pdf.

Malik, Arun S. 1990. "Markets for Pollution Control When Firms are Noncompliant. *Journal of Environmental Economics and Management* 18 (2): 97–106.

Mizrach, Bruce. 2010. "Integration of the Global Carbon Markets." http://ssrn.com/ abstract=1542871.

Montero, Juan-Pablo. 1999. "Voluntary Compliance with Market-Based Environmental Policy: Evidence from the U.S. Acid Rain Program." *Journal of Political Economy* 107 (5): 998–1033.

———. 2002. "Prices versus Quantities with Incomplete Enforcement." *Journal of Public Economics* 85 (3): 435–54.

Sigman, Hilary, and Howard F. Chang. 2011. "The Effect of Allowing Pollution Offsets with Imperfect Enforcement." *American Economic Review* 101 (3): 268–72.

Silva, Emilson C. D., and Xie Zhu. 2008. "Global Trading of Carbon Dioxide Permits with Noncompliant Polluters." *International Tax and Public Finance* 15 (2): 430–59.

Stranlund, John K., Carlos A. Chavez, and Barry C. Field. 2002. "Enforcing Emissions Trading Programs: Theory, Practice, and Performance." *Policy Studies Journal* 30 (3): 343–61.

Stranlund, John K., and Kanwalroop Kathy Dhanda. 1999. "Endogenous Monitoring and Enforcement of a Transferable Emissions Permit System." *Journal of Environmental Economics and Management* 38 (3): 267–82.

Thompson, Barton H. Jr. 2000. "The Continuing Innovation of Citizen Enforcement." *University of Illinois Law Review* 2000 (1): 185–236.
UK Department of Energy and Climate Change. 2009. *Report on 2008 EU Emissions Trading System Emissions Data.* London: DECC.
US Environmental Protection Agency. 2009. "EPA Analysis of the Waxman-Markey Discussion Draft: The American Clean Energy and Security Act of 2009." http://www.epa.gov/climatechange/economics/economicanalyses.html.

Comment   Severin Borenstein

Hilary Sigman does an excellent job of presenting both a compelling theoretical argument and some interesting data on the impact of enforcement and detection in tradeable pollution permit markets. The conclusion that extending the market to areas with lower detection rates could actually raise compliance rates is particularly thought-provoking. For me, it provoked thoughts about optimal combinations or separations of markets. In particular, while it might make sense to include uncovered polluters in an existing market even if it is more difficult to detect cheating among the new participants, I believe it can also make sense to establish separate markets for participants with differential detection probabilities.

Consider an exisiting emissions market in which the probability of detection, d_1, and the fine for failing to purchase sufficient permits, f_1, are such that there is perfect compliance among all emitters. For the purpose of this intuitive discussion, assume that enforcement costs are zero, and detection rates are purely exogenous. Assume that the equilibrium permit price in that market is p. Now consider a second set of emitters who, in aggregate, have exactly the same abatement cost curve as in the first market, but may have a different probability of being detected if they purchase fewer emission permits than their actual emissions, d_2, could differ from d_1. The fine for detection is the same in both markets, $f_2 = f_1 = f$. There are (at least) three possible treatments of this second set of emitters: (a) include them in the existing emissions market, (b) establish a separate emissions market for the second set of emitters, or (c) do not regulate the second set of emitters at all. With zero enforcement costs, option (b) clearly dominates option (c). The comparison of options (a) and (b) is more interesting, however.

Consider expanding the permit market to include the second market while

Severin Borenstein is the E. T. Grether Professor of Business Economics and Public Policy at the Haas School of Business, University of California, Berkeley; codirector of the Energy Institute at Haas; director of the University of California Energy Institute; and a research associate of the National Bureau of Economic Research.

My thanks to Jim Bushnell, Stephen Holland, and Meredith Fowlie for helpful discussions. For acknowledgments, sources of research support, and disclosure of the author's material financial relationships, if any, please see http://www.nber.org/chapters/c12141.ack.

simultaneously giving permits to all participants in the second market equal to their zero-price (premarket) emissions. With $d_2 = d_1$, this would be a Pareto improvement with no change in the level of emissions and a decrease in abatement costs as some of the abatement is undertaken by low-cost abaters in the second market who displace higher-cost abaters in the first market. Near the other extreme, with d_2 near zero, bringing in the second market would lead to virtually no actual abatement in either market. It would all be falsely claimed abatement by members of the second market, and p would drop to near zero. If abatement policy had been undertaken in the first market because it was welfare improving, then expanding to the second market would lower welfare.

For d_2 sufficiently close to d_1, bringing in the second market will be a welfare improvement, but for d_2 sufficiently less than d_1, it will not be. In the latter case, with the exception of $d_2 = 0$, it would still be valuable to set up a second separate market for the participants in market 2. If $d_2 > 0$, but very small, then for any pool of permits in market 2 even somewhat below the market's zero-price output, the equilibrium price of a permit in market 2 would have to be $d_2 f$, which is the expected avoided fine from owning a permit. Essentially, this is a tax of f with a very low probability of enforcement. It would cause the lowest-cost abatement in market 2 to occur, though the quantity could be measured by the regulator only through some sampling procedure because all emitters would claim they are in compliance. That quantity could displace an identical amount of abatement in market 1— which has a higher marginal abatement cost of p—and result in the same total amount of abatement at lower total cost. This would not be as efficient as combining the markets if they each had full compliance, but it would still be more efficient than ignoring market 2.

My goal in this very simplified model is to suggest that differences in detection and compliance rates can lead to optimal pooling or separating of permit markets for the same pollutant.[1] In fact, there is probably a detection rate difference, $|d_1 - d_2|$, below which markets should be merged and above which they should be treated separately. This argument is separate from and complementary to Sigman's point that incorporating abatement in the second market can lower the price for participants in the first market and, thus, increase their incentive to comply.

A complete analysis of optimal separation or integration of emissions markets would also have to include recognition that the monitoring costs will differ between markets, as Sigman does in studying optimal expansion of the market. Another practical cost of expanding the market, which a complete analysis would have to recognize, is the cost of determining property rights. While economic models often take property rights as exogenous,

---

1. This is somewhat analogous to the issue of hotspots, where abatement in different markets is of different expected *value*.

that is far from true in practice. With negative externalities, of course, it is nearly always the case that the activity is unpriced not because the property right has been clearly allocated to the polluter, but because it has not been clearly allocated at all. The costs of arriving at acceptable processes for determining property rights for new market participants (i.e., baselines from which abatement is measured) and of making whatever measurements are necessary to apply those processes are formidable. These have proven to be extremely difficult problems even within the developed world for easy-to-measure fossil-fuel combustion emissions. For the much-less-understood counterfactuals on which baselines are determined for new industries or more complex greenhouse gas (GHG) sinks or sources, determining property rights seems even more challenging, as I have suggested in my chapter (chapter 6) in this volume.

How Can Policy Encourage Economically Sensible Climate Adaptation?

V. Kerry Smith

14.1 Introduction

There is broad consensus among scientists that the climatic services, such as what the public might associate with local weather patterns, will change due to the accumulation of greenhouse gases (GHGs). Action on a US climate policy, regardless of what it turns out to be, will not stop the change due to past activities. As a result, adaptation is now viewed as an important focus for new policies along with those aimed at reducing GHGs.

In these discussions, adaptation is described as the adjustments in natural or human systems that exploit the beneficial opportunities and moderate the negative effects of any changes arising due to the altered climate system.[1] Several maintained assumptions are taken as given in nearly all discussions of climate adaptation. First, it is assumed that there is a key role for government and that *anticipatory* action is essential. Second, the discussions maintain that the experts know what to do. A mix of physical and natural infra-

V. Kerry Smith is the Regents' Professor, W. P. Carey Professor of Economics, and Distinguished Sustainability Scientist at the Global Institute of Sustainability at Arizona State University; a university fellow at Resources for the Future; and a research associate of the National Bureau of Economic Research.

Thanks are due to Allen Klaiber, Nicolai Kuminoff, Sheila Olmstead, and Carlos Valcarcel Wolloh for suggestions on this research and to both Don Fullerton and Erin Mansur for very constructive comments on an earlier draft. In addition, Erin helped me to remove some mistakes in that earlier draft. Any remaining errors are mine. Thanks are also due to Natalie Cardita and Jon Valentine for preparing several versions of this chapter. For acknowledgments, sources of research support, and disclosure of the author's material financial relationships, if any, please see http://www.nber.org/chapters/c12160.ack.

1. This definition is consistent with what is used by the Intergovernmental Panel on Climate Change (IPCC).

structure investments, coordinated by government, is generally presented as the best adaptive response to expected changes in the climate system. Finally, it is assumed that reliance on ex post responses, by either consumers or firms, will magnify the damages experienced from climate change. Numerous examples could be used to document this summary. The National Research Council's (2010) *Adapting to the Climate Change,* a newly released report that is likely to be influential, is one of them. It offers ten recommendations for adaptation. None of them considers using economic incentives as part of climate adaptation policy. There is nearly a complete reliance on information programs and government action.

This chapter is about the design of adaptation policies that rely on economic incentives. It begins, following Mendelsohn (2000), by asking why anticipatory adaptation is believed to be an efficient response. After that, it discusses current pricing policies for the private goods that households and firms can be expected to use as substitutes in adjusting to the natural services that are altered by climate change. Electricity for heating and cooling and water from public (and private) centralized water systems are both examples of the types of substitutes used to respond to regional changes in temperature and precipitation. Changes in the price structure for these commodities may well make sense independent of anticipatory adaptation policy, especially because current pricing assumes changes in the service reliability standards with different levels of interruption and associated price discounts are not policy options.

My analysis "dusts off" an early framework used in considering pricing structures with uncertain demand. After reviewing the basic model, the analysis discusses alternative ways a natural substitute might be introduced. Four conclusions follow from this analytical model. First, the pricing and capacity choices for substitute services will depend on how the natural capacity is assumed to contribute to the services supporting people's activities. Second, decisions to augment the capacity for climate substitutes, in response to a decline in natural capacity or changes in demand uncertainty, cannot be considered independent of the pricing policy. Third, and equally important, when produced capacity of the substitute is selected ex ante and its price is not easily adjusted, the optimal decisions depend on the rationing rule for allocating the available supply during periods of excess demand. When prices do not adjust easily, short-run variation in excess demand conditions must be managed. Rules defining who is served under these conditions translate into changes in the reliability of service. Thus, a practical implication of these simplified models is to suggest that policy consider pricing service reliability. These price schedules could be designed to change from year to year as expectations for natural conditions that would affect demands for climate substitutes change.

Finally, there is an indirect implication of incentive-based adaptation for climate mitigation policy. The terms of access to services that substitute for natural climate conditions affect the value of climate mitigation. Borenstein (2005) makes a related point using a specific example—suggesting that dynamic pricing can increase the value of investments in residential solar power in some regions. This conclusion follows because the renewable power can displace the highest-cost substitute at exactly the times that power is needed.

The next section outlines an economic perspective on the reasons for intervention to promote climate adaptation and summarizes Carlton's (1977) version of a model to describe optimal pricing and capacity decisions with stochastic demand. The model is used as a template to consider two issues: (a) the effects of the conditions of access on the "ideal" pricing and capacity choices; and (b) the implications of alternative ways of characterizing climate services in models of the demands for substitutes.

14.2   Climate Adaptation Policies and Substitutes

14.2.1   Context

If the external conditions governing temperature and precipitation in a location change exogenously, we usually assume the people and firms affected by the change will adjust when it makes sense for them to do so. Of course, those involved have to be able to distinguish a permanent change from "normal" variability in their local environment. In the climate adaptation literature, these types of actions are labeled as autonomous adaptations (see Fankhauser, Smith, and Tol 1999). Most climate policy recommendations call for *anticipatory* adaptation, which amounts to doing things in advance of the changes that are expected. Mendelsohn (2000) has questioned the need for these advance interventions. His arguments are the traditional ones we expect from economists. That is, if there is a market failure or incomplete information, then the first best response is usually to correct the source of the failure. Actions taken assuming the failures will persist may be inefficient.

In the real world, some market failures are the result of practical compromises. Pricing policies for electricity and water reflect past metering technologies (and are changing slowly) as well as the regulations governing the reliability of these services. For example, we realize that the incremental costs of delivering another kilowatt hour of power depends on the overall demands imposed from the full system of users at each time. These total demands vary with the location, the season, the days of the week, and the hours of the day. Initially it was impractical to have residential electric meters that provided this temporal resolution. In addition, meters had to be read by people.

Today it is not only possible to vary the recording systems for power, but the readings can be collected remotely. Usage could also be controlled remotely. Residential devices with these controls may well be cost-effective in many areas independent of whether the price schedules are changed or service is controlled remotely. The savings in manpower reading meters may be sufficient to justify the change.

This example helps to explain the source of a failure in pricing schemes. Initially, metering technology could not accommodate prices that adjusted to changes in the costs of service. In addition, the firms providing the service were regulated. To adjust prices in many areas, these firms must seek permission from a regulatory commission. This is broadly true for electricity and true in many areas for residential water supplies as well.

Firms providing these goods face uncertain demands and varying costs of meeting a reliability mandate. Current practice imposes the risks created by the differential costs of meeting varying system demands (and prices that don't readily adjust) on the suppliers. Significant changes in either the variability of demand or the costs of providing service will alter the nature of these risks. Changes in local weather conditions due to climate change could be one source for such a shift. As a result, it may be efficient to reconsider the predefined pricing contracts and reliability mandates. To illustrate the economic rationale for this suggestion, the next section reviews a class of models that has been used to describe socially optimal pricing and capacity decisions under demand uncertainty. These models assume the social objective function is to maximize the expected consumer surplus from the service.

14.2.2   Pricing and Capacity Planning

Over forty years ago, a series of papers considered situations where a firm (or a stylized description of a policymaker) faced a stochastic demand and had to select the production capacity and a single price for output.[2] The intended application was to motivate a reconsideration of pricing policies for resources with these attributes. An important by-product of the research was a conclusion that these choices can depend on the conditions of access to the resource when demand exceeds capacity and prices do not adjust. My analysis begins with what I believe was the last major paper in this sequence of past research by Carlton (1977). His model assumes the random component of demand scales the quantity demanded at each price. This paper finds that selecting a price and capacity to maximize expected consumer surplus would, under some conditions of access to the service, imply an "optimal" price *above* long-run marginal costs. The assumed terms of access when demand exceeds available capacity also affect the prospects for profits (or losses). Thus, they affect the need for taxes or subsidies to assure reliable provision of service.

Demand is a function of prices and defined as the product of two terms, $x(p)$ and u. $x(p)$ could be considered a per capita demand; p is the price of

2. The initial research was developed by Brown and Johnson (1969). A subsequent comment by Visscher (1973) raised the issue of how the excess demand would be allocated among different demanders.

service; and u a positive, random variable with distribution function $F(u)$. u could be interpreted as a measure of the number of customers. Capacity is planned as multiples of unit demand under "normal" conditions. Capacity is given by $k = s \cdot x(p)$. So when $u > s$, then with a fixed price that is set in advance, not all customers can be served. Assuming p and s are selected before the size of u is known, then the policymaker must also consider rules to determine which consumers will have their demand satisfied.

Once decisions about capacity and price are made, the conditions of access (or rationing schemes) will influence what "counts" in defining the expected consumer surplus. Price does not play a role in clearing the market. Few markets allow instantaneous price adjustment. However, the assumption of no price adjustment is especially relevant to the issue of climate adaptation. This conclusion follows because climate's substitute services have historically been provided in situations with limited price adjustment. For example, consider the cases of "time of use" pricing or increasing price block structures for water. These policies do not allow prices to adjust as the amount demanded changes. Rather these structures amount to replacing constant *prices* with constant *price schedules*.

To illustrate the logic of the model, consider the simple graph presented in figure 14.1. Price is measured on the vertical axis and total quantity demanded on the horizontal. With multiplicative uncertainty, the variability in u pivots the demand function about the choke price, given by the point A. At the time s and p must be selected, the planner does not know what the

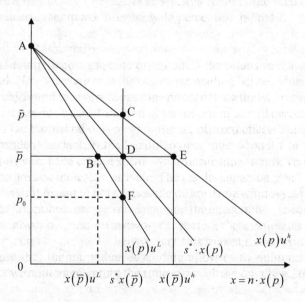

Fig. 14.1　Illustration of the effects of stochastic demand with ex ante price and capacity decisions

aggregate demand will be. To begin this summary, consider first the case of planning when efficient rationing is assumed to govern situations when demand exceeds available capacity. Three cases need to be distinguished to describe all possibilities: (a) demand matches exactly the planned capacity; in this case the diagram represents demand as $s^* \cdot x(p)$; (b) demand is less than planned capacity or $x(p) \cdot u^L$ in the figure, and (c) demand exceeds the planned capacity, given by $x(p) \cdot u^h$ in the figure. If the value for the capacity that maximizes the expected surplus is s^* multiples of demand at the optimal price of $\bar{p}$, or $s^*x(\bar{p})$, then the realized consumer surplus is $AD\bar{p}$. If we assume b is the constant (per unit), variable cost of producing the output, and β is the constant (per unit), long-run cost of capacity, then the need for a subsidy will depend on how revenue $(\bar{p}s^*x(\bar{p}))$ compares with $bs^*x(\bar{p})$ in the short run and $(b + \beta)s^*x(\bar{p})$ in the long run. The demand possibilities in figure 14.1, aside from the exact match with planned capacity, represent two (i.e., $x(p) \cdot u^L$ and $x(p) \cdot u^h$) of an infinite array of possible demands. The model assumes the policymaker focuses on the expected value of the aggregate consumer surplus net of costs. If demand is less than capacity (i.e., $x(\bar{p}) \cdot u^L$), at $\bar{p}$, then consumer surplus will be $AB\bar{p}$, and we consider $bx(\bar{p}) \cdot u^L$ versus $\bar{p}x(\bar{p})u^L$ to determine the need for subsidies in the short run. The contribution to net benefits is the consumer surplus plus revenue, $(A\bar{p}B + \bar{p}x(\bar{p})u^L)$, less the variable $(bx(\bar{p})u^L)$ and fixed costs $(\beta sx(\bar{p}))$. At $\bar{p}$, all consumers with willingness to pay represented along the demand curve from A to B want to consume the service, and there is sufficient capacity to accommodate them. Indeed, if the price could be adjusted, more users could be accommodated because aggregate demand is less than the capacity. When price effectively rations use, as it does in this example, then benefits are defined assuming those with highest willingness to pay are served first. Other consumers are not "counted." At the selected price, $\bar{p}$, they would not purchase the good.

The issue of other rationing schemes arises when the aggregate demand at the price, $\bar{p}$, exceeds capacity. This is case (c). All the consumers represented along the demand curve $x(p) \cdot u^h$ from A to E would be willing to pay at least $\bar{p}$. However, only $s^*x(\bar{p})$ of this total demand can be served. Price does not screen out users consistent with the predefined capacity of $s^*x(\bar{p})$. If price cannot be raised, then someone must decide who among the consumers represented from A to E gets access to the service. Efficient rationing assumes those with the highest willingness to pay, or the segment from A to C, are the customers to be served. Random rationing assumes anyone from A to E has *an equal chance* of service.

The point of this earlier literature is to recognize that the definition for the access conditions, or the rationing rule when demand exceeds capacity, influences how the policymaker would select both the ex ante price and the amount of capacity. The rationing rules define who "counts" in the objective function. Equations (1) and (2) specify the objective functions for these two cases (S_E for the expected surplus with efficient rationing and S_R for random rationing).

(1) $$S_E = \int_0^s u\left[\int_0^{x(p)} x^{-1}(q)dq - bx(p)\right]dF(u)$$

$$+ \int_s^\infty u\left[\int_0^{x(p)} x^{-1}(q)dq - bx(p)\right]dF(u)$$

$$- \int_s^\infty u\left[\int_{(s/u)x(p)}^{x(p)} x^{-1}(q)dq - \left(1-\frac{s}{u}\right)bx(p)\right]dF(u) - \beta sx(p)$$

(2) $$S_R = \int_0^s u\left[\int_0^{x(p)} x^{-1}(q)dq - bx(p)\right]dF(u)$$

$$+ \int_x^\infty u\cdot\frac{s}{u}\left[\int_0^{x(p)} x^{-1}(q)dq - bx(p)\right]dF(u) - \beta sx(p)$$

In these specifications, $x^{-1}(q)$ is the inverse demand function for $x(p)$ with q the quantity demanded at a price of p (i.e., $q = x(p)$). Both objective functions describe ex ante choices of p and s. As such, they describe what counts when demand is less than $sx(p)$ and when it exceeds $sx(p)$ for every possible value of p and s, the choice variables. Equation (1) could be written more compactly. This more detailed form is used because it helps to illustrate the issues to be considered in extending the model to include a natural supply.

The first term in equation (1) provides the contribution to expected surplus if demand is less than selected capacity at any selected price. The second term *overstates* the contribution to expected surplus for demand in excess of capacity. In terms of figure 14.1 it would count all of the surplus along the demand to point E. In fact, at $\bar{p}$ only $s^*x(\bar{p})$ units of demand can be served. So we need *two* corrections that are represented in the third term. First, we remove the extra surplus (illustrated by $s^*x(\bar{p})CEx(\bar{p})u^h$ in figure 14.1) and correct the variable cost embedded in the second term. The term, $(1-s/u)bx(p)$, removes the cost used in the second term and includes variable cost for only those units actually sold, $bsx(p)$. As the more compact version of the objective function in equation (2) illustrates, this amount is all that can be counted for a capacity price selection with random rationing. Moreover, in this case, we attach to each unit of consumption the "average" surplus over the full range of users that would "like to" have the ability to use the service at price $\bar{p}$. The last term in equations (1) and (2) is the cost of a selected capacity. This long-run cost does not change with the rationing schemes.

Table 14.1 summarizes the implications for capacity and price selections under the two objective functions and rationing schemes. The capacity/price pair for the objective function associated with efficient rationing summarizes the results from Brown and Johnson (1969; with a somewhat different

Table 14.1   Capacity and pricing with demand uncertainty

	Pricing	Capacity choice
Efficient rationing	$p = b$	$\int_s^\infty \bar{p}(u)dF(u) = \beta + b(1 - F(s))$
Random rationing	$p = b + \dfrac{\beta}{(1/s)\int_0^s udF(u) + (1 - F(s))}$	$s = F^{-1}\left(\dfrac{(cs/x(p)) - b - \beta}{(cs/x(p)) - b} \right)$

Notes: These results are derived maximizing expected consumer surplus using equations (1) and (2) in the text; $F^{-1}(.)$ refers to the inverse of the distribution function $F(u)$; $\bar{p}$ is defined implicitly based as the price required to assure the unit quantity demand would equal the proportional reduction required so that $ux(\bar{p}) = sx(p)$. Thus $\bar{p} = x^{-1}[(s/u)x(p)] = x^{-1}[(s/u)x(b)]$ when p is set equal to b.

specification for capacity) and those for random rationing are taken from Carlton (1977). Clearly, the selection of an "optimal" price (p) and capacity (s) pair depends on how access conditions are determined in periods of excess demand.

It is not easy to compare the capacity choices under efficient and random rationing. Direct results depend on what we assume for $x(p)$ and $F(u)$. s is defined implicitly by equality between the truncated expected consumer surplus of the marginal user who is not served (less corresponding operating costs), $(\int_s^\infty \bar{p}dF(u) - b(1 - F(s)))$ and the marginal capacity cost. With random rationing, capacity depends on the relative size of consumer surplus per unit demanded net of both unit variable and capacity costs compared to consumer surplus per unit net of the variable cost. With efficient rationing, prices would be set *below* long-run marginal costs, while with random rationing, they would be greater than long-run marginal costs.

14.2.3   Adding Natural Supply

To relate these results to incentive-based policies for climate adaptation, we need to describe how the private goods substitute for climate services. Assume, for simplicity, that x is a perfect substitute for some climate service. If the level of natural service provided by climate is initially η, then each person's demand for a substitute is conditional to the amount of η available. If η represents the aggregate services to everyone, and climate change eliminates these natural services, then the market demand for the substitute would shift out by η (parallel to $x(p) \cdot u$). If we assume natural services are specific to each individual user, then $(x(p) - \eta)u$ is the market demand. In this case, natural supply reduces needs for x but could accentuate the variability in the aggregate demand for x. The introduction of these natural services into the formal model in the simplest case (where natural supply affects aggregate demand) is similar to adding natural capacity. It influences how we define excess demand (the upper limits of the first integral and the lower limit of the second and third in equation (1) and in a similar fashion the two integrals in equation [2]). As a result, it influences the effects of assumptions about rationing.[3] The natural supply would not influence the cost of the substitutes. Nonetheless, the comparison of price and capacity choices with the two rationing schemes would be altered.

Relaxing the assumption of perfect substitution between x and η is another variation that would further change the results. Alternatively, we could also assume the amount natural services affect the unit demands for x. This formulation would change the slope and position of $x(p)$. Finally, we could assume that u and η are not independent random variables. In this case, a joint distribution for these two random variables needs to be defined, and the problem becomes more complex.

One does not need to display all the algebra for these cases to conclude that pricing and capacity decisions would change in all of them. Thus, regardless of how we treat natural supply, anticipatory adaptation must consider both the pricing and the conditions of access to services provided by the planned substitutes for climate services at the same time as capacity planning takes place.

The incremental value of policies that would alter natural capacity also depends on adaptation policy. Access conditions determine the value of capacity as demonstrated in table 14.1. The lesson from this algebra is adaptation planning will implicitly (or explicitly) incorporate rules for allocating supply when all cannot be served. With a permanent change in the climate regime at some locations, these allocation rules serve to redefine reliability conditions for the substitute services. A more direct way of providing incentives to substitute for productive capacity would be by using pricing schemes that share the risks between suppliers and demanders of these substitutes. These price structures can also be described as methods for including the reliability of service as part of a nonlinear price schedule. In the model, these possibilities are represented through the rationing alternatives. In a more realistic setting, consumers would select among plans for service that define prices and the ability of a centralized control to remove service at particular intervals. These terms could vary with season, time of day, or whatever. They might be more complex for some substitutes than others.

Today, they are feasible policy alternatives due to the changes in our ability to meter and inform consumers of their patterns of use. It is certainly possible to envision a consumer-friendly device that would track the changes in usage and switch off electric appliances (i.e., heat pumps, refrigerators, and so forth) for short periods. It is also possible to envision remote systems a consumer might use to monitor home usage and conditions. In the case of water as a substitute for climate services, this type of continuous adjustment seems unlikely. Nonetheless, price signals that varied by season and

---

3. For random rationing, it could influence how we average consumer surplus but not the costs of capacity produced.

year based on climate along with decentralized storage could be options that policymakers and customers might consider.

14.3   Weather and Water

Climate change will alter local weather conditions. People and firms adjust by using substitutes. This chapter's analysis of this process envisioned changes to a system that already acknowledged stochastic demand for these substitutes and pricing conventions that do not allow markets to alter prices as the demand-supply imbalance changes. As a result, the effects of new uncertainties on this system and the design of revised policies depend upon what is assumed about the interrelationships between uncertainties in the supplies of climate services and the stochastic demand for substitutes. Can we treat the two as approximately independent? Or are there reasons to believe the demand for substitutes changes when the natural services they displace are also more variable? The previous section posed these as alternative model specifications.

A detailed answer for the cases of electricity and water is not possible. It is difficult to estimate the demands for these substitute goods under any set of conditions. This task is confounded by a variety of issues: increasing block rate pricing structures, limited price variation, incomplete metering of use (especially for outdoor uses in the case of water), and a variety of other challenges. Instead, this section summarizes some recent empirical research on residential water demand in the urban Southwest that suggests independence would not be a good assumption. It suggests that the nature of the residential demand for water changes with *seasonal* levels of precipitation. As a result, models that treat the uncertainty in water demand and the response of water consumed to price as independent of the uncertainty in the climate system would understate the complexity of the problem.

Table 14.2 summarizes some of the estimates for the price elasticity of demand for water by residential users in Phoenix taken from Klaiber et al. (2010). These results were developed by exploiting two types of changes in water prices for Phoenix households. In each of these years, the Phoenix water department varied residential water customers' rates between winter and summer. There was also a gradual transition in marginal prices and a change in the threshold consumption level (in the block structure) for higher marginal prices from 600 to 1,000 cubic feet consumed between winter and summer. Finally, over time, the level of the marginal prices by block and month also changed to reflect cost increases.

The estimation strategy matched records by month for years experiencing cost increases and evaluated the *change* in the quantity thresholds that define the 10th, 25th, 50th, 75th, and 90th percentiles for residential customers in each census block group served by the Phoenix water department. Summer and winter months were considered separately. As a result, each consumption group did not move between the blocks associated with dif-

Table 14.2   **Price elasticity for residential water demand**

Percentile	2003–2000 (normal/normal)			2002–2000 (normal/dry)		
	Overall	Winter	Summer	Overall	Winter	Summer
10	−1.068	−0.528	−0.959	−0.296	−0.758	−0.362
	(−27.78)	(−3.9)	(−15.22)	(−7.37)	(−7.92)	(−4.54)
25	−0.899	−0.215	−0.823	−0.143	−0.627	−0.335
	(−37.19)	(−2.17)	(−20.34)	(−5.54)	(−10.03)	(−6.28)
50	−0.743	−0.061	−0.652	−0.99	−0.524	−0.307
	(−40.13)	(−0.71)	(−22.25)	(−5.16)	(−11.05)	(−7.87)
75	−0.625	−0.075	−0.537	−0.003	−0.438	−0.195
	(−35.21)	(−0.91)	(−19.42)	(−0.15)	(−9.67)	(−4.71)
90	−0.528	*	−0.437	*	−0.428	−0.138
	(−27.38)		(−14.94)		(−6.27)	(−2.99)

Source: Klaiber et al. (2010).
Notes: The numbers in parentheses are asymptotic Z-statistics, treating the price difference, price, and quantity at their sample means as constants for estimating the variance of the estimated price elasticity.
*Positive and statistically insignificant.

ferent marginal prices. Thus, the endogeneity of price due to "choosing" a consumption block does not need to be considered. The customers in each consumption group experienced the same price change due to changes in the rates for each block over time.[4]

The effects of natural supply variability can be seen through the difference in price elasticity estimates implied for different pairings of the years used in the models. Consumption in 2000 is compared with 2002 and 2003 in forming the quantity differences used to estimate the first difference model. The average annual precipitation (as well as in average days with measurable rain) in 2002 was less than half the level experienced in 2000 and 2003. The estimates for price elasticities in winter and summer indicate quite distinct changes when pairing two normal years as compared to the pairing of a normal and a dry year.[5] For the normal/dry combination, summer demand is much *less* responsive to price changes compared to the estimates derived

4. Erin Mansur noted that the increasing block pricing structure implies that all marginal prices enter demand under uncertainty. Our analysis is a short-run model that examines changes in matched months for the typical household as the marginal price for a pricing block changes over time. This change is separate from the seasonal change winter to summer and is not part of the block structure. It reflects increases over time in marginal prices due to cost increases and would not be anticipated by households. The Olmstead, Hanemann, and Stavins (2007) result relates to a given increasing block structure and the movements within that block structure that take place due to uncertain needs for water. Our analysis holds constant the price block for consumption and considers how use changes over time as marginal price for *that* block changes.

5. By pairing the consumption at a block-group level, we control for demographics and landscape conditions. The models include temperature and precipitation controls for changes in minimum temperature and precipitation in the months paired to estimate the differences in quantity demanded for the paired years.

using changes between two normal years. By contrast, the winter demand for a normal/dry combination is *more* responsive to price than when two normal years are used to estimate the price response.

While these results are for residential water demand in one city, it is important to note that it is the first evidence of a response in monthly demand to differences in seasonal conditions, after controlling for differences in both the monthly temperatures and the monthly precipitation in the two years. It is consistent with an early stated preference study by Howe and Smith (1994). This study offered a change in the likelihood, on an annual basis, of a standard annual shortage event.[6] They found that the level of baseline reliability of the water system and average water expenditure in each of three Colorado towns influenced the choices their respondents from those towns would make to policies explained as being intended to enhance reliability.

If the demand results reported here hold up in other studies, they suggest that the stochastic nature of water demand itself may change with factors influencing natural sources of climate-related services. That is, one might speculate that climate change would not only alter the amount of water demanded as a substitute for natural sources but the price responsiveness might also change. This finding would imply larger price changes may be needed to induce greater conservation and that prices might need to depend on seasonal conditions. This conclusion parallels the Howe and Smith (1994) finding that the value of reliability depends both on the costs of water and the extent to which natural supply makes water shortfalls a more common event. Changing prices as these conditions are anticipated would offer a parallel to the more complex pricing systems described earlier for electricity.

14.4 Implications

Climate adaptation is not synonymous with augmenting the capacities of systems that provide substitutes for the climate services. Changes in pricing can reduce the demands for the services of substitutes (especially during times when demand is high) and can signal the potential for higher, long-term end-user costs for those with higher levels of use. Household commitments to power and water using devices change both the level of demand and the ability of the overall system to respond to climate changes. To the extent new price systems change the incentives households face as they make such power and water using commitments, and alter the level or the efficiency of these commitments, we might describe them as altering effective capacity

---

6. This was defined as a drought of sufficient severity and duration that residential outdoor water use would be restricted to three hours every third day for the months of July, August, and September.

of the system to meet households' needs with variation in long-run natural conditions.[7] Some types of demand are reduced or displaced. As a result, a smaller capacity can meet the revised demand pattern with less likelihood of shortfalls. This interpretation is commonly used in the demand response literature associated with pricing schemes for electricity. It has not been connected in formal models with discussions of climate adaptation.[8]

This chapter has used the early literature on pricing and capacity decisions in the presence of demand uncertainty to describe how an economic analysis of capacity planning, as a response to climate change, cannot be undertaken independent of considering how substitute services are priced. In addition, with inflexible prices, the rules used to determine who is served when demand exceeds supply will be important to both capacity and price choices. Considering the design of price schedules as part of anticipatory adaptation would imply that prices for a wide range of activities serving as substitutes for climate services might be considered. These types of changes offer the potential to create incentives that can feed back to influence both the pace of climate change and the demands for the services facilitating adaptation.

References

Borenstein, Severin. 2005. "Valuing the Time-Varying Electricity Production of Solar Photovoltaic Cells." Center for the Study of Energy Markets, University of California Institute Working Paper no. 142, March.
Brown, Gardner M. Jr., and M. Bruce Johnson. 1969. "Public Utility Pricing and Output Under Risk." *American Economic Review* 59:119–28.
Carlton, Dennis W. 1977. "Peak Load Pricing with Stochastic Demand." *American Economic Review* 67 (3): 1006–10.
Earle, Robert, Edward P. Kahn, and Edo Macan. 2009. "Measuring the Capacity Impacts of Demand Response." *Electricity Journal* 22 (6): 47–58.
Fankhauser, Samuel, Joel B. Smith, and Richard S. J. Tol. 1999. "Weathering Climate Change: Some Simple Rules to Guide Adaptation Decisions." *Ecological Economics* 30:67–78.
Howe, Charles W., and Mark G. Smith. 1994. "The Value of Water Supply Reliability in Urban Water Systems." *Journal of Environmental Economics and Management* 26:19–30.
Klaiber, H. Allen, V. Kerry Smith, Michael Kaminsky, and Aaron Strong. 2010. "Measuring Residential Water Demand under 'Data Duress'." Arizona State University, Unpublished Manuscript.
Mendelsohn, Robert. 2000. "Efficient Adaptation to Climate Change." *Climatic Change* 45:583–600.
National Research Council. 2010. *Adapting to the Impacts of Climate Change.* Amer-

---

7. Price schedules that smooth demand reduce the need for capacity to meet a peak, and, in this sense, function like added capacity. See Earle, Kahn, and Macan (2009).
8. Smith (2010) discusses this connection but does not attempt to develop a formal analysis.

ica's Climate Choices: Panel on Adapting to the Impacts of Climate Change. Prepublication copy, May. Washington, DC: National Academies Press.

Olmstead, Sheila, W. Michael Hanemann, and Robert N. Stavins. 2007. "Water Demand under Alternative Price Structures." *Journal of Environmental Economics and Management* 54 (2): 181–98.

Smith, V. Kerry. 2010. "Pre-Positioned Policy as Public Adaptation to Climate Change." Resources for the Future, Issue Brief no. 10-07, June.

Visscher, Michael. 1973. "Welfare Maximizing Price and Output with Stochastic Demand: Comment." *American Economic Review* 63:224–29.

Comment   Erin T. Mansur

From one point of view, climate adaptation can be thought of as a series of responses to supply and demand shocks. From this perspective, a well-functioning economy determines the socially optimal response. In other words, if markets are perfectly competitive—whereby all market failures of externalities, market power, imperfect information, and so on have been addressed—then the economy will adapt to market shocks in an efficient manner. Thus, the role of government is not to impose the outcome (for example, by subsidizing farmers to use more heat-tolerant crops or requiring power companies to construct more dams for hydropower capacity) but rather to facilitate well-functioning markets.

Thus, correcting failures in those markets most sensitive to climatic change becomes the focus of market-based adaptation policy. In particular, Smith looks at consumer pricing of two goods that are especially likely to become increasingly scarce, water and power, due to supply and demand shocks, respectively. These goods are expensive to store and have volatile supply and demand, respectively. Dynamic, or real-time, pricing of such goods would be a possible response. We observe this type of pricing in other markets with similar characteristics, such as hotels and airplane flights. However, utilities have been restricted (in part, because of regulation but also, at least historically for electricity, because of technological limitations). Thus, a single price, or price schedule, has been used without correcting for volatile supply and demand. Climate change is expected to increase the importance of peak load pricing in both water and power.

Smith begins by modifying a model on peak load pricing from Carlton (1977). Carlton and others noted the importance of allocation rules when prices do not clear the market. In some cases, there will be excess demand and, without variable prices in the short run, the good may still be allocated

Erin T. Mansur is associate professor of economics at Dartmouth College and a research associate of the National Bureau of Economic Research.

For acknowledgments, sources of research support, and disclosure of the author's material financial relationships, if any, please see http://www.nber.org/chapters/c12161.ack.

to those willing to pay the most for it (i.e., efficient), or it may be randomly assigned.

The main focus of Smith's chapter is on how climate change will affect these optimal decisions. In particular, climate change will affect the supply of water and the demand for electricity in a stochastic manner. This additional source of variation complicates the objective function and needs to be taken into account when thinking about optimal pricing and capacity decisions. The chapter does this by adding natural supply to this discussion of capacity: $\phi = \phi(s, n)$.

This is a useful modification for water and also for power if we think of demand shocks as negawatts. Note that much of the discussion of demand side management programs also includes demand as part of the "supply" function. Nonetheless, a more direct treatment of this uncertainty may be to include it in demand, u. However, climate change may have a direct effect on supply in regions with a significant amount of hydropower.

The chapter discusses two important features: ϕ may be nonlinear; and shocks to natural supply could be correlated with with demand shocks, u. For water, less precipitation will likely increase people's willingness to pay for utility-provided water (for watering lawns) and also decrease the utility's ability to supply water as its reservoirs will likely have less in them. This negative correlation will exacerbate the welfare loss from incomplete pricing (namely the loss that would be avoided by real-time pricing). While this correlation has not yet been included in the model, I think that this would be an interesting extension of the current chapter.

Smith suggests that this correlation may be an important characteristic of actual water demand. In particular, Klaiber et al. (2010) estimate water demand using data from households in Phoenix. For each census block and month, they calculate the quantity consumed at the 10th, 25th, 50th, 75th, and 90th percentiles. They then look at the change in consumption for that calendar-month, census block percentile group from the base year (2000) to another year (2002 or 2003). Averaging across census blocks and summer/winter months, they find several results that are consistent with those found in Mansur and Olmstead (2010): larger consumers are less elastic (Mansur and Olmstead find consumers with greater income are less elastic and purchase more water); and summer elasticity is greater than that in the winter (Mansur and Olmstead find outdoor demand is more elastic than indoor demand and makes up a larger share of total demand in the summer). Klaiber et al. (2010) then compare price changes from a normal to a normal year versus prices changes from a normal to a dry year. They find that summer demand is less elastic in the dry year. However, somewhat surprisingly, they then find that winter demand is more elastic in the dry year.

On identification, Klaiber et al. (2010) argue that ordinary least squares (OLS) is unbiased. In general, OLS estimates of a cross section of households facing increasing block pricing will result in biased, possibly positive, estimates of demand elasticity (e.g., Olmstead 2009). However, Klaiber et al. (2010) are mostly identifying demand response from changes in prices over time. They argue that OLS will be consistent as none of their consumer groups changed from the low price block to the high price block, or vice versa, when prices changed over time. However, Olmstead, Hanemann, and Stavins (2007) note that *all* prices enter into a household's demand function given uncertainty. This suggests that more complicated estimation strategies that account for nonlinear pricing may result in different estimates. In particular, demand elasticity estimates for those households near the block pricing kink point may be the most biased. Olmstead (2009) finds that the structural model of water demand and two stage least squares result in similar estimates for her sample, so the bias in Klaiber et al. (2010) may be small.

References

Carlton, Dennis W. 1977. "Peak Load Pricing with Stochastic Demand." *American Economic Review* 67 (3): 1006–10.
Klaiber, H. Allen, V. Kerry Smith, Michael Kaminsky, and Aaron Strong. 2010. "Measuring Residential Water Demand under 'Data Duress'." Arizona State University, Unpublished Manuscript.
Mansur, Erin T., and Sheila M. Olmstead. 2010. "The Value of Scarce Water: Measuring the Inefficiency of Municipal Regulations." Yale, Unpublished Manuscript.
Olmstead, Sheila M. 2009. "Reduced-form vs. Structural Models of Water Demand under Non-linear Prices." *Journal of Business and Economic Statistics* 27 (1): 84–94.
Olmstead, Sheila M., W. Michael Hanemann, and Robert N. Stavins. 2007. "Water Demand under Alternative Price Structures." *Journal of Environmental Economics and Management* 54 (2): 181–98.

15

Setting the Initial Time-Profile of Climate Policy
The Economics of Environmental Policy Phase-Ins

Roberton C. Williams III

This chapter considers the question of under what circumstances a new environmental regulation should "phase in" gradually over time, starting with an initially lax regulation and then gradually tightening, rather than being immediately implemented at full force. Phase-ins are a very common—perhaps ubiquitous—feature of new environmental regulations and can greatly influence the near-term costs and benefits of policy.

This differs from the broader, longer-term question of whether regulation should tighten over time. There are other reasons why it might be efficient for regulation to gradually tighten (e.g., if incomes are rising and, thus, the willingness to pay for a cleaner environment is also rising). The key distinction is that with a phase-in, the reason for gradually tightening over time is because the policy is new. A natural argument for a phase-in is that it provides time for individuals and firms to adjust to the new policy. Therefore, much of this chapter focuses on the role of adjustment costs.

Prior work on the broader issue of the optimal time-profile of climate policy has indirectly addressed the issue of phase-ins. For example, Wigley, Richels, and Edmonds (1996) show that because of capital adjustment costs, the least-cost path to achieve a given atmospheric concentration of CO_2 departs only gradually from the business-as-usual path, thus implicitly suggesting some sort of phased-in policy. And a substantial literature focuses on

Roberton C. Williams III is associate professor of agricultural and resource economics at the University of Maryland, a senior fellow of Resources for the Future, and a research associate of the National Bureau of Economic Research.

For their helpful comments and suggestions, I thank Don Fullerton, Stephen Holland, and participants in the NBER Design and Implementation of US Climate Policy Conference. For acknowledgments, sources of research support, and disclosure of the author's material financial relationships, if any, please see http://www.nber.org/chapters/c12144.ack.

the question of whether learning by doing accelerates or slows the optimal pace of carbon abatement (e.g., Goulder and Mathai [2000] or Manne and Richels 2004), a question that implicitly relates to phase-ins. However, none of these papers specifically considers the phase-in question or separates out this question from other influences on the optimal abatement path. And to my knowledge, no prior work in the environmental literature even implicitly addresses the phase-in question in a general context or in any specific context other than carbon abatement, even though phase-ins have been included in many other environmental regulations.[1]

This topic is also closely related to the broader literature on policy transitions. Kaplow (2003) addresses the general issue of transitions in legal rules. It includes a very brief discussion of regulation of newly discovered externalities that argues for retroactive application of environmental taxes (which is effectively the opposite of a phase-in) because it gives polluters an incentive to reduce emissions even before policy is announced.[2] Perhaps the most widely studied transition issue is the effect of switching from taxing income to taxing consumption, which is quite different from the environmental phase-in issue but, nonetheless, shares some similarities in that the way new rules affect existing capital can have important incentive and distributional effects.[3]

This chapter uses an analytical dynamic model to consider the phase-in question in a general environmental regulation context and then discusses implications of that model in the specific context of climate policy. The chapter shows that while adjustment costs provide a strong efficiency argument for phasing in a quantity-based regulation (or allowing intertemporal flexibility that creates the equivalent of a phase-in), this argument does not apply for price-based regulation. Indeed, in many cases, it will be more efficient to do just the opposite, setting an initially very high emissions price that then gradually falls over time. This difference in results comes not from any fundamental difference between price and quantity policies, but simply from a difference in how one defines whether the policy is phased in or not: under either policy, the efficient quantity of abatement rises over time, while the efficient price stays constant or even falls. However, other considerations, such as distributional concerns or monitoring and enforcement issues, may still argue for a gradual phase-in even for a price-based policy.

The next section of this chapter presents a simple analytical dynamic model of environmental regulation and uses that model to address the

1. Montero (2000) addresses a different issue: the optimal design of a trading program that allows otherwise unregulated sources to opt-in to the program. Such opt-in provisions have been included in early phases of a number of emissions trading programs (most notably the US SO_2 trading program).

2. I thank an anonymous referee for pointing out this paper.

3. A few examples of papers that focus specifically on transition issues are Bradford (1996), Kaplow (2008), and Sarkar and Zodrow (1993), but nearly every paper on consumption taxation discusses transition issues at least briefly.

phase-in question. The following section considers possible extensions to that model that might provide a further rationale for a gradual phase-in of a new regulation. A final section concludes and discusses implications for policy.

15.1   A Simple Model

This section introduces a simple analytical dynamic model of environmental regulation and uses that model to address the question of under what circumstances an environmental policy should be phased in gradually rather than immediately implemented at full force. A key element of this problem is that capital cannot instantly adjust in response to policy. This provides the main argument for phasing in policy: a gradual phase-in avoids making existing capital prematurely obsolete and allows time to build up a stock of less-polluting capital.

To incorporate this issue, production follows:

$$(1) \qquad\qquad Y_t = F(H_t, E_t),$$

where Y is output of a pollution-intensive good, H is the stock of pollution-intensive capital, and E is the pollution emissions rate. The production function is concave and twice-differentiable. In addition, pollution and capital are complements, so $\partial^2 F / \partial H \partial E > 0$.

For simplicity, this model explicitly considers only a single type of capital, and production of only one good. This is probably best understood as a partial-equilibrium model, with the single good representing an aggregate of output from all pollution-intensive industries. A model with two distinct types of capital, one polluting and one nonpolluting (or less-polluting), or with production of both pollution-intensive and nonpollution intensive goods, would be more complex but would yield fundamentally the same results.

Capital depreciates at the rate δ, and, thus, the rate of change of the capital stock is given by:

$$(2) \qquad\qquad \dot{H}_t = I_t - \delta H_t,$$

where I is the rate of investment (or disinvestment, if negative). The cost of investment is given by:

$$(3) \qquad\qquad C(I_t),$$

which is strictly convex and twice-differentiable. This function includes the cost of the capital itself (which will be negative if I is negative), plus any adjustment cost. The profit-maximization problem of a representative firm is then given by:

$$(4) \qquad\qquad \max_{E,I} \int_0^\infty [p_t F(H_t, E_t) - C(I_t) - \tau_t E_t] e^{-rt} dt,$$

subject to the capital transition equation (2), where p is the price of output, τ is the emissions tax rate or emissions permit price, and r is the discount rate. The first-order condition for the emissions rate is then:

$$(5) \qquad p_t \frac{\partial F}{\partial E_t} = \tau_t,$$

which equates the marginal value product of emissions with the emissions tax rate. The first-order condition for investment is:

$$(6) \qquad \frac{\partial C}{\partial I_t} = \lambda_t,$$

which sets the marginal cost of capital equal to its current-value shadow price, λ. The costate equation gives the rate of change of λ as:

$$(7) \qquad \dot{\lambda}_t = (r + \delta)\lambda_t - \frac{p \partial F}{\partial H_t}.$$

The intuition for this equation is that the return on capital (its marginal value product plus the change in its shadow price) must equal the cost of holding capital (the discount rate plus the depreciation rate, times the shadow price).

The next subsection considers regulation of a flow pollutant (one for which pollution damage is caused entirely by the current flow of emissions). The following subsection then considers regulation of a stock pollutant (one for which damage is caused by the accumulated stock of emissions, as is the case for greenhouse gases).

15.1.1   Regulation of a Flow Pollutant

In the flow pollutant case, pollution damage will be given by the function $D(E_t)$, which is increasing, convex, and twice-differentiable. The regulator's problem is given by:

$$(8) \qquad \max_{\tau} \int_0^{\infty} [p_t F(H_t, E_t) - C(I_t) - D(E_t)] e^{-rt} dt,$$

which is very similar to the firm's problem (4), except that in the regulator's objective, the cost of pollution is the pollution damage done, whereas the analogous term in the firm's objective is the emissions tax paid. The regulator's first-order condition for the emissions tax rate is:

$$(9) \qquad p_t \frac{\partial F}{\partial E_t} = \frac{\partial D}{\partial E_t},$$

which sets the marginal value product of emissions equal to the marginal damage. The regulator can achieve this by setting the emissions tax rate equal to marginal damage:

$$(10) \qquad \tau_t = \frac{\partial D}{\partial E_t},$$

which causes the firm's first-order condition (5) to be equivalent to the regulator's first-order condition (9). If the pollution tax is set equal to marginal damage at all points in time, then the firm's first-order condition for investment (6) and costate equation (7) will also be equivalent to the analogous equations for the regulator. Just as in a simple static model, the optimal emissions tax simply equals the marginal damage from emissions.

What does this imply for phase-ins? First, consider the case in which the marginal damage from pollution is constant (i.e., damage is linear in emissions). The optimal emissions tax, equal to marginal damage, will then also be constant. Thus, in this case, it is not optimal to phase in the emissions price: the optimal path has the emissions price go immediately to its fully phased-in level and stay constant at that level.

However, the optimal time path for emissions in this case does involve a phase-in. Imposing a constant emissions price causes an immediate drop in emissions. Because capital and emissions are complements, that drop in emissions causes a corresponding drop in the marginal product of capital, which, in turn, causes the shadow price of capital to fall, leading to a reduction in investment. That drop in investment means that the quantity of capital will gradually fall, with a corresponding gradual fall in the emissions rate (again, because capital and emissions are complements), eventually converging to a new steady state with lower levels of capital and emissions.[4]

Thus, the optimal policy doesn't phase in the emissions price but does phase in the emissions quantity. Regardless of how quickly or slowly capital can adjust, setting the emissions price equal to marginal damage internalizes the externality and, thus, leads to the efficient level of emissions. This might lead to earlier retirement of polluting capital and to higher costs than would a gradual increase in the emissions price, but, if so, then retiring that polluting capital earlier and incurring higher costs is efficient. But because the capital stock takes time to adjust, the level of emissions reductions implied by any given emissions price will rise over time, thus gradually phasing in the emissions reductions. Another way of thinking about this is to view it as a higher price elasticity of emissions in the long run than in the short run, so the same emissions price will lead to a greater emissions reduction in the long run than in the short run.

This could be implemented with a phased-in permit program, where the annual allocation of permits gradually drops over time. But calculating the appropriate phase-in rate would be very challenging for regulators because calculating the optimal path for emissions requires knowing how quickly

---

4. Note that if emissions and capital were substitutes, as would be the case for abatement capital, the optimal path would still entail a gradual drop in emissions. The chain of reasoning is the same as for the complements case, except that the signs of the changes in the shadow price of capital, investment rate, and quantity of capital are all reversed. A similar logic would apply in a model with both polluting and nonpolluting capital: along the optimal path, the level of polluting capital would fall, and the level of nonpolluting capital would rise.

polluting capital will be required. A much simpler policy would set an emissions tax with no phase-in. Similarly, a permit program that allows banking and borrowing (such that permit prices are equalized across time periods) would provide a constant price for emissions without requiring regulators to determine the appropriate path for a phase-in.

Now consider the case in which marginal damage is increasing in the level of emissions. As just shown, a constant emissions price implies a gradually falling level of emissions over time. But if marginal damage is increasing in emissions, a gradually falling level of emissions implies that marginal damage will also be gradually falling. Therefore, the optimal path must entail an emissions price that initially jumps to a level above its long-run level and then gradually falls over time as the capital stock adjusts. In this case, not only does the optimal path not entail a phase-in of the emissions price, but it actually implies the opposite.

Going one step further, consider a case with threshold damages: marginal damage is very low up to some threshold level of emissions and very high beyond that threshold. In this case, the optimal policy will hold the level of emissions right at that threshold. This will require a very high initial emissions price that then gradually falls over time. Thus, in this case, the quantity of emissions jumps immediately to its long-run level, without any phase-in, while the emissions price phases-in in reverse, starting high and then falling over time.

15.1.2   Regulation of a Stock Pollutant

Now consider the case of a stock pollutant such as greenhouse gases. Let P_t represent the stock of pollution and $D(P_t)$ the damage caused. In this case, the regulator's problem is:

$$(11) \qquad \max_\tau \int_0^\infty [p_t F(H_t, E_t) - C(I_t) - D(P_t)] e^{-rt} dt,$$

subject to the same capital transition equation (2) and to a transition equation for the pollution stock, given by:

$$(12) \qquad \dot{P_t} = E_t - \eta P_t,$$

where η is the natural rate of decay of the pollution stock. The regulator's first-order condition for the emissions price is now:

$$(13) \qquad p_t \frac{\partial F}{\partial E_t} = \mu_t,$$

where μ_t is the shadow price of emissions at time t. The costate equation for μ_t is:

$$(14) \qquad \dot{\mu_t} = (r + \eta)\mu_t - \frac{\partial D}{\partial P_t},$$

which can be solved to give:

$$(15) \qquad \mu_t = \int_0^\infty \frac{\partial D}{\partial P_{t+i}} e^{-(r+\eta)i} di.$$

Just as in the flow pollutant case, the first-order condition equates the marginal benefit from emissions with the marginal damage from emissions, which, in this case, is the discounted value of future pollution damage caused by a marginal unit of emissions at time t.

As in the flow pollutant case, it is helpful first to consider the case in which marginal pollution damage is constant. In this case, the shadow price on emissions (μ_t) will also be constant (as can be seen by examining equation [14] or [15]), and, therefore, the optimal emissions price will be constant. The intuition is that the optimal emissions price will equal the discounted flow of damages caused by a marginal unit of emissions, and, if marginal damage is constant, then that discounted flow of future damages will also be constant over time. Just as in the flow pollutant case, the optimal path has the emissions price jump immediately to its long-run level and then stay constant, while emissions gradually fall over time—in other words, the emissions price is not phased-in, but the emissions quantity is.

For the case in which marginal pollution damage is increasing in the stock of pollution, the results are again similar to the analogous results for a flow pollutant. A constant emissions price would imply a gradual fall in emissions, which, in this case, implies a gradual fall in μ_t (again, this can be seen by examining equation [14] or [15]). Thus, the optimal path must entail an emissions price that initially jumps to a level above its long-run level and then gradually falls over time. This effect will be much less pronounced than it would be for a flow pollutant (because a gradual fall in $\partial D/\partial P_t$ implies a much slower fall in μ_t), but it nonetheless demonstrates the same pattern, which is the opposite of the usual phase-in.

This argument assumes that the stock of pollution is at a steady state prior to the introduction of any regulation. This is not the case for carbon dioxide, for which the stock of pollution is currently rising rapidly. In such a case, the shadow price of emissions will follow a path similar to that of the pollution stock, initially rising, possibly overshooting its long-run level, and then eventually converging to a steady state. This resembles a phase-in because of the initially rising optimal emissions price, but arises for different reasons. In this case, the emissions price is initially rising because the current stock of emissions is below the postpolicy long-run steady state level, whereas a phase-in would be a case where the emissions price gradually rises because the policy is newly introduced. This distinction is important because it is not generally the case that the prepolicy stock of pollution will be below the postpolicy long-run steady state: the opposite could easily be true for other pollutants and would be true for carbon if we wait longer before taking action or if we were planning more aggressive action.

15.2   Possible Alternative Justifications for Phase-Ins

The previous section's results show that capital adjustment costs imply that phasing in the quantity of emissions reductions is optimal but that phasing in the emissions price is not: the optimal emissions price jumps immediately to a level at or above its long-term level, without any phase-in period. Nonetheless, many environmental regulations have gradually phased in both the quantity and price of emissions. Phase 1 of the US SO_2 trading program covered only a small fraction of the pollution sources that were covered in Phase 2, so for those not covered by Phase 1, the emissions price they faced was clearly higher in Phase 2. And even for those sources covered during Phase 1, the emissions caps in Phase 2 were enough tighter to imply a higher emissions price. Similarly, under the European Union (EU) Emissions Trading System for carbon, the second-phase caps were enough tighter than the caps during the first phase to imply a substantially higher permit price. Moreover, environmental regulations are almost always announced well before they are to take effect, which also represents a phase-in.

Were the initial phases of those programs inefficiently designed, or do other factors provide some justification for phasing in both the quantity and price of emissions? This section discusses two such extensions to the model: distributional concerns and monitoring and enforcement issues.

15.2.1   Distributional Considerations

The model in section 15.1 assumed that the regulator is setting policy to maximize efficiency. In practice, however, distributional considerations are often at least as important as efficiency, and policy decisions frequently represent a compromise between these two factors.

Suppose that, in addition to maximizing efficiency, the regulator would also like to limit the cost imposed on firms (or more generally, on the owners of pollution-intensive capital). Under some circumstances, this additional goal could imply a gradual phase-in of the emissions price as well as the emissions quantity.[5]

Consider an extreme case as an illustrative example: suppose that emissions and polluting capital are perfect complements in production (i.e., the production function [1] is Leontief), polluting capital cannot be liquidated once it is installed (i.e., $C(I_t) \geq 0$ for $I_t < 0$), and the efficiency-maximizing

---

5. This is similar to Feldstein's (1976) argument for announcing a tax reform well in advance of the date it will take effect, which is less efficient than having it take effect immediately but may still be worthwhile for distributional reasons. Zodrow (1985) considered this argument in a dynamic model with capital adjustment costs and found that whether it justifies some form of phase-in depends on the magnitude of capital adjustment costs. The optimal capital tax problem that Feldstein and Zodrow consider differs substantially from the optimal environmental tax problem considered here, but, nonetheless, it seems likely that a similar result would hold here.

emissions price is not high enough to cause capital to be idled but is more than enough to stop new investment. Thus, on the optimal path, pollution-intensive production continues, but the stock of polluting capital is allowed to depreciate over time, eventually converging to zero.

In this case, announcing that an emissions tax will be imposed at some future date but not imposing any tax before that date can have the same effect on investment and emissions as immediately imposing a tax. If the future date is not too distant (sufficiently near to cause the shadow value of capital to drop below the marginal cost of investment), then, in either case, investment will stop immediately, and emissions will fall gradually as the capital stock depreciates. Thus, the efficiency consequences of these two policies are identical, but waiting to impose the tax reduces the cost to capital owners. If the regulator puts more weight on the cost to capital owners than on government revenue, then waiting to impose the tax would be optimal.

In less extreme cases, phasing in the emissions price will have some efficiency cost, but that cost could still be outweighed by distributional considerations. However, in these cases, an emissions price phase-in would still be a second-best policy: the regulator could achieve the same distributional outcome at lower efficiency cost by immediately imposing an emissions price equal to marginal damage and providing a compensating transfer (which could take the form of emissions permit allocations or inframarginal exemptions from an emissions tax) to owners of polluting capital. Only if such transfers aren't possible would an emissions price phase-in be optimal.

15.2.2   Monitoring and Enforcement

The model in section 15.1 also ignores issues of emissions monitoring and enforcement of regulations. Incorporating such issues might provide another argument for phase-ins. Suppose that the regulatory agency has a limited capacity for monitoring and enforcement and that increasing that capacity will take time (one could view this as accumulating "enforcement capital"). In such a case, it could be optimal initially to regulate only a relatively small set of polluters, those who would be expected to achieve relatively large reductions in emissions at relatively low cost. Then the set of regulated firms could be expanded over time as the regulatory agency's enforcement capacity grows. The resulting phase-in policy would look much like the phase-in of the US SO_2 trading program, which started by regulating a relatively small number of large and particularly pollution-intensive plants in Phase 1 and then expanded to include smaller and less-polluting plants in Phase 2.

Again, though, it is not at all clear that such a policy is genuinely optimal. It might well be more efficient for the regulation to cover all polluters immediately but for limited enforcement resources to be directed primarily (though not exclusively) toward the largest polluters.

15.3   Conclusions

This chapter has shown that capital adjustment costs provide an efficiency justification for a gradual phase-in of the quantity of emissions reductions under a new environmental regulation. But this argument does not hold for phase-ins of the emissions price. Indeed, the optimal policy is just the opposite—the emissions price immediately jumps to a point above its long-run level and then gradually declines to that long-run level over time—for any case in which marginal pollution damage is increasing in the quantity of emissions.

This result calls into question the approach taken with many environmental regulations, which have gradual phase-ins of both the quantity of emissions reductions and the emissions price. Given this chapter's simple and highly stylized model, it certainly cannot rule out the possibility that there are other considerations that would justify such phase-ins, and further work to explore such possible justifications would be valuable. But these results do suggest that policymakers should consider a more aggressive emissions price path in the initial implementation of a new regulation.

References

Bradford, David. 1996. "Consumption Taxes: Some Fundamental Transition Issues." In *Frontiers of Tax Reform,* edited by Michael Boskin, 123–50. Stanford, CA: Hoover Institution Press.

Feldstein, Martin. 1976. "On the Theory of Tax Reform." *Journal of Public Economics* 6:77–104.

Goulder, Lawrence, and Koshy Mathai. 2000. "Optimal CO_2 Abatement in the Presence of Induced Technological Change." *Journal of Environmental Economics and Management* 39:1–38.

Kaplow, Louis. 2003. "Transition Policy: A Conceptual Framework." *Journal of Contemporary Legal Issues* 13:161.

———. 2008. "Capital Levies and Transition to a Consumption Tax." In *Institutional Foundations of Public Finance: Economic and Legal Perspectives,* edited by Alan Auerbach and Daniel Shaviro, 112–46. Cambridge, MA: Harvard University Press.

Manne, Alan, and Richard Richels. 2004. "The Impact of Learning-by-Doing on the Timing and Costs of CO_2 Abatement." *Energy Economics* 26:603–19.

Montero, Juan-Pablo. 2000. "Optimal Design of a Phase-In Emissions Trading Program." *Journal of Public Economics* 75:273–91.

Sarkar, Shounak, and George Zodrow. 1993. "Transitional Issues in Moving to a Direct Consumption Tax." *National Tax Journal* 46 (3): 359–76.

Wigley, T., R. Richels, and J. Edmonds. 1996. "Economic and Environmental Choices in the Stabilization of Atmospheric CO_2 Concentrations." *Nature* 379:240–43.

Zodrow, George. 1985. "Optimal Tax Reform in the Presence of Adjustment Costs." *Journal of Public Economics* 27:211–30.

Comment   Stephen P. Holland

How can we regulate carbon emissions from factories and power plants, many of which were built years or decades ago? One potential solution is to "phase-in" any regulations so that businesses have time to anticipate and prepare for the regulations.

Williams's chapter analyzes whether and how environmental regulations should be phased in when capital adjustment is costly. In the context of the model, the chapter shows that quantity regulations—such as cap and trade—generate the greatest social welfare when they are phased in over time (become more stringent). However, on the same criterion, price instruments should *not* be phased in and possibly should be set initially at a higher level and reduced over time.

The chapter makes the intuition for these slightly counterintuitive results quite clear. As capital equipment adjusts over time, it becomes cheaper to attain a given level of emissions. The optimal environmental regulation takes advantage of this either by tightening quantity regulations or by relaxing price regulations.

Given this surprising result, Williams's chapter nicely discusses several reasons outside the model for phasing in environmental taxes. In particular, distributional concerns or the development of regulatory capacity may be reasons for phasing in regulations. Moreover, the chapter describes an additional reason for phasing in a carbon tax: increasing marginal damages with a stock pollutant. Although this case receives only brief mention, it is probably the most relevant case for current climate policy.

To understand this rationale for phasing in a carbon tax, consider equation (15) of Williams's chapter:

$$(15) \qquad \mu_t = \int_0^\infty \frac{\partial D}{\partial P_{t+i}} e^{-(r+\eta)i} di.$$

When damages are a function $D(P_t)$ of the stock of the pollutant (e.g., the atmospheric concentration of carbon), this equation shows that the optimal carbon tax should equal the sum of the present value of marginal damages.[1] If marginal damages are constant, then this equation implies that the optimal carbon tax should be constant, that is, should not be phased in.

But are marginal damages constant? While damages from global warming are notoriously tricky to specify, it is likely that marginal damages are

Stephen P. Holland is associate professor of economics at the University of North Carolina at Greensboro and a research associate of the National Bureau of Economic Research.

For acknowledgments, sources of research support, and disclosure of the author's material financial relationships, if any, please see http://www.nber.org/chapters/c12145.ack.

1. For simplicity, I discuss the equivalent discrete version of Williams's equation (15), in which the integral is a summation.

increasing.[2] In this case, Williams states that the optimal carbon tax gradually *falls* over time. However, this rests on the unrealistic assumption that climate policy reduces the atmospheric concentration of carbon from one steady state to another.

In the more realistic case where climate policy simply slows the growth of the atmospheric concentration of carbon, the model's implications are quite different. If the atmospheric concentration of carbon is growing, then marginal damages are increasing over time. As shown in equation (15), the optimal carbon tax in the first year is then the sum of the present values of future marginal damages beginning with the first year. Similarly, the optimal carbon tax in the second year is the sum of the present values of future marginal damages beginning with the second year. Because each year the marginal damages are higher than in the preceding year, the sum of their present values beginning with the second year is higher. Thus, the optimal carbon tax in the second year is higher than in the first year; that is, the optimal carbon tax is phased in.

As Williams points out, this phase-in arises due to increasing marginal damages rather than from adjustment in the capital stock. Because increasing marginal damages and a rising atmospheric concentration of carbon imply a rising carbon tax, but capital adjustment costs imply a falling carbon tax, theory alone cannot tell whether an optimal carbon tax should be phased in.

2. See Martin Weitzman "What is the 'Damages Function' for Global Warming—and What Difference Might It Make?" forthcoming in *Climate Change Economics*.

IV

Sector-Specific Issues

Urban Policy Effects on Carbon Mitigation

Matthew E. Kahn

16.1 Introduction

Suppose that a household was choosing between living in Houston or San Francisco. In each city, what would this household's annual carbon footprint be? Glaeser and Kahn (2010) estimate that a standardized household would create 12.5 extra tons of carbon dioxide per year if it moved to Houston rather than moving to San Francisco. In Houston, the same household drives more, lives in a bigger home, uses more residential electricity—electricity that is generated by power plants with a higher carbon emissions factor. Using data from the year 2000 across sixty-four major metropolitan areas, Glaeser and Kahn (2010) document that Pittsburgh is the city with the median residential household footprint of 28.3 tons of carbon dioxide per year, while San Francisco is the third "greenest" metropolitan area, and Houston is the third "brownest" metro area. This cross-sectional descriptive work creates a benchmark for comparing cities' household carbon emissions from transportation, electricity consumption, and home heating, at a point in time and tracking city trends over time.

Given that greenhouse gas emissions are a global externality, households are unlikely to internalize the carbon consequences of moving to a city such as Houston rather than San Francisco. Why is San Francisco "greener" than Houston? San Francisco is blessed with a temperate climate. Northern California's electric utilities emit less greenhouse gas emissions than their Texas

Matthew E. Kahn is professor in the Institute of the Environment, the Department of Economics, and the Department of Public Policy at the University of California, Los Angeles, and a research associate of the National Bureau of Economic Research.

I thank Varun Mehra, Frank Wolak, and Catherine Wolfram for useful comments. For acknowledgments, sources of research support, and disclosure of the author's material financial relationships, if any, please see http://www.nber.org/chapters/c12128.ack.

counterparts. Due to its amenities, land prices are higher in San Francisco, and its residents live in smaller homes than their Houston counterparts. Land prices are highest downtown, and this encourages economy activity to be highly compact. Such population density in San Francisco encourages households to live a walking, "new urbanist" life. This low-carbon lifestyle would be rare in sprawling Houston.

Over the last 100 years, people and jobs have been moving away from center cities. The average person who lived in a metropolitan area lived 9.8 miles from the city center in 1970, and this distance grew to 13.2 miles by the year 2000. While privately beneficial, this trend has helped to exacerbate the challenge of mitigating greenhouse gas production. Suburbanites drive more and live in larger homes that require more heating and cooling than their urban counterparts.

This chapter uses three different data sets to document that households who live in center cities drive less, use public transit more, and consume less electricity than observationally similar households who live in the suburbs. This center city/suburban differential is largest in the Northeast's monocentric cities such as New York City. Given that households choose where to live, this differential could be caused by both residential self-selection and a true causal "treatment effect" of urban living. If these correlations are due to a treatment effect, then any pro-center city policy, such as policies directed toward reducing inner-city crime, is likely to reduce a metropolitan area's carbon footprint.

16.2 Urban Transportation

16.2.1 Miles Driven as a Function of Urban Form

The Department of Transportation has recently released the 2009 National Household Transportation Survey (NHTS) micro data.[1] This micro data set is distinctive because it reports household vehicle mileage for a large representative sample of households. Using a special version of the data set that has census tract identifiers, I restrict the sample to households living within thirty-five miles of a major city center. For each household, I observe which metropolitan area it lives in, its distance to the city center, and its distance to the closest rail transit station using data from Baum-Snow and Kahn (2005). I estimate ordinary least squares (OLS) regressions for household i living in census tract 1 in metropolitan area j using observations on over 92,000 households based on equation (1):

(1) $\text{Miles}_{ilj} = \text{MSA}_j + B_1 \cdot \text{Demographics}_i + B_2 \cdot \text{Urban Form}_{ilj} + U_{ilj}$

In this regression, the dependent variable is the household's total miles driven in the last year. I trim the dependent variable and set the dependent

---

1. See http://nhts.ornl.gov/download.shtml#2009.

variable equal to the 99th percentile of the empirical distribution for observations in the top percentile. In table 16.1 columns (1) and (2) are identical except that the results in column (2) include metropolitan area fixed effects. In these regressions, I control for household income, the number of people in the household, and the age of the head of the household. Controlling for these demographic factors, my primary interest is to measure the association between a household's total miles driving and its location within the metropolitan area, the population density where it lives, and where it lives within one mile of a rail transit line. The results are roughly similar with and without metropolitan area fixed effects. As shown in table 16.1 distance from the city center is positively correlated with miles driving. Moving a household from the 25th percentile of the distance to the city center distribution to the 75th percentile of this distribution is associated with driving an extra 1,300 miles per year. A household's census tract's population density is negatively correlated with miles driving. Moving a household from the 25th percentile of the population density distribution to the 75th percentile of this distribution is associated with a driving 2,400 fewer miles per year. In both regressions, proximity to a rail transit line has a negative effect on

Table 16.1   Household miles driven as a function of urban form

	(1)	(2)
Log(Distance to city center)	1239.66	1140.41
	[217.000]***	[254.272]***
Log(Census tract density)	−1154.22	−1066.32
	[131.712]***	[150.854]***
Within 1 mile of rail transit	−1941.15	−1906.90
	[643.066]***	[743.939]**
Household size	4997.63	5051.96
	[224.129]***	[209.260]***
Age of head of household	−77.72	−77.71
	[12.073]***	[11.589]***
Constant	6822.21	4751.63
	[2752.336]**	[3365.890]
Observations	92,597	92,597
R^2	0.26	0.27
Geographical fixed effect	No	Metro
Household income fixed effect	Yes	Yes

Notes: The unit of analysis is the household. The dependent variable is the household's annual mileage. The sample includes all households in the 2009 National Household Transportation Survey (NHTS) who live in a census tract whose centroid is within thirty-five miles of a city center. Standard errors are reported in brackets. The standard errors are clustered by census tract. The omitted category is a household who lives more than a mile from the closest rail transit line. Dummy variables for the household's income category are included in the regressions, but their coefficients are suppressed. Miles driven has a mean of 17,925 and a standard deviation of 19,061.

***Significant at the 1 percent level.

**Significant at the 5 percent level.

driving. All else equal, households who live within one mile of a rail transit line drive 1,900 fewer miles per year than households who live more than one mile from a rail transit line. This correlation must reflect both sorting effects (selection) and a treatment effect due to easy access to fast public transit.

These correlations are suggestive about the role that urban policy plays in encouraging driving less. If households were randomly assigned to homes, then OLS estimates of equation (1) would be of immediate use to policymakers in determining how urban policies affect an important part of the household carbon footprint (miles driven). But we know that households self-select where they want to live. Liberal environmentalists are likely to self-select and choose to live in the high density areas, close to center city, and close to subway stations (Kahn and Morris 2009). An active research agenda in urban planning examines the importance of attitudes, beliefs, and preferences in determining residential location choice and travel behavior (e.g., Cao, Handy, and Mokhtarian 2006; Cao, Mokhtarian, and Handy 2007; Krizek 2003).

16.2.2   Public Transit Use from 1970 to 2000

In previous research, I have examined how worker public transit varies across cities and over time for cities that have expanded their rail transit systems (Baum-Snow and Kahn 2005). This work has public policy implications because rail transit construction is a favorite urban policy for encouraging center city growth and compact urban development.

Communities differ with respect to their distance to the central business district (CBD) and their distance to rail transit stations. Rail transit is a fast means for commuting to the city center. As discussed in Glaeser and Kahn (2001, 2004), a fundamental challenge for urban policy in battling climate change is that jobs continue to suburbanize. When people work in the suburbs, they do not use public transit to commute there. But, in cities such as New York City with a vibrant center city core, public transit remains an important commuting mode.

To examine how rail transit access affects commute mode choice, I use census tract level data for forty-two major metropolitan areas. I examine how proximity to the CBD and proximity to rail transit correlates with public transit use. I use a geocoded census tract panel data set from 1970, 1980, 1990, and 2000 and observe public transit use by workers (the percent of the census tract who commute using public transit), while restricting the sample to tracts within thirty-five miles of a major city's CBD.

In table 16.2, the dependent variable is a tract's public transit use share. Each column reports a separate regression. I include all metropolitan areas that are within thirty-five miles of a city center that has a rail transit system. Controlling for metropolitan area fixed effects, several facts emerge. Relative to the omitted category (1970), the propensity to commute by public transit has declined each decade. The propensity to commute by public transit

Table 16.2   **Public transit use and urban form from 1970 to 2000**

	Tract share of workers commuting using public transit		
	(1)	(2)	(3)
Log(Tract distance to CBD)	–0.119	–0.109	
	[0.001]***	[0.001]***	
1(Within 1 mile of rail station)	0.043	0.038	0.023
	[0.002]***	[0.002]***	[0.001]***
1980 year dummy	–0.024	–0.022	–0.011
	[0.001]***	[0.001]***	[0.000]***
1990 year dummy	–0.032	–0.033	–0.021
	[0.001]***	[0.001]***	[0.000]***
2000 year dummy	–0.032	–0.034	–0.021
	[0.001]***	[0.001]***	[0.000]***
Tract share college graduate		0	
		[0.002]	
Tract share black		0.148	
		[0.002]***	
Constant	1.329	1.204	0.154
	[0.006]***	[0.005]***	[0.000]***
Observations	74,076	74,076	74,076
R^2	0.665	0.701	0.969
Fixed effects	Metro	Metro	Tract

Notes: This table reports three ordinary least squares regressions. The omitted category is a 1970 census tract whose centroid is more than one mile from the closest rail transit station. In columns (1) to (3), the sample includes all census tracts whose centroid is within thirty-five miles of a central business district (CBD) in a city with a rail transit system. Standard errors are reported in brackets. The dependent variable has a mean of .16 and a standard deviation of .19.

***Significant at the 1 percent level.

declines with distance from the city center and increases if a tract's centroid is within one mile of a rail transit station. These results are robust to controlling for tract demographics such as the share black and the share college graduate (see column [2]).

16.3   Differences in Residential Energy Consumption between Urban and Suburban Households

In this section, I examine household residential energy consumption for center city versus suburban residents, and I compare how energy consumption varies across geographical regions of the nation. I use micro data from the 2005 Residential Energy Consumption Survey (RECS) for households who live in urban and suburban areas.[2] While the RECS data does not pro-

2. See http://www.eia.doe.gov/emeu/recs/recspubuse05/pubuse05.html.

vide information on the exact metropolitan area where a household lives, it does provide information on the household's census division, cooling degree days where the home is located, and the household's urban versus suburban status. I estimate versions of equation (2) for household i living in census division j.

$$(2) \qquad Y_{ij} = \text{Division}_j + B_1 \cdot \text{Demographics}_i + B_2 \cdot \text{Suburb}_{ij} + U_{ij}$$

Controlling for a household's income and the number of people in the home, why would electricity consumption differ for suburban versus urban households? In a classic monocentric city, all jobs are located in the CBD. Land prices decline with distance from the city center to compensate households for longer commutes. If land prices are lower in the suburbs, then suburban homes will be larger, and the households who live there will consume more energy. As documented by Glaeser and Kahn (2001, 2004), employment has been suburbanizing but major center cities such as New York City and others in the Northeast continue to be major employment centers. I predict that suburbanites in regions where a large share of employment continues to be downtown (and, thus, the monocentric model has more predictive power) will feature a larger center city/suburb energy consumption differential.

The energy consumption regressions are reported in table 16.3. In column (1), the dependent variable is the log of household electricity consumption. Controlling for the household's income and demographics, the average suburbanite outside of the Northeast region consumes 10 percent more electricity than the average urbanite. In the Northeast region, this differential is much larger. The average suburbanite in the Northeast consumes 51 percent more electricity than the average urbanite. The divisional dummies show the spatial variation in energy consumption with the East South Central division having the highest residential electricity consumption.[3] Column (2) shows a similar Northeast urban/suburban differential for natural gas consumption. This suburban/center city differential could simply reflect selection rather than a treatment effect. In column (3), the dependent variable is a dummy that equals one if the household lives in a single detached home. Single detached homes are likely to be larger than apartments in multifamily housing. Larger homes require more electricity. The probability that a household lives in single family home is higher in the suburbs and much higher (23.5 percentage points) in the Northeast suburbs. This evidence supports the claim that the urban/suburban electricity consumption gap is larger in monocentric cities because of the within metropolitan area differences in the housing stock. In a metropolitan area with a uniform distribution of employment and no spatially differentiated amenities, I would

---

3. These states include Alabama, Kentucky, Mississippi, and Tennessee.

Table 16.3   **Urban form and residential electricity consumption**

	(1) Log(electricity)	(2) Log(natural gas)	(3) Single family home
Suburb	0.099	0.038	0.05
	[0.027]***	[0.039]	[0.019]***
Suburb*Northeast region	0.306	0.348	0.235
	[0.063]***	[0.093]***	[0.044]***
Persons in household	0.143	0.102	0.062
	[0.008]***	[0.012]***	[0.006]***
Age of head of household	0.002	0.007	0.007
	[0.001]**	[0.001]***	[0.001]***
Middle Atlantic	−0.078	−0.321	−0.035
	[0.066]	[0.103]***	[0.047]
East North Central	0.335	0.279	0.26
	[0.063]***	[0.098]***	[0.045]***
West North Central	0.403	0.082	0.261
	[0.074]***	[0.112]	[0.052]***
South Atlanta	0.567	−0.071	0.213
	[0.066]***	[0.106]	[0.046]***
East South Central	0.822	0.09	0.331
	[0.077]***	[0.125]	[0.054]***
West South Central	0.539	−0.042	0.295
	[0.073]***	[0.118]	[0.051]***
Mountain	0.313	−0.085	0.302
	[0.074]***	[0.113]	[0.053]***
Pacific	0.034	−0.571	0.123
	[0.062]	[0.098]***	[0.044]***
Cooling degree days in 1000s	0.14	−0.242	−0.023
(base 65)	[0.014]***	[0.026]***	[0.010]**
Constant	9.234	10.722	0.021
	[0.072]***	[0.115]***	[0.051]
Observations	2,602	1,892	2,602
R^2	0.421	0.273	0.267
Housing income fixed effect	Yes	Yes	Yes

Notes: Standard errors are reported in brackets. The omitted category is an urban household in the New England Census Division. The dependent variable in column (3) is a dummy variable that equals one if the household lives in a single family home.
***Significant at the 1 percent level.
**Significant at the 5 percent level.

not expect to observe spatial differences in residential energy consumption because the price of land would be the same throughout the area.

16.4   Conclusion

The macro debate about the costs and benefits of adopting carbon pricing has not discussed how carbon mitigation incentives will affect competition between center cities and their suburbs or between low-carbon cities such as San Francisco and high-carbon cities such as Houston. The low-carbon city is a city that is compact and dense and offers fast, frequent public transit that helps people commute to downtown.

Across regions, there are large differences in the household carbon footprint. Glaeser and Kahn (2010) find that cities in California, such as Los Angeles, San Diego, and San Francisco, have the smallest residential carbon footprint, while cities in the South, such as Memphis, Oklahoma City, Houston, and Nashville, have the largest carbon footprint. Their results suggest that housing development policies that lower barriers to development in California's coastal cities would "green" the overall national average. Housing economists have ranked cities with respect to the stringency of their antigrowth policies. The low-carbon cities identified by Glaeser and Kahn tend to be the same cities that have high land use regulation (Glaeser, Gyourko, and Saks 2005; Quigley and Raphael 2005). Glaeser and Kahn (2010) argue that housing regulation does not cause a low carbon footprint. They argue that the antigrowth cities are green because of their local climate conditions and their relatively low electric utility emissions factor (using natural gas rather than coal). This claim merits future research.

If carbon legislation is passed soon, the residents of the low-carbon cities will face less of a tax burden, and this will be capitalized into local land prices. Previous incidence studies have focused on geography and income categories but not the center city/suburbs dimension (Hassett, Mathur, and Metcalf 2007). The introduction of a significant carbon tax may help to reverse a fifty-year trend in the suburbanization of households and firms (Glaeser and Kahn 2004).

Such a tax could reduce the current carbon gap between cities such as San Francisco and Houston. In the long run, in the presence of such a tax, Houston's transport infrastructure, residential building stock, and portfolio of electric utilities might resemble San Francisco's. Urban economists have tried to use the "natural experiment" of the Organization of Petroleum Exporting Countries (OPEC) oil shocks to examine whether high gas prices encourage densification (Muth 1984). These short-run shocks did not increase the demand for center city living in the 1970s.

This chapter's evidence suggests that center city residents do produce less carbon emissions than their suburban counterparts. A productive future line of research could use panel data to disentangle whether the observed correlation between center city living and the low-carbon lifestyle represents a self-selection effect or a true causal effect.[4] If future research substantiates the causal role of center city living, then this raises the public policy issue of how

---

4. See Eid et al. (2008) for a recent study that uses panel data to attempt to estimate the impact of urban form on obesity. By observing weight changes for migrants from center cities to suburbs (and vice-versa), they reject the claim that "sprawl is making us fat."

do we encourage more households to live downtown? Policies that improve the center city's quality of life and local public goods bundle can achieve this goal. The vibrancy of a downtown can be spurred by fighting crime and by improving urban public schools (Berry Cullen and Levitt 1999).

References

Baum-Snow, Nathaniel, and Matthew E. Kahn. 2005. "The Effects of Urban Rail Transit Expansion: Evidence from Sixteen Cities from 1970 to 2000." *Brookings-Wharton Papers on Urban Affairs,* edited by Gary Burtless and Janet Rothenberg Pack, 147–206. Washington, DC: Brookings Institution.

Berry Cullen, Julie, and Steven D. Levitt. 1999. "Crime, Urban Flight, and the Consequences for Cities." *Review of Economics and Statistics* 81 (2): 159–69.

Cao, X., S. L. Handy, and P. L. Mokhtarian. 2006. "The Influences of the Built Environment and Residential Self-Selection on Pedestrian Behavior: Evidence from Austin, TX." *Transportation* 33 (1): 1–20.

Cao, X., P. L. Mokhtarian, and S. L. Handy. 2007. "Do Changes in Neighborhood Characteristics Lead to Changes in Travel Behavior? A Structural Equations Modeling Approach." *Transportation* 34 (5): 535–56.

Eid, Jean, Henry Overman, Diego Puga, and Matthew Turner. 2008. "Fat City: Questioning the Relationship between Urban Sprawl and Obesity." *Journal of Urban Economics* 63 (2): 385–404.

Glaeser, Edward L., Joseph Gyourko, and Raven Saks. 2005. "Why is Manhattan So Expensive? Regulation and the Rise in House Prices." *Journal of Law and Economics* 48 (2): 331–70.

Glaeser, Edward, and Matthew E. Kahn. 2001. "Decentralized Employment and the Transformation of the American City." *Brookings-Wharton Papers on Urban Affairs,* edited by Janet Peck and Gary Burtless, 1–47. 2. Washington, DC: Brookings Institution.

———. 2004. "Sprawl and Urban Growth." In *Handbook of Urban Economics.* Vol. 4, edited by Vernon Henderson and J. Thisse, 2481–2527. Amsterdam: North-Holland Press.

———. 2010. "The Greenness of Cities: Carbon Dioxide Emissions and Urban Development." *Journal of Urban Economics* 67 (3): 404–18.

Hassett, Kevin, Aparna Mathur, and Gilbert Metcalf. 2007. "The Incidence of a U.S. Carbon Tax: A Lifetime and Regional Analysis." NBER Working Paper no. 13554. Cambridge, MA: National Bureau of Economic Research.

Kahn, Matthew E., and Eric Morris. 2009. "Walking the Walk: Do Green Beliefs Translate into Green Travel Behavior." *Journal of the American Planning Association* 75 (4): 389–405.

Krizek, Kevin. 2003. "Residential Relocation and Changes in Urban Travel: Does Neighborhood-Scale Urban Form Matter?" *Journal of the American Planning Association* 69 (3): 265–81.

Muth, Richard. 1984. "Energy Prices and Urban Decentralization." In *Energy Costs, Urban Development and Housing,* edited by Anthony Downs and Katharine L. Bradbury, 85–104. Washington, DC: Brookings Institution.

Quigley, John M., and Stephen Raphael. 2005. "Regulation and the High Cost of Housing in California." *American Economic Review* 95 (2): 323–128.

Comment   Christopher R. Knittel

Reductions in greenhouse gas (GHG) emissions typically focus on increased use of lower GHG technologies that already exist, such as increases in insulation and shifts to higher mileage vehicles, and the advent new technologies, such as more efficient air conditioning units and vehicles. The chapter by Matthew Kahn adds to our understanding of a third mechanism for GHG reductions: shifts in where economic activity takes place.

Understanding this mechanism can have large implications for how we regulate GHG emissions and the social costs associated with those regulations.

The main question that Kahn wants to answer is essentially the following: suppose we moved everyone living in Houston to San Francisco; how would their carbon footprint change? Reductions are likely to come from this move for a number of reasons. First, California electricity is generated from cleaner sources. Second, San Francisco is more walking-friendly than Houston. And third, there is an income effect given the higher land prices.

Kahn uses a variety of data sources to answer this question. Using these data, he documents that households living closer to the center of cities drive less, rely on public transit more, and consume less electricity than those households living in the suburbs. These effects are largest in the Northeast's "monocentric" cities.

To analyze how household location affects miles driven, Professor Kahn uses the 2009 National Household Transportation Survey, which surveys a large number of households and reports household location and miles driven. Kahn then regresses miles driven on the distance the household is from the center of the city (in logs), the population density of the household's census tract (in logs), a dummy for whether the household is within rail transit, household size, the age of the head of household, and household income. The results suggest that the correlations between miles driven and distance from center of the city, population density, and distance from a rail transit system are large.

Kahn then uses census-tract-level data to analyze how the share of public transit use correlates with distance from the center of a city and proximity to a rail transit station. In these empirical models, Kahn controls for the decade the data are taken from, the tract's share of college graduates, and the tract's share of African Americans. In two specifications, he includes metropolitan

Christopher R. Knittel is the William Barton Rogers Professor of Energy Economics at the Sloan School of Management, Massachusetts Institute of Technology, and a research associate of the National Bureau of Economic Research.

For acknowledgments, sources of research support, and disclosure of the author's material financial relationships, if any, please see http://www.nber.org/chapters/c12129.ack.

area fixed effects, and, in a third, he includes tract fixed effects. The results, again, are quite intuitive. He finds that the further the tract is from the city's center, the lower is public transit usage, and those tracts with a rail station within one mile have higher public transit usage. This last correlation exists even when he uses within-tract changes in the variable. That is, if we take the same tract in two different time periods, one where there is not a transit station, the other where there is, transit usage is higher, on average, in the period where there is a transit system.

Finally, Kahn looks at electricity and natural gas usage. Here Kahn regresses the log of electricity and natural gas usage on a dummy for whether the household is in the suburbs, the number of household members, the age of the head of household, dummies for eight regions, and the number of cooling degree days. Kahn finds that a suburban household is associated roughly 10 percent greater electricity usage and 4 percent higher natural gas usage. These effects quadruple in size in the country's Northeast region!

To summarize, Kahn provides compelling evidence that living closer to the center of the city and public transit are correlated with lower energy use, both in terms of transportation and home energy use. This is an important set of results and, I hope, sparks further research in this area. The elephant in the room, entirely visible to Professor Kahn, is whether these results represent correlations or are they causal relationships. That is, for the latter, if we were to pick up the Knittel family, who lives in the suburbs, and move them to the center of Sacramento, would we observe the same changes in energy usage as represented by Kahn's statistical analysis?

The results in Kahn's chapter can be viewed as upper bounds on these effects, highlighting the importance of his analysis. Had the correlations not been as large, policymakers might have concluded that land use policies are unlikely to lead to large greenhouse gas reductions. The size of the estimated correlations leaves open the door for these policy instruments. Whether policymakers should go through the door requires more analysis. For one, it may be the case that those households living further from the city differ from city dwellers for reasons other than simple geography. They may prefer larger vehicles, cooler in-home temperatures in the summer, larger homes, and so on. All of these other factors are not controlled for in Kahn's analysis. Kahn understands this, but the data limitations are severe; it will take more time and more data to be able to control for these factors. Second, Kahn's analysis only speaks to the external-benefit side of land use policies. We still don't know how costly it is to the Knittels to "force" them from their lakefront home in the suburbs to the city center.

As is often the case in Kahn's research, this chapter is sure to launch a stream of important papers on this topic. Kahn has established an important set of initial results. Future work will continue to refine these estimates insofar as they are causal linkages between where people live and their energy use.

Is Agricultural Production Becoming More or Less Sensitive to Extreme Heat?
Evidence from US Corn and Soybean Yields

Michael J. Roberts and Wolfram Schlenker

17.1 The Role of Agriculture in the US Economy

The share of employment in the agricultural sector in the United States has been continuously declining. About half of the occupations in the 1870 census were classified as agriculture, and a significant share of the workforce in the manufacturing and service sector were related to agriculture.[1] In the 2000 census, only 1.9 percent of the workforce was employed in agriculture, forestry, fishing, hunting, or mining. Employment in the agricultural sector decreased by 1.8 percent per year in the postwar period 1947 to 1985, while agriculture exhibited one of the highest postwar productivity growth rates of 1.6 percent per year, only surpassed by communications (Jorgenson and Gollop 1992).

Growth in agricultural productivity is shown in figure 17.1, which displays yields for corn, soybeans, and wheat for the years 1866 to 2009.[2] It shows yearly outcomes as well as trend lines. Before World War II, yields were stable over time, and production increases were driven by expansion of the growing area into the Western United States. Following World War II, growth switched from the extensive to the intensive margin: output per acre increased significantly due to new seed varieties and increased use of fertil-

Michael J. Roberts is associate professor of agricultural and resource economics at North Carolina State University. Wolfram Schlenker is assistant professor of economics at Columbia University and a faculty research fellow of the National Bureau of Economic Research.

We would like to thank James Bushnell and Catherine Wolfram for helpful comments. This material is based upon work supported in part by the Department of Energy under grant no. DE-FG02-08ER64640 and the National Science Foundation under grant no. 0962559. For acknowledgments, sources of research support, and disclosure of the authors' material financial relationships, if any, please see http://www.nber.org/chapters/c12162.ack.

1. See Table 26: 6 out of 12.5 million occupations were classified as agriculture.
2. See www.nass.usda.gov. The time series for soybeans does not start until 1924.

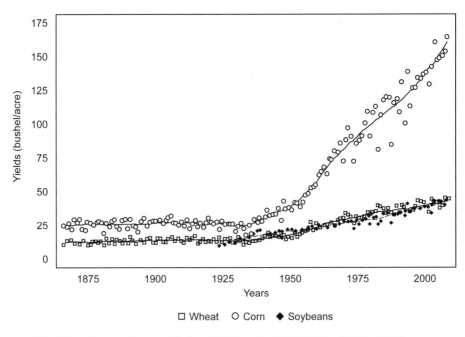

Fig. 17.1   Corn, soybean, and wheat yields in the United States (1866–2009)

Notes: Graphs shows yields over time (1866–2009 for corn and wheat and 1924–2009 for soybeans). Yearly observations are shown as scatter plot and a nonparametric trend line is added (Epanechnikov kernel with a bandwidth of ten years).

izer, while the growing area stabilized and even declined slightly.[3] In addition to yield growth, productivity was enhanced with steadily improving farm equipment, which allowed each farmer to manage increasingly larger growing areas.

On a global scale, supply growth outpaced demand growth causing commodity prices to decline in real terms over the twentieth century. Today, at least in relatively developed nations, agriculture's share of gross domestic product (GDP) is small. Estimates vary depending on how much of food processing and distribution is included in the calculation. In the United States, it is comparable to its employment share, that is, about 2 percent.

Given the small share of GDP that is attributable to agriculture in the United States, some people have argued that climate change does not pose a significant threat. There are, however, three reasons why changing climate conditions might still be economically meaningful. First, while agriculture constitutes a small share of GDP, it accounts for a large share of consumer surplus. Demand for agricultural goods is highly inelastic (Roberts and

3. The exception is soybeans, a relatively new crop that is grown in rotation with corn, which showed area increase throughout the twentieth century.

Schlenker 2010). A shortage of food has the potential to cause large price increases, as was evident in the fourfold price increase in staple commodity prices between 2005 and 2008. Second, agricultural production depends directly on weather fluctuations and is more susceptible to changing climatic conditions than other sectors of the economy. In contrast, most manufacturing today occurs within buildings, thereby insulating the process from weather fluctuations unless extreme events keep inputs or the workforce from reaching the plant. Third, agriculture in the United States is important because it constitutes a large share of global production. Corn, rice, soybeans, and wheat comprise about 75 percent of the caloric consumption of humans (Cassman 1999). The United States share of caloric production among these four commodities has been relatively constant at around 23 percent for the last forty years. This share is about twice as large as Saudi Arabia's share of world oil production (13 percent of world total, US Energy Information Administration).[4] Given its sheer size, any impact on US agricultural production can have large repercussion on world food markets.[5] And for less-developed nations, food expenditures comprise a much larger share of national income.

17.2   The Effect of a Changing Climate on Agricultural Output

Economic studies have used both cross-sectional and panel data to empirically study potential effects of climate change on agriculture. Cross-sectional studies typically associate climate with land values (Mendelsohn, Nordhaus, and Shaw 1994, Schlenker, Fisher, and Hanemann 2006), while panel studies link agricultural output to year-to-year weather fluctuations (Deschênes and Greenstone 2007; Schlenker and Roberts 2009). Each of these approaches has strengths and weaknesses. Cross-sectional differences in current climate can capture how farmers adapt to permanent difference in climate, yet they can be susceptible to omitted variables and specification biases. On the other hand, year-to-year weather fluctuations are plausibly random and exogenous to farm decision making but cannot account for long-run adaptation and, thus, may over- or underestimate effects of climate change depending on whether the set of possible adaptation choices are larger in the long term or short term. Both approaches are incapable of capturing equilibrium effects; that is, they effectively assume all prices remain constant. We first summarize earlier findings in section 17.2.1. We

4. See http://www.eia.doe.gov/emeu/cabs/Saudi_Arabia/Oil.html.
5. The United States is generally predicted to experience larger temperature increases than the global average. Despite nonuniform temperature increases, Battisti and Naylor (2009) observe that equatorial regions have a greater likelihood of experiencing temperatures that are outside the historic range because historic weather variation has also been lower. This makes statistical identification of weather and climate effects for off-equatorial zones more plausible because larger historic variations can be used to estimate a model and predicted climate change impacts do not require large out-of-sample extrapolations.

present the average effect of various weather variables in section 17.2.2 and examine the evolution of the key variables over time in section 17.2.3.

17.2.1   The Importance of Extreme Heat on Agricultural Output—Earlier Results

In previous work, we found similar relationships between corn or soybean yields and temperatures using three distinct sources of identification: (a) a fifty-six-year panel of yields from 1950 to 2005; (b) a cross-section linking average yields to average weather outcomes; (c) and a time series linking annual yields to annual weather outcomes (Schlenker and Roberts 2009). Temperature effects were modeled using a flexible functional form: yields are increasing in temperature up to 29°C (84°F) for corn and 30°C (86°F) for soybeans, but further temperature increases are harmful to yields. The ideal growing condition would be a constant temperature of 84°F for corn and 86°F for soybeans. Deviations from this optimal temperature result in approximately linear yield reductions. This linearity is captured by the concept of degree days, which are the number of degrees above a baseline, summed over all days for the growing season. For example, a temperature of 34°C with a baseline of 30°C would result in four degree days, while all temperatures below 30°C would results in zero degree days.

In the following, we use the data and optimal bounds from Schlenker and Roberts (2009), that is, all counties east of 100 degree longitude, an approximate boundary between the irrigated west and the dryland east.[6] The exception is Florida, which is excluded as most counties are highly irrigated.

17.2.2   Average Relationship between Weather and Yields

The baseline model is[7]

$$(1) \qquad y_{it} = \alpha_i + \beta_1 h_{it} + \beta_2 m_{it} + \beta_3 p_{it} + \beta_4 p_{it}^2 + t_s + t_s^2 + \varepsilon_{it},$$

where y_{it} are log corn or soybean yields in county i in year t, α_i is a county fixed effect, m_{it} captures moderate temperatures (degree days 10–29°C for corn and degree days 10–30°C for soybeans), h_{it} extreme heat (degree above 29°C for corn and degree days above 30°C for soybeans), p_{it} precipitation (in cm), and t_s and t_s^2 are state-specific quadratic time trends.

Results from estimation equation (1) are reported in table 17.1. The first column of the table uses only the measure of extreme heat and omits all other weather variables. The second column adds additional weather variables for moderate heat and precipitation. The third column uses area-weights (amount of cropland in each county) in the regression, while the fourth column uses all counties from the United States as opposed to just the Eastern subset. While previous research has argued that the response

---

6. The 100 degree meridian roughly cuts Texas in half in the east-west dimension.
7. In all regressions but column (5) of table 17.1, the errors are clustered by state.

Table 17.1   **The effect of weather on corn and soybean yields**

	(1)	(2)	(3)	(4)	(5)	(6)
			Corn			
Extreme heat	–0.594***	–0.637***	–0.746***	–0.700***	–0.639***	–0.410***
	(0.054)	(0.070)	(0.048)	(0.056)	(0.114)	(0.135)
Moderate heat		0.031***	0.043***	0.040***	0.034*	0.002
		(0.007)	(0.007)	(0.007)	(0.018)	(0.020)
Precipitation		1.031***	1.681***	1.48***	3.963**	3.544
		(0.217)	(0.338)	(0.345)	(1.658)	(2.555)
Precipitation squared		–0.008***	–0.015***	–0.013***	–0.039***	–0.035
		(0.002)	(0.003)	(0.003)	(0.015)	(0.021)
Observations	105,981	105,981	105,981	120,995	56	2,275
			Soybeans			
Extreme heat	–0.541***	–0.581***	–0.586***	–0.583***	–0.395***	–0.380***
	(0.031)	(0.029)	(0.040)	(0.039)	(0.089)	(0.129)
Moderate heat		0.040***	0.040***	0.039***	0.022	0.005
		(0.005)	(0.007)	(0.007)	(0.013)	(0.012)
Precipitation		1.222***	1.275***	1.244***	1.224	0.864
		(0.166)	(0.206)	(0.200)	(1.517)	(1.69)
Precipitation squared		–0.009***	–0.010***	–0.010***	–0.011***	–0.009
		(0.001)	(0.002)	(0.002)	(0.013)	(0.013)
Observations	82,385	82,385	82,385	85,225	56	2,078
Subset	East	East	East	US	East	East
Area-weighted	No	No	Yes	Yes	No	No

Notes: Table regresses log yields on weather variables for the months March–August. All coefficients are multiplied by 100, so they roughly report the effect of each variable in percent. Extreme heat is measured by degree days above 29°C for corn and degree days above 30°C for soybeans. Moderate heat is measured by degree days between 10°C and 29°C for corn, and 10 and 30°C for soybeans. Precipitation is the season total and measured in centimeters. Columns (1) through (4) use a panel of yields, while column (5) uses the time series (average yield and weather in a year) and column (6) uses the cross-section (average yield and weather in a county). Errors are clustered in STATA at the state level.

***Significant at the 1 percent level.

function is different in irrigated areas, adding them in a pooled model has limited effects as the irrigated crop area is small and, therefore, receives little weight.

Column (5) uses only the time series in the identification where both the dependent variable y_t and exogenous weather variables are averaged over all counties in the sample and cropland area-weights are used:

(2) $$y_t = \alpha + \beta_1 h_t + \beta_2 m_t + \beta_3 p_t + \beta_4 p_t^2 + t + t^2 + \varepsilon_t$$

Column (6) considers only the cross-section and uses the model:

(3) $$y_i = \alpha_i + \beta_1 h_i + \beta_2 m_i + \beta_3 p_i + \beta_4 p_i^2 + \varepsilon_i,$$

where y_i is the average difference of the yield in a county from the overall log yield in that year; that is, $y_i = 1/T\, \Sigma_t(y_{it} - y_t)$, and the weather variables are averages over the fifty-six years.

All coefficients have the expected sign: deviations from optimal precipitation levels (both too little and too much) are harmful. An increase in moderate heat (shifting from cold temperatures to moderate temperatures) is beneficial, while an increase in extreme heat is detrimental. Climate change is predicted to move the temperature distribution upward. Shifting the lower (cooler) part of temperature distribution toward the optimum temperature of 29°C or 30°C is beneficial, but the effects are dwarfed by the damaging effects of more frequent temperatures above the optimum. The dominating factor that drives predicted yield impacts is the measure of extreme heat as (a) the magnitude of the coefficient in the first row of the corn and soybeans panels of table 17.1 are large, and (b) the measure of extreme heat is predicted to increase significantly in higher latitudes as described in the previous section. Each twenty-four-hour exposure to each degree above 29°C for corn and 30°C for soybeans lowers annual yields by 0.4 to 0.7 percent. In other words, if a plant is exposed twenty-four hours to 40°C, yields decrease by about 5 percent.

The most striking feature of these results is that they are similar whether we use the time series in column (5) or the cross-section in column (6). While the former measures the effect of year-to-year weather shocks to which farmers can only partially adapt as they are realized after the plants have been sown, the latter compares how places with different average growing conditions have adapted to them. It seems quite unlikely that unobserved factors might confound the cross-sectional relationship in a manner that causes it to spuriously match the same nonlinear relationship observed in the time series.

17.2.3   The Evolution of Heat and Drought Tolerance over Time

Given the high sensitivity of yields to extreme temperatures, one might wonder whether there has been progress in developing plants that are more resistant to extreme heat over time. Average yields improved greatly over our fifty-six-year time frame. But what happened to heat tolerance? In Roberts and Schlenker (2011), we examined the evolution of the weather-yield relationship over the entire twentieth century in the state of Indiana, including such extreme events as the Dust Bowl in the 1930s. Here, we focus on the more recent past (1950–2005), but instead use a richer geographic coverage. To answer this question, we generalize specifications reported in column (2) of table 17.1 by allowing the coefficients to vary over time. The model is:

$$(4) \qquad y_{it} = \alpha_i + f_h(t)h_{it} + f_m(t)m_{it} + f_p(t)p_{it} + f_{p2}(t)p_{it}^2 + t_s + t_s^2 + \varepsilon_{it},$$

where $f(t)$ includes a constant plus restricted cubic splines of time.[8] The results for $f_h(t)$ are shown in the top row of figures 17.2 and 17.3, while the results for $f_{p2}(t)$ are shown in the bottom row. Recall that the average

8. We use 5 knots as the baseline but obtain comparable figures if we use 7 knots instead.

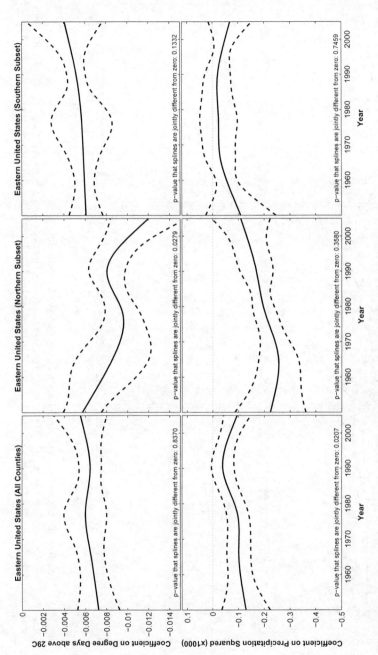

Fig. 17.2   The effect of extreme heat and precipitation on corn yields (1950–2005)

Notes: Graphs display the effect of extreme heat (top row) and precipitation (bottom row) on corn yields over time. The left column uses all counties east of the 100 degree meridian (except for Florida), while the second and third column use only the northern and southern subsets, respectively.

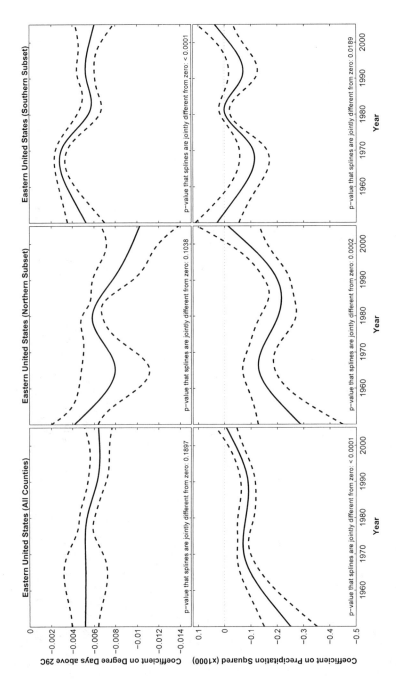

Fig. 17.3 The effect of extreme heat and precipitation on soybean yields (1950–2005)

Notes: Graphs display the effect of extreme heat (top row) and precipitation (bottom row) on soybean yields over time. The left column uses all counties east of the 100 degree meridian (except for Florida), while the second and third column use only the northern and southern subsets, respectively.

coefficients were given in column (2) of table 17.1. These figures plot how two of these coefficients evolve over time.[9] Each figure shows the results for all counties east of the 100 degree meridian (except Florida) in the left column, while the second and third column limit the sample to the cooler northern and warmer southern subset of states, respectively.[10] Each graph also lists the *p*-value for an *F*-test that the splines are jointly different from zero, implying that the coefficient of interest evolves over time. The only significant change in heat tolerance can be detected for northern subsample for corn and southern subsample for soybeans. Both suggest that heat tolerance *decreased* over time. On the other hand, the coefficient on the squared precipitation term, which measures the reduction in yield as precipitation deviates from the optimal level, is significant in most cases and generally shows an upward trend, which indicates crop yields have become less sensitive to fluctuations in precipitation.

17.2.4   Policy Implications

Why do corn and soybean show large improvements in average yields and better resistance to precipitation fluctuations yet show no improvement or even a worsening in heat tolerance? One reason might be that greater heat tolerance comes partially at the expense of reduced yield potential. Wahid et al. (2007) emphasize that "acquiring thermotolerance is an active process by which amounts of plant resources are diverted to structural and functional maintenance to escape damages caused by heat stress" (199). Presumably, those resources would otherwise go into seed formation and greater yields in the event of less extreme weather.

At the same time, heat tolerance and drought tolerance are inherently intermingled because water requirements increase with temperature. A plant wilts because it did not receive enough water for a given level of heat, or it received too much heat for a given amount of water. While plants require more water when temperatures go up, historic weather data in the United States has shown the opposite association: there is a negative correlation between extreme heat and precipitation in the fifty-six-year time series in table 17.1 as evaporation following rainfall results in cooling. This helps to explain the highly damaging effect of extreme heat in the historic time series: water requirements increase with extreme heat, yet water availability generally decreases as it only gets very hot once the soils are dry. But some effects of extreme heat act separately from water availability. For example,

9. Results for the remaining two coefficients on moderate heat and precipitation are available on request.
10. Northern states are Illinois, Indiana, Iowa, Michigan, Minnesota, New Jersey, New York, North Dakota, Ohio, Pennsylvania, South Dakota, and Wisconsin. Southern states are Alabama, Arkansas, Georgia, Louisiana, Mississippi, North Carolina, Oklahoma, South Carolina, Tennessee, and Texas.

regardless of water availability, corn does not flower if temperatures are too high.

Breeding new crop varieties is a long-term process. Alston, Pardey, and Roseboom (1992) review the history of crop research, which traditionally has been publicly funded, especially in developed countries. The economic rationale for a public role in basic research is that there are positive spillovers from the spread of ideas, so private companies do not necessarily have the right incentives to breed the most socially beneficial crops as they cannot reap all the benefits. In the United States, the Morrill Act and the Hatch Act created Land Grant universities in the second half of the nineteenth century with a mission to teach and study agriculture and created a cooperative extension service to interact with farmers. On an international level, the Consultative Group on International Agricultural Research (CGIAR) has several research centers around the world designed to improve yields of plants that are native to a region. Norman Borlaug, a previous director at CGIAR's International Maize and Wheat Improvement Center in Mexico received the Nobel Peace Prize for his work in improving yields and avoiding starvation. The Bill and Melinda Gates Foundation set out to become a CGIAR center and assist in reforming the agency as recent articles have highlighted that the budget for various CGIAR centers, for example, the International Rice Research Institute, have been cut significantly as world production outpaced demand and led to a downward drift in prices until 2005.[11]

Alston, Pardey, and Roseboom (1992) note that recently private companies have taken over a larger share of research and development. One reason for growing interest from private companies is that bioengineering can allow seed companies generate more productive seed that do not reproduce, thereby enhancing excludability.[12] Herbicide resistance, a key trait in the first generation of commercial genetically modified crops, also complements other products owned by seed companies. Some biotechnology companies have reported success in developing new strains with increased drought tolerance, yet critics have argued that such success has been reported before but did not materialize in the field.[13] There is little documented evidence on increased heat tolerance that is not counterbalanced by reduced yield potential.

Recent food price spikes have shown the most detrimental consequences of reduced supply fall on people living in poorest countries who can no longer afford food consumption when prices start to rise. The problem is

11. See "World's Poor Pay Price as Crop Research Is Cut," *New York Times,* May 18, 2008.

12. Seed companies might still have problems capturing all the rent from innovation if a drastic innovation significantly lowers production cost (Gallini and Wright 1990).

13. See "Drought Resistance Is the Goal, but Methods Differ." *New York Times,* October 22, 2008.

not necessarily a shortage of caloric production, but large income dispari-
ties. The taste for meat in developed countries, which requires many more
calories in production, might price poor people out of the market and lead
to famine.

On the other hand, many remain optimistic about the potential of genetically modified crops, which might usher in a new era of innovation that
breaks historical the trade-off between heat tolerance and yield potential.
To date, most commercially successful genetically modified crops resist
pests or herbicides. But more ambitious efforts exist to develop plants
that manufacture their own nitrogen fertilizer and possess more nutrients.
These innovations, among others, would be especially beneficial to poor
countries. While public funding of basic research has diminished, private
donations from charities like the Gates Foundation or by profit-driven
companies like Monsanto might replace these funds. But given public good
attributes of research, there remain important questions about the extent to
which private incentives to fund basic research align with potential social
welfare.

17.3   Conclusions

This chapter considers how changing climatic conditions may affect agricultural output and how heat tolerance has evolved over time. Changing
climate conditions, specifically the increased frequency of extreme heat, has
the potential to significantly decrease yields of staple crops in the United
States that form an important basis of caloric consumption. Because the
United States is by far the world's largest producer and exporter of commodity calories, this has the potential to impact world prices of staple food
commodities. One big question is whether advances in biotechnology might
increase heat tolerance enough to make crops more resistant to extreme
heat. On the upside, crops have become less susceptible to precipitation fluctuations. On the downside, the recent trend has been toward varieties with
higher average yields with unchanged or even *more* sensitive to extreme heat.
This suggests that the kind of technological change needed to cope with a
warming climate would be historically unprecedented. The changes needed
differ markedly from the kind of technological changes that has brought
about the green revolution—a three- to fourfold increase in yields that has
occurred since World War II.

Comparative advantages will obviously change with the climate. Thus,
the most natural and least-cost form of adaptation would seem to involve
simply changing the locations where agricultural activities take place. However, the best soil is found in currently moderate temperate zones, thereby
limiting the potential to shift to higher latitudes. How all the shifts will add
up globally remains highly uncertain.

References

Alston, Julian M., Philip G. Pardey, and Johannes Roseboom. 1992. "Financing Agricultural Research: International Investment Patterns and Policy Perspectives." *World Development* 26 (6): 1057–71.

Battisti, David S., and Rosamond L. Naylor. 2009. "Historical Warnings of Future Food Insecurity with Unprecedented Seasonal Heat." *Science* 323 (5911): 240–44.

Cassman, Kenneth G. 1999. "Ecological Intensification of Cereal Production Systems: Yield Potential, Soil Quality, and Precision Agriculture." *Proceedings of the National Academy of Sciences* 96 (11): 5952–59.

Deschênes, Olivier, and Michael Greenstone. 2007. "The Economic Impacts of Climate Change: Evidence from Agricultural Output and Random Fluctuations in Weather." *American Economic Review* 97 (1): 354–85.

Gallini, Nancy T., and Brian D. Wright. 1990. "Technology Transfer under Asymmetric Information." *Rand Journal of Economics* 21 (1): 147–60.

Jorgenson, Dale W., and Frank M. Gollop. 1992. "Productivity Growth in U.S. Agriculture: A Postwar Perspective." *American Journal of Agricultural Economics* 74 (3): 745–50.

Mendelsohn, Robert, William D. Nordhaus, and Daigee Shaw. 1994. "The Impact of Global Warming on Agriculture: A Ricardian Analysis." *American Economic Review* 84 (4): 753–71.

Roberts, Michael J., and Wolfram Schlenker. 2011. "The Evolution of Heat Tolerance of Corn: Implications for Climate Change." In *The Economics of Climate Change: Adaptations Past and Present,* edited by Gary D. Libecap and Richard H. Steckel. Chicago: University of Chicago Press.

———. 2010. "Identifying Supply and Demand Elasticities of Agricultural Commodities: Implications for the US Ethanol Mandate." NBER Working Paper no. 15921. Cambridge, MA: National Bureau of Economic Research.

Schlenker, Wolfram, Anthony C. Fisher, and W. Michael Hanemann. 2006. "The Impact of Global Warming on U.S. Agriculture: An Econometric Analysis of Optimal Growing Conditions." *Review of Economics and Statistics* 88 (1): 113–25.

Schlenker, Wolfram, and Michael J. Roberts. 2009. "Nonlinear Temperature Effects Indicate Severe Damages to U.S. Crop Yields under Climate Change." *Proceedings of the National Academy of Sciences* 106 (37): 15594–98.

Wahid, A., S. Gelani, M. Ashraf, and M. R. Foolad. 2007. "Heat Tolerance in Plants: An Overview." *Environmental and Experimental Botany* 61 (3): 199–223.

Comment   James B. Bushnell

This intriguing chapter explores two important central questions regarding climate change and the agricultural sector. First, how will changing climate conditions effect the productivity of the sector? Second, how will the sec-

James B. Bushnell is associate professor of economics at the University of California, Davis, and a research associate of the National Bureau of Economic Research.

For acknowledgments, sources of research support, and disclosure of the author's material financial relationships, if any, please see http://www.nber.org/chapters/c12163.ack.

tor be able to adapt to mitigate the impacts of climate change? Climate change threatens the continuation of the remarkable gains in agricultural productivity that have been experienced over the last fifty years at a time when continued productivity increases are projected to be needed to satisfy the demand for calories from a growing global population. The diversion of cropland to bioenergy production or land-use related forms of carbon sequestration will put further stress on the need for continued gains. This further emphasizes the need to understand the relationship between future climate conditions and the production capabilities of the sector.

Among the many changes predicted by climate models is the increase in the variability in weather conditions, resulting, for example, in an increase in extreme weather events that might be experienced during the growing season in agricultural areas. The important contribution of the work of Roberts and Schlenker (R&S) has been to estimate the relationship between weather and yields more flexibly. This flexibility is necessary to tease out the potential impacts of these extreme weather incidences, which can be dramatically different than the potential impacts predicted by average weather changes. For example, an increase of a tenth of a degree in temperature spread evenly over a four-month growing season may have negligible effects. If that same average, 122 degree-day increase were instead experienced as ten days with a 12.2 degree increase, it can have dramatically more severe implications for yields. In other work, R&S have explored this relationship using very flexible specifications. Here they apply previous insights to classify seasons according to the number of "moderate" or "extreme" (above 29° or 30°C) days.

Focusing on corn and soybeans, they find that yields are very sensitive to extreme heat events. This general finding is concerning as climate change is widely believed to increase the number of extreme heat days in key US growing regions. Most provocative is the evidence that, for corn, this sensitivity to extreme temperatures appears to have been increasing over the last fifty years. In other words, while yields have grown substantially, the sensitivity to extreme heat may have grown with them.

One possible implication is these two trends are not coincidental, that the remarkable gains in yields has come at the cost of increased sensitivity to heat. If this interpretation is correct, and past evidence can be extrapolated into a future of more extreme weather conditions, then the authors' findings imply that the agricultural sector may have to give up some of its productivity gains in order to address increasing volatile weather conditions.

It is worth noting that a sense of increased weather vulnerability is not consistent with much of the conventional wisdom in the corn belt. Part of this seeming contradiction could be attributable to the fact that the midwest has experienced relatively mild growing seasons in past decades. Another source of difference could be the industry and academic focus on *drought* as opposed to heat. While closely related, R&S contend that the effects of

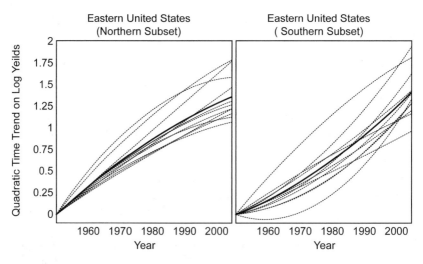

Fig. 17C.1   Time trends in corn yields

high heat can be separated from those of low moisture. By contrast, Yu and Babcock study the joint interaction of high heat and low rainfall and find that US corn and soybeans have become more tolerant of drought.[1]

There remains the curious and distinct difference in corn yield response between the northern and southern states in the sample of R&S. While heat tolerance appears to decline in the northern states for corn, there is no discernable trend in the south. One possibility might be that this is further evidence of a yield-versus-tolerance trade-off. Corn yields are (and always have been) quite a bit higher in the north. However, improvements in yields appear to be relatively consistent for both regions. Figure 17C.1 plots the quadratic time trends in yields produced from equation (4) in R&S. While there is some state to state variation, trends northern yields do not appear to be much different from those in southern states.

Implications for Adaption and Mitigation

Roberts and Schlenker also document another trend in the industry relating to research on seeds and productivity. Much of the research relating to agricultural productivity has migrated from publicly supported institutions to private firms. In the context of the results of this chapter, such a trend would be of concern if there were a mismatch between crop attributes that yield societal benefits and those that produce private economic gains. At first glance, these incentives seem reasonably aligned. Both society and individual producers would like to see an increase in expected yields although there

---

1. See Tian Yu and Bruce Babock, "Are U.S. Corn and Soybeans Becoming More Drought Tolerant?" *American Journal of Agricultural Economics* (forthcoming).

Table 17C.1   **Changes in planted corn acreage by region (millions)**

	Corn			Soybeans		
Region	1996	2004	Change	1996	2004	Change
North	42.71	43.81	1.10	31.88	39.64	0.78
South	4.47	3.78	−0.70	9.74	9.39	−0.35

Source: USDA Economic Research Service.

may be differences in the relative tolerance of risk associated with those expected gains.

Such a misalignment in risk preferences, which really would lie at the heart of private incentives to focus on the mean, as opposed to the variance, of yields, could be a side effect of U.S. crop insurance and disaster relief policies. Farmers have access to insurance on yields and revenues through government subsidized crop insurance programs and also receive periodic disaster relief payments in the face of extreme drought and other natural disasters.[2] It is worth noting that enrollment in crop insurance programs is much higher in the "northern" states that R&S identify as experiencing increased vulnerability to heat. Their findings may be evidence of moral hazard effects of these programs.

From 1996 to 2004, planted corn acreage increased by over 1 million acres in the northern states, while *decreasing* by over half a million acres in the south (table 17C.1). It could be that planting was retreating from areas that would be more sensitive to heat in the south, while expanding into such areas in the north. For this to be a factor in the results of R&S, however, this expansion into "heat sensitive" soil would have to be within-county. Further, similar trends in planted acreage played out in soybeans over this period (table 17C.1), with no apparent impact on heat sensitivity in either the south or the north.

---

2. Crop insurance was intended to replace disaster relief, but in general this has not happened, creating a potential "double indemnity" for effected producers. See J. Glauber, "Double Indemnity: Crop Insurance and the Failure of U.S. Agricultural Disaster Relief Policy," in *Agricultural Policy for the U.S. Farm Bill and Beyond.* Washington, DC: American Enterprise Institute, 2007.

Carbon Prices and Automobile Greenhouse Gas Emissions
The Extensive and Intensive Margins

Christopher R. Knittel and Ryan Sandler

18.1 Introduction

The transportation sector accounts for nearly one-third of the United States' greenhouse gas emissions. While over the past number of decades, policymakers have avoided directly pricing the externalities from vehicles, both in terms of global and more local pollutants and Corporate Average Fuel Standards have changed little since the mid-1980s, there is now considerable interest in reducing greenhouse gas emissions form the transportation sector. Many have argued that the unique features of the sector imply that pricing mechanisms would have little effect on emissions.

This chapter analyzes how pricing carbon through either a cap-and-trade system or carbon tax might affect greenhouse gas emissions from the transportation sector. Pricing carbon can influence emissions from the transportation sector in a number of ways. On the firm side, a positive carbon price incentivizes firms to reduce lifecycle emissions from liquid fuels either through the refining process or by switching to fuels that have a lower carbon content. Pricing carbon also incentivizes automobile manufacturers to change their product mix. On the consumer side, pricing carbon differentiates fuels by their carbon content, so consumers have an incentive to switch

Christopher R. Knittel is the William Barton Rogers Professor of Energy Economics at the Sloan School of Management, Massachusetts Institute of Technology, and a research associate of the National Bureau of Economic Research. Ryan Sandler is a PhD candidate in the Department of Economics at the University of California, Davis.

We thank Matthew Kotchen and Catherine Wolfram for helpful comments. We gratefully acknowledge financial support from the University of California EEE. Knittel gratefully acknowledges financial support from the Energy Institute at Haas. For acknowledgments, sources of research support, and disclosure of the authors' material financial relationships, if any, please see http://www.nber.org/chapters/c12134.ack.

to cleaner gasoline or alternative fuels. Consumers also have an incentive to drive more efficiently and to keep their vehicles operating more efficiently. The scrapping decisions of consumers are also affected. High-mileage vehicles may stay on the roader longer as they become relatively more valuable, while low-mileage vehicles may exit faster. New vehicle decisions are also likely to change as consumers switch to more fuel efficient vehicles. Finally, driving habits and trip decisions may also be affected, reducing the number of miles vehicles are driven.

Existing work has focused on the extensive margin by estimating how changes in gas prices affect what new cars people purchase and how scrappage decisions change. Using a single unique data source, we focus on two influences: scrappage decisions and vehicle miles traveled. We both summarize recent empirical work and present new results.

We bring together a number of unique data sets. The first is the universe of test records for California's emissions inspection and maintenance program, so-called smog tests, for the period of 1996 to 2010. California requires vehicles older than six years to receive biennial testing. In addition, testing occurs each time a vehicle changes ownership and randomly for a small share of vehicles. Among other things, the inspection data report odometer readings, which we use to measure vehicle miles traveled between tests. To measure greenhouse gas emissions, we link these data to Environmental Protection Agency (EPA) fuel economy ratings. In addition, the data are linked to Energy Information Administration (EIA) gas prices for the same years.

Our work builds on a recent literature analyzing how changes in gasoline prices influence consumer behavior. On the extensive margin, Busse, Knittel, and Zettelmeyer (2009) study purchase decisions and dealer pricing decisions using transaction-level data for both new and used vehicles. They find that increased gasoline prices influence both which vehicles consumers buy and the prices they pay for them in both the new and used vehicle markets. Furthermore, market shares are most influenced in the new vehicle market, while prices are most affected in the used market. Using model-level registration data for twenty metropolitan statistical areas (MSAs), Li, Timmins, and von Haefen (2009) find higher gas prices affect both new vehicle purchase decisions and used vehicle scrapping decisions. Hughes, Knittel, and Sperling (2008) estimate the short-run elasticity for gasoline and find that, over time, elasticities have fallen.

Most of these papers use monthly variation in gasoline prices for their empirical results. Therefore, their results are short run in nature. One of the contributions of this chapter is that because of the richness of our data, in particular a large number of individual decisions that take place throughout the year, we are able to estimate how changes in gas prices affect decisions that take place within two-year intervals. That is, our empirical models estimate how miles driven decisions over a two-year period are affected by changes in the average gas price throughout this entire period. We are able to estimate this longer-run elasticity because we observe many two-year intervals within our sample. To estimate a "two-year elasticity" with aggregate data would require a much larger time series than we require.

The chapter proceeds as follows. Section 18.2 discusses the empirical setting. The data are discussed in section 18.3. Section 18.4 provides graphical support for the two channels, while section 18.5 presents the empirical models and results. Finally, section 18.6 concludes the chapter.

18.2   Empirical Setting

Our empirical setting is California. Our primary data source is the universe of test data from California's Smog Check program from 1996 to 2010. California implemented its first inspection and maintenance program in 1984 in response to the 1977 Clean Air Act amendments. The initial incarnation of the Smog Check program relied purely on a decentralized system of privately run, state-licensed inspection stations and was plagued by cheating and lax inspections. Although the agreement between California and the federal EPA promised reductions in hydrocarbon and carbon monoxide emissions of more the 25 percent, estimates of actual reductions of the early Smog Check program range from zero to half that amount (Glazer, Klein, and Lave 1995).

The 1990 Clean Air Act amendments required states to implement an enhanced inspection and maintenance program in areas with serious to extreme nonattainment of ozone limits. Several of California's urban areas fell into this category, and, in 1994, a redesigned inspection program was passed by California's legislature after reaching a compromise with the EPA. The program was updated in 1997 to address consumer complaints and fully implemented by 1998. Among other improvements, California's new program introduced a system of centralized "test-only" stations and an electronic transmission system for inspection reports.[1] Today, more than a million smog checks take place each month.

An automobile appears in our data for a number of reasons. First, vehicles that are older than four years old must pass a smog test within ninety days of any change in ownership. Second, in parts of the state (details in the following), an emissions inspection is required every other year as a prerequisite for renewing the registration on the vehicle for vehicles that are seven years old or older. Third, a test is required if a vehicle moves from out of state. Vehicles that fail an inspection must be repaired and receive another inspection before they can be registered and driven in the state. There is also

---

1. For more detailed background, see http://www.arb.ca.gov/msprog/smogcheck/july00/ if.pdf.

a group of exempt vehicles. These are vehicles of 1975 model year or older, hybrid and electric vehicles, motorcycles, diesel powered vehicles, and large trucks powered by natural gas.

Since 1998, the state has been divided into three inspection regimes (recently expanded to four), the boundaries of which roughly correspond to the jurisdiction of the Regional Air Quality Management Districts. "Enhanced" regions, designated because they fail to meet state or federal standards for carbon monoxide (CO) and ozone, fall under the most restrictive regime. All of the state's major urban centers are in Enhanced areas, including the greater Los Angeles, San Francisco, and San Diego metropolitan areas. Vehicles that are seven years or older that are registered to an address in an Enhanced area must pass a biennial smog check in order to be registered, and they must take the more rigorous Acceleration Simulation Mode (ASM) test, which involves the use of a dynamometer. In addition, a randomly selected 2 percent sample of all vehicles in these areas is directed to have their smog checks at so-called test-only stations, which are not allowed to make repairs. High emitter profile vehicles are also directed to test-only stations, as are vehicles that are flagged as "gross polluters" (this occurs when a vehicle fails an inspection with twice the legal limit of one or more pollutant in its emissions). More recently, some "Partial Enhanced" areas have been added, where a biennial ASM test is required, but no vehicles are directed to test-only stations.

Areas with air pollution that does not exceed legal limits fall under the Basic regime. Cars in a Basic area must have biennial smog checks as part of registration, but they are allowed to take the more lax Two Speed Idle (TSI) test, and no vehicles are directed to test-only stations. The least restrictive regime, consisting of rural mountain and desert counties in the east and north of the state, is known as the Change of Ownership area. As the name suggests, inspections in these areas are only required upon change of ownership; no biennial smog check is required. (See table 18.1.)

18.3   Data

Our data come from the Bureau of Automotive Repair (BAR) and are the universe of smog tests from 1996 to 2010 and report the location of the test, the vehicle's vehicle identification number (VIN), odometer reading, the reason for the test, and test results. We decode the VIN to obtain the vehicles' make, model, engine and transmission. Using this, we match the vehicles to EPA data on fuel economy. Because the VIN decoding only holds for vehicles made after 1981, our data are restricted to these models, although to date we have only matched the EPA data for model years 1984 to the present. We also restrict our sample to 1998 and beyond given the large changes that occurred in 1997. This yields roughly 120 million observations. For the analysis in this chapter, we use a random 10 percent sample.

Table 18.1   **Means of greenhouse gas emission-related variables**

Year	Count	MPG	VMT/day	Gasoline price	Cents/mile
1998	661,729	23.92	33.42	1.52	7.31
1999	512,168	23.98	34.62	1.75	7.74
2000	869,975	23.78	30.84	2.09	8.45
2001	791,347	23.77	31.95	1.95	8.48
2002	788,716	23.56	29.40	1.82	8.49
2003	809,615	23.45	30.30	2.04	9.30
2004	1,119,371	23.29	28.07	2.38	10.59
2005	833,477	23.30	28.88	2.71	12.02
2006	953,961	23.17	27.23	3.02	13.33
2007	877,855	23.16	27.32	3.10	14.27
2008	959,873	23.05	25.84	3.71	14.73
2009	799,774	23.06	26.27	2.45	14.37

For biennial tests, we construct the average gasoline price between the two test data using the EIA's national average prices.

18.4   Initial Evidence

Before discussing the econometric models and results, we provide graphical evidence suggesting that increasing fuel prices affects both the intensive and extensive margins.

18.4.1   Extensive Margin

Changes in the extensive margin will manifest themselves in changing the mix of vehicles that are registered through both scrappage and new vehicle sales. We present evidence of both. Figure 18.1 plots both gas prices and the average fuel economy of newly registered vehicles within one year of the current year. While the Smog Check program does not require dealers to test new vehicles, tested vehicles within one year of the current year are likely to correlate well with new vehicle sales as we are capturing changes in ownership and vehicles moving into California that are one or two years old.[2]

From 1998 to 2004, there was a steady decrease in the fuel economy of new vehicles registered in California. This corresponds to the increase in sport utility vehicle (SUV) sales and a period of relatively low gasoline prices. As gasoline prices rose, however, this trend reversed. Remarkably, the trend again reversed as gasoline prices began to fall in 2008. We take this as evidence, consistent with Busse, Knittel, and Zettelmeyer (2009), that new vehicle sales respond to gasoline prices. Indeed, this figure extends their analysis to include the drop in gasoline prices beginning in 2008.

2. The graphs in this section smooth the series using a lowess smoothed line with a bandwidth equal to four months.

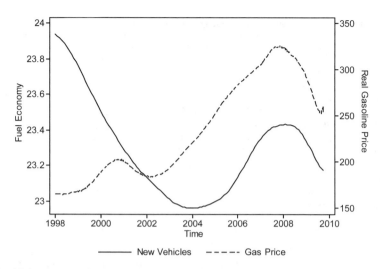

Fig. 18.1   Average fuel economy of new vehicles registered in California

As evidence that scrappage rates respond to gasoline prices, we plot the average fuel economy of vehicles of a specific model year over time. If the scrappage rates of vehicles of a specific vintage are independent of a vehicle's fuel economy or gasoline prices, then the average fuel economy of a particular model year over time will be constant. There is reason to believe, however, that less fuel efficient vehicles have lower hazard rates because trucks typically last longer than passenger cars.[3]

Figure 18.2 plots the average fuel economy of vehicles with model years of 1984, 1986, 1988, and 1990 being tested as part of either the random or biennial test programs, as well as gasoline prices. The model years are old enough to be at risk of scrappage and required biennial smog checks in each year of our data. All four model years, early in the sample, show a general decreasing trend in fuel economy, consistent with the higher durability of low fuel economy vehicles. This trend continues even as gasoline prices begin to rise in 2003. However, this trend appears to break and in three of the four cases reverse the higher are gasoline prices.

18.4.2   Intensive Margin

We present preliminary evidence that gasoline prices affect the intensive margin by plotting monthly gas prices and the average miles driven (VMT) daily within a year (figure 18.3). The figure suggests that VMT rose from 1998 to 1999 and then began a steady decline. This corresponds to the period

---

3. Therefore, all else equal, we might expect the average fuel economy of a given model year to fall over time. See, for example, Lu (2006), who finds different scrappage rates for cars and trucks.

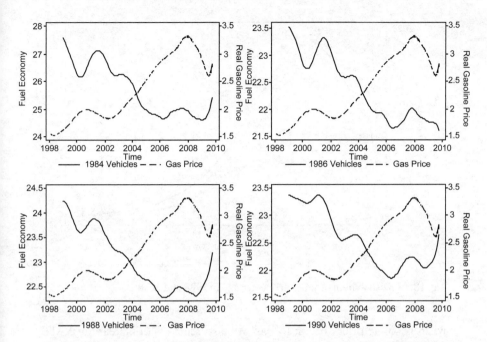

Fig. 18.2 Average fuel economy for vehicles with model years of 1984, 1986, 1988, and 1990 over time

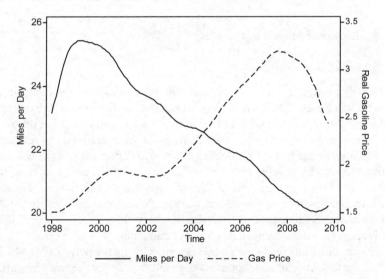

Fig. 18.3 Average miles driven per day and gasoline prices

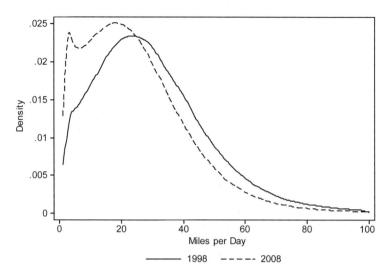

Fig. 18.4   Distribution of miles driven per day in 1998 and 2008

when gasoline prices began to rise. We also see a small increase in VMT during 2009, which corresponds to the decrease in gasoline prices, albeit with some lag. Figure 18.4 plots the distribution of VMT in 1998 and 2008. The figure suggests an entire shift in the distribution over this time period.

18.5   Empirical Models and Results

18.5.1   Extensive Margin

Our first empirical model estimates the hazard rate of the decision to scrap a vehicle as a function of the cost per mile of the vehicle. We define a vehicle as being scrapped if it had a biennial smog test in year X but does not have another smog test by year $X + 3$. We estimate a stratified discrete time Cox proportional hazard model. The stratified model allows the baseline hazard to vary by groups, where we define groups by either the make of the vehicle or the specific make/model/engine/drivetrain/transmission. We also include a sixth-order polynomial in the vehicle's odometer, whether the vehicle previously failed a smog test, whether it was flagged as a gross polluter, and separate vintage fixed effects for cars and trucks.

The key covariate is a vehicle's cost per mile during the period after a biennial smog test. We calculate the average gasoline price for the two years after the vehicle took the test and divide this by its fuel economy rating. As gasoline prices increase, the cost of operating all vehicles increase. All else equal, this will tend to increase the hazard rate for all vehicles. However, there is also a more general equilibrium effect. Busse, Knittel, and Zettelmeyer (2009) find that prices for fuel efficient vehicles increase as gasoline prices increases. This implies that as gasoline prices increase, the "continuation value" of fuel efficient vehicles might also increase, despite their increase in usage costs. This may reduce the scrapping rates of these vehicles. For this reason, we separate the effect of change in cost per mile by fuel efficiency quartile.

While our data are rich, one shortcoming of our data is that a vehicle can exit our data for a number of reasons. For one, it might be retired while still capable of being used; the decision of interest. Second, it could have been scrapped as a result of an accident. Third, it might move to a county that does not require smog tests. Finally, it might leave the state. These final three reasons for exit present a difficulty for us. Insofar as they are correlated with gas prices, we will tend to over- or under-predict the effect of changes in gas prices on scrappage decisions. In the following, we discuss empirical evidence that suggests our results are likely lower bounds on how increases in gasoline prices increase fleet fuel economy by changing the scrappage decisions of owners of existing vehicles.

Table 18.2 reports the results. Model 1 estimates the scrappage of vehicles that are over fifteen years old and includes only the dollars per mile of the vehicle as the key regressor along with a dummy for whether the vehicle is a truck, the polynomial in the odometer reading, year fixed effects, vintage fixed effects, and make fixed effects. Model 2 splits the effect of dollars per mile by the whether the vehicle falls in the first, second, third, or fourth fuel economy quartile.[4] Because these coefficients represent changes in the baseline hazard, coefficients smaller than one represent decreases in the probability of scrappage when the respective variable increases.

Both models 1 and 2 suggest that increases in the per mile cost of driving reduces the chances a vehicle is scrapped. Specifically, model 1 suggests that a five cent increase in the cost per mile of driving reduces the chances all vehicles are scrapped by roughly 12 percent. When we split this effect by fuel economy quartile, there is some evidence that this effect is strongest for the top three fuel economy quartiles. That is, low fuel efficient vehicles become relatively more likely to be scrapped.

There are at least two explanations for why higher gas prices may lower the likelihood a vehicle is scrapped. First, there may be an income effect, in the sense that as gas prices increase, vehicles stay on the road longer as consumers have less disposable income to buy newer vehicles. Second, and somewhat related, the propensity to sell a vehicle might fall as gas prices increase. As noted in the preceding, because in our data "scrappage" will also capture vehicles moving from counties that require smog checks to those that don't, as such as the number of transactions fall, this will be expressed as a reduction in scrappage. We find evidence of this when looking at the probability

---

4. We define the quartiles across the entire sample, but the results are robust to defining them within year as well.

Table 18.2   Probability of exit as a function of gasoline prices—Cox proportional hazard model

	Model 1	Model 2	Model 3	Model 4
Dollars per mile	0.882**		0.872**	
	(0.005)		(0.024)	
Dollars per mile * MPG quartile 1		0.862**		1.119**
		(0.007)		(0.034)
Dollars per mile * MPG quartile 2		0.841**		0.892**
		(0.009)		(0.028)
Dollars per mile * MPG quartile 3		0.837**		0.592**
		(0.010)		(0.021)
Dollars per mile * MPG quartile 4		0.840**		0.314**
		(0.012)		(0.014)
Truck	0.778**	0.776**	0.815**	
	(0.005)	(0.005)	(0.041)	
Odometer	0.936	0.936	0.927	0.925
	(0.054)	(0.054)	(0.057)	(0.057)
Odometer2	1.560**	1.557**	1.607**	1.607**
	(0.076)	(0.076)	(0.084)	(0.084)
Odometer3	0.816**	0.817**	0.806**	0.806**
	(0.015)	(0.015)	(0.016)	(0.016)
Odometer4	1.037**	1.037**	1.039**	1.039**
	(0.003)	(0.003)	(0.004)	(0.004)
Odometer5	0.997**	0.997**	0.997**	0.997**
	(0.000)	(0.000)	(0.000)	(0.000)
Odometer6	1.000**	1.000**	1.000**	1.000**
	(0.000)	(0.000)	(0.000)	(0.000)
Gross polluter	1.127*	1.126**	1.099**	1.103**
	(0.016)	(0.016)	(0.017)	(0.017)
Failed in the past	1.491**	1.488**	1.409**	1.383**
	(0.016)	(0.016)	(0.016)	(0.016)
Time trend	1.001+	1.000	1.021**	1.020**
	(0.000)	(0.000)	(0.000)	(0.000)
Time trend squared	1.000**	1.000**	1.000**	1.000**
	(0.000)	(0.000)	(0.000)	(0.000)
Vintage fixed effects	Yes	Yes	Yes	Yes
Stratified on make	Yes	Yes	—	—
Stratified on VIN prefix	No	No	Yes	Yes
Observations	676,321	676,322	676,323	676,324

**Significant at the 5 percent level.

of being "scrapped" of vehicles between six and nine years old. For these vehicles, as gas prices increase, their probability of being scrapped, where scrappage likely reflects the chances of being sold to a county that does not require a smog test, falls. Given this, we view these estimates as lower bounds on how gas prices affects the *retirement* decisions of our at-risk category.

The first two models only allow the baseline hazard to depend on make, vintage, and whether the vehicle is a car or truck. Within a make, there is a lot of heterogeneity in terms of the longevity of a vehicle within manufacturer and model year. To account for this, the last two models stratify by make, model, model year, engine, and transmission.

Model 3 constrains the dollar per mile effect to be constant across fuel efficiencies. The results are similar to model 1. However, when we allow the cost effect to differ by fuel efficiency quartile, we find much larger heterogeneity. In addition, despite the likelihood that the level of these effects are likely biased downward, we find that as gas prices increase the lowest fuel efficiency quartile vehicles are *more* likely to be scrapped. While the highest fuel efficient vehicles are much less likely to be scrapped compared to model 2.

We note that the right-hand-side variable is the dollars per mile; therefore, the fact that we find heterogeneity across fuel efficiency quartiles implies that to incentivize a low fuel efficiency vehicle to exit requires a smaller change in gasoline prices than a high fuel efficiency vehicle. To put these estimates into perspective, the average fuel efficiency of a bottom quartile vehicle is 16.7 miles per gallon (MPG). A one dollar increase in gasoline prices increases the cost per mile of these vehicles by roughly six cents. Therefore, because our coefficients are scaled for a five cent change in the cost per mile, a one dollar increase in gas prices increases the chance a bottom quartile vehicle is scrapped by approximately 15 percent.[5] The average fuel efficiency of a fourth quartile vehicle is 30.3 MPG, implying a one dollar increase in gas prices increases the cost per mile of these vehicles by 3.3 cents. Therefore, a one dollar increase in gas prices reduces the chances a top-quartile vehicle is scrapped by roughly 45 percentage points.

18.5.2   Intensive Margin

We next estimate how gasoline prices affect the intensive margin. To do this, we calculate the change in the odometer reading between biennial tests for each vehicle and the average gasoline prices during the two years between tests. This leaves roughly 3.6 million observations in our 10 percent sample.

As with the hazard model, we vary the set of fixed effects included. The key independent variable is the log dollars per mile. Model 1 in table 18.3 includes just year fixed effects, within-year linear time trends, vintage fixed effects, and a truck indicator variable. The results suggest a VMT elasticity of 0.399. It is important to note that while we are using within-year variation in gasoline prices, because we are estimating the effect of a 1 percent change in gasoline prices over the entire two-year period, these estimates represent fairly long-run elasticity. We believe that this makes these results unique in the sense that the individual-level data allow us to estimate long-run elasticities without aggregating the time series of the data (e.g., this would

---

5. Again, the Cox model is not linear, but for small changes around coefficients scaled for a five cent change, the error is likely small.

Table 18.3   Vehicle miles traveled, dollars per mile, and fuel economy

	(1) Model 1	(2) Model 2	(3) Model 3	(4) Model 4	(5) Model 5	(6) Model 6
ln(DPM)	−0.399**	−0.140**		−0.224**	−0.257**	
	(0.068)	(0.040)		(0.052)	(0.055)	
ln(DPM) * MPG Q1			−0.286**			−0.310**
			(0.024)			(0.059)
ln(DPM) * MPG Q2			−0.245**			−0.288**
			(0.024)			(0.058)
ln(DPM) * MPG Q3			−0.179**			−0.237**
			(0.024)			(0.059)
ln(DPM) * MPG Q4			−0.154**			−0.198**
			(0.022)			(0.066)
Truck	0.125**	0.097*	0.100**	0.038		
	(0.042)	(0.048)	(0.010)	(0.075)		
Year fixed effects	Yes	Yes	Yes	Yes	Yes	Yes
Within-year time trends	Yes	Yes	Yes	Yes	Yes	Yes
Vintage fixed effects	Yes	Yes	Yes	Yes	Yes	Yes
Demographics	Yes	Yes	Yes	Yes	Yes	Yes
Make fixed effects	No	Yes	Yes	No	No	No
Vin prefix fixed effects	No	No	No	Yes	No	No
Vehicle fixed effects	No	No	No	No	Yes	Yes
Observations	3,640,436	3,640,436	3,640,436	3,640,436	3,640,436	3,640,436
R^2	0.141	0.159	0.160	0.083	0.110	0.110

**Significant at the 5 percent level.

be infeasible using average yearly California consumption over a two-year period). Because of this, these estimates are larger than recent estimates of short-run elasticities (e.g., Hughes, Knittel, and Sperling 2008). Model 2 adds manufacturer fixed effects to model 1, lowering the elasticity substantially to 0.140. Model 3 allows the elasticity to vary by fuel efficiency quartile and finds that the least fuel efficient vehicles have an elasticity that is about 1.9 times larger than that of the most fuel efficient vehicles.

Model 4 includes VIN prefix (equivalent to make/model/engine/model year) fixed effects, and model 5 adds vehicle fixed effects. The average elasticities increase a little but are largely stable once vehicle make is controlled for. Model 6 allows the elasticity to vary by fuel efficiency quartile with vehicle fixed effects, and we find the heterogeneity persists. The top-quartile vehicles' elasticity is 1.6 times smaller than that of the bottom-quartile vehicles. One potential explanation for this heterogeneity is that we observe within-household substitution from the fuel inefficient vehicles to the fuel efficient vehicles. We are exploring this in current work.[6] Another potential explanation is that a given change in gasoline prices implies a larger change in the cost per mile for fuel inefficient vehicles. But we note that the inde-

6. This is consistent with the household bargaining that took place for one of the authors.

pendent variable here is price per model, and, in any event, this should not matter for models 5 and 6, where we use only variation *within* vehicle.

18.6 Conclusions

This chapter estimates how changes in gasoline prices affect both the extensive and intensive margins of automobile use. We find significant effects on scrapping decisions, new vehicle purchase decisions, and miles traveled. The results highlight the variety of avenues through which carbon pricing policies may affect emissions from the transportation sector.

References

Busse, Meghan, Christopher R. Knittel, and Florian Zettelmeyer. 2009. "Pain at the Pump: The Differential Effect of Gasoline Prices on New and Used Automobile Markets." NBER Working Paper no. 15590. Cambridge, MA: National Bureau of Economic Research.

Glazer, Amihai, Daniel B. Klein, and Charles Lave. 1995. "Clean on Paper, Dirty on the Road: Troubles with California's Smog Check." *Journal of Transport Economics and Policy* 29 (1): 85–92.

Hughes, Jonathan E., Christopher R. Knittel, and Daniel Sperling. 2008. "Evidence of a Shift in the Short-Run Price Elasticity of Gasoline Demand." *Energy Journal* 29 (1).

Li, Shanjun, Christopher Timmins, and Roger H. von Haefen. 2009. "How Do Gasoline Prices Affect Fleet Fuel Economy?" *American Economic Journal: Economic Policy* 1 (2).

Lu, S. 2006. *Vehicle Survivability and Travel Mileage Schedules.* Working Paper DOT HS 809 952, NHTSA Technical Report.

Comment   Matthew J. Kotchen

Most economists agree that establishing a price on carbon should be a central component of any climate policy in the United States. This could occur explicitly through a carbon tax or implicitly through a cap-and-trade mechanism. The chapter by Chris Knittel and Ryan Sandler (hereafter KS) helps us understand how a carbon price that affects the price of gasoline is likely to impact decisions about automobile use. In particular, KS consider how the price of gasoline affects decisions about when to scrap a vehicle and how

Matthew J. Kotchen is associate professor of environmental economics and policy at Yale University and faculty research fellow of the National Bureau of Economic Research.

For acknowledgments, sources of research support, and disclosure of the author's material financial relationships, if any, please see http://www.nber.org/chapters/c12135.ack.

much to drive. Understanding these effects is important because the transportation sector accounts for a third of all US greenhouse gas emissions.

One of the noteworthy features of the chapter is simply the data set KS were able to obtain. Using data from California's Smog Check program, they have detailed information about the characteristics of all vehicles registered in the state, when each vehicle is inspected, and how many miles it traveled between inspections. While the complete data set includes roughly 120 million observations, the reported results are based on a random sample of only 10 percent of the data. We already learn much from their analysis, but the KS chapter is only the beginning of new and important insights that we can expect from these data.

Regarding the decision about whether to scrap a vehicle, KS use as a proxy instances when a vehicle ceases to show up for its required smog check. They find that when gasoline prices are higher, vehicles with relatively low fuel efficiency are more likely to be scrapped, while those with relatively high fuel efficiency are less likely to be scrapped. These results are quite intuitive when one thinks about which vehicles will be associated with more "pain at the pump" from higher gasoline prices. From a policy perspective, the results draw attention to the fact that changes in the price of gasoline will affect not only *how* people drive, but also *what* they drive. There exists a growing body of evidence on how fuel prices affect new car purchases, but KS provide new evidence on what happens with respect to peoples' decisions about when to scrap their existing cars.

When it comes to driving, KS find that higher gasoline prices decrease the number of vehicle miles traveled. This, of course, is what one would expect with a downward sloping demand curve. Specifically, they estimate a price elasticity of 0.44. This implies that a 1 percent increase in the price of gasoline results in just under a 1/2 percent decrease in the number of vehicle miles traveled. Also consistent with microeconomic theory, their elasticity estimate is larger than existing estimates in the literature for short-run price elasticities. An important contribution of the KS chapter, therefore, is that it provides evidence on a more long-run price elasticity, which is more relevant for purposes of evaluating the potential impact of climate policy on the transportation sector.

With the KS results in hand, we can begin to think about questions for further research that will help us understand the interaction between climate policy and the transportation sector. What happens, for example, when vehicles are scrapped? If they are exported out of state or out of the country, the effect on greenhouse gas emissions is not clear. Moreover, given the estimated price elasticities, what kind of price increases would be necessary to achieve emission targets of the type that have been specified in recent climate bills at the state and federal level? While researchers and policymakers await answers to these more general questions, the KS chapter provides new and important evidence on a crucial step in the process.

Evaluating the Slow Adoption of Energy Efficient Investments
Are Renters Less Likely to Have Energy Efficient Appliances?

Lucas W. Davis

19.1 Introduction

While public discussion of H.R. 2454 (the Waxman-Markey Bill) has focused on the cap-and-trade program that would be established for carbon emissions, the bill also includes provisions that would tighten energy efficiency standards for consumer appliances. Appliance standards have been used in the United States since the 1970s, and currently standards are in place for dozens of different appliance types. There is an important trade-off inherent with standards. A standard truncates the market, removing goods that are preferred by some buyers. This cost must be balanced against potential benefits. In particular, supporters of standards argue that they help address a number of market failures that would not be addressed by a cap-and-trade program alone.

One frequently discussed example is the landlord-tenant problem. Many studies have pointed out that landlords may buy cheap inefficient appliances when their tenants pay the utility bill. Although investments in energy efficient appliances could, in theory, be passed on in the form of higher rents, it may be difficult for landlords to effectively convey information about

Lucas W. Davis is assistant professor of economic analysis and policy at the Haas School of Business, University of California, Berkeley; a faculty affiliate at the Energy Institute at Haas; and a faculty research fellow of the National Bureau of Economic Research.

I thank Severin Borenstein, Howard Chong, Olivier Deschênes, David Levine, Aaron Swoboda, Catherine Wolfram, Yves Zenou, and seminar participants at the University of Illinois College of Business, the UC Berkeley Green Buildings Conference, RAND, and the California Energy Commission for helpful comments. Support from the Energy Institute at Haas is gratefully acknowledged. For acknowledgments, sources of research support, and disclosure of the author's material financial relationships, if any, please see http://www.nber.org/chapters/c12130.ack.

the efficiency characteristics of appliances. Landlords have an incentive to inform tenants about energy efficient appliances. However, it may be difficult for tenants to evaluate these claims because most tenants are not experienced in evaluating the energy efficiency of appliances. Moreover, old energy bills are typically of limited value in evaluating claims from landlords because appliance utilization varies across households.

The landlord-tenant problem and other principal-agent problems are important to consider when designing carbon policy. Cap-and-trade programs work by increasing the price of energy, causing agents to internalize the social damages from their choices. Principal-agent problems reduce the effectiveness of this approach because the person experiencing these increased prices may not be the same person who is making decisions about energy use. For example, landlords may continue to purchase inefficient appliances even as their tenants' energy bills increase. In short, it may not be enough to simply put a price on carbon and the presence of principal-agent problems in addition to environmental externalities may justify combining appliance standards with cap and trade.

The landlord-tenant problem has been widely discussed in the literature (see, e.g., Blumstein, Krieg, and Schipper 1980; Fisher and Rothkopf 1989; Jaffe and Stavins 1994; Nadel 2002; and Gillingham, Newell, and Palmer 2009), but its practical importance has yet to be determined empirically. Understanding the mechanisms that explain this behavior and the magnitude of the distortion is important for determining how to most effectively target policies.

This chapter compares appliance ownership patterns between homeowners and renters using household-level data from a nationally representative survey, the Residential Energy Consumption Survey. The results show that renters are significantly less likely to report having energy efficient refrigerators, clothes washers, and dishwashers. Differences are large in magnitude and remain after controlling for household income, demographics, energy prices, weather, and other controls. The results imply nationwide an annual increase in energy consumption of approximately nine trillion btus, equivalent to 165,000 tons of carbon emissions annually.

The chapter focuses on a set of appliances that together represent about one-fourth of energy consumption in rental housing units.[1] There is reason to believe, however, that the other three-fourths (mostly heating and cooling) is also subject to the landlord-tenant problem. The agency issues with building energy efficiency may actually be worse than with appliances. Although it is relatively easy to verify that a dishwasher is energy efficient, it requires considerably more expertise to verify investments in, for example, roof insulation or heating and cooling ductwork. Given pending legislation

---

1. See U.S. Department of Energy (DOE), 2005 Residential Energy Consumption Survey, "Total Energy Consumption, Expenditures, and Intensities," table US12.

aimed at weatherization, an important priority for future work is to examine directly this broader class of energy efficient investments.

The chapter proceeds as follows. Section 19.2 provides relevant background information about energy efficiency standards in the United States and describes the data. Section 19.3 describes the estimating equation used to test for differences in appliance ownership patterns between homeowners and renters. Results are presented and discussed. Section 19.4 calculates the total energy consumption, expenditure, and carbon emissions implied by the estimates, and section 19.5 concludes.

19.2   Background and Data

Under the Energy Policy and Conservation Act of 1975 the U.S. Department of Energy (DOE) is required to establish energy efficiency standards for refrigerators, room air conditioners, clothes washers, dishwashers, and a broad class of additional residential appliances. Standards are periodically revised as warranted by technological improvements. Most recently, the Energy Policy Act of 2005, the Energy Independence and Security Act of 2007, and H.R. 2454 (the Waxman-Markey Bill) include provisions regarding energy efficiency standards for residential appliances.[2]

Since 1992, the Department of Energy in cooperation with the Environmental Protection Agency has, in addition, maintained a set of more stringent standards called "Energy Star" standards. Appliances exceeding these standards are among the most energy efficient in a particular class and receive an Energy Star label that is prominently displayed on the appliance at the time of purchase. Participation in the Energy Star program is voluntary though in practice all appliance manufacturers choose to participate. Similar programs are used in Australia, Canada, Japan, New Zealand, Taiwan, and the European Union. In addition, many utilities offer rebates for households that purchase Energy Star appliances, and the DOE recently committed $300 million in funding for rebates for qualified Energy Star appliances.[3]

This chapter examines the saturation of Energy Star appliances using household-level data from the 2005 Residential Energy Consumption Survey (RECS), a nationally representative in-home survey conducted approximately every five years by the DOE. The RECS provides detailed information about the appliances used in the home as well as information about the demographic characteristics of the household, the housing unit itself, weather characteristics, and energy prices. In addition, RECS reports state

2. See Nadel (2002) and US Department of Energy (2009), "Code of Federal Regulations, Energy Conservation Program for Consumer Products, Energy and Water Conservation Standards and Their Effective Dates, 430.32" for more information about appliance efficiency standards in the United States.

3. See Department of Energy, "Secretary Chu Announces Nearly $300 Million Rebate Program to Encourage Purchases of energy efficient Appliances," Press Release, July 14, 2009.

of residence for households living in New York, California, Florida, and Texas, and census division for all other households. The RECS is a national area-probability sample survey, and RECS sampling weights are used throughout the analysis.

The RECS also provides detailed information on who pays for utilities. The main results exclude households whose utilities are included in the rent. In the 2005 RECS sample, this includes 13.4 percent of all renters (4.2 percent of all households). These households do not pay directly for energy and, thus, tend to use their appliances more intensively.[4] In addition, the incentives for the adoption of energy efficient technologies are very different. Paying utilities themselves, landlords in these housing units have more incentive to invest in energy efficient appliances.

Beginning in 2005, households in the RECS were asked whether their major appliances were Energy Star.[5] These questions are somewhat unusual. Although many surveys ask about appliance ownership (e.g., American Community Survey), nationally representative surveys typically do not elicit information about energy efficiency. The question was asked for refrigerators, dishwashers, room air conditioners, and clothes washers, and households were shown an Energy Star label when answering the question. Households with appliances more than ten years old were assumed not to have Energy Star appliances and were not asked the question.

With any self-reported information, there is reason to be concerned about accuracy.[6] Perhaps most problematic for this analysis, it would seem reasonable to believe that homeowners may be better informed than renters about whether their appliances are Energy Star. This could provide an alternative explanation for the finding that homeowners are more likely to report having Energy Star appliances. In light of these concerns, the following analysis also examines two alternative measures of energy efficiency. Results are

4. Levinson and Niemann (2004) use RECS data to test whether energy use is higher in apartments where utilities are included in the rent. Controlling for observable characteristics of households, they find that tenants in apartments where utilities are included set their thermostats between one and three degrees (Fahrenheit) warmer during winter months when they are not at home.

5. Earlier RECS surveys do not ask about appliance energy efficiency. The 2001 RECS does include a question about whether your clothes washer is front loading or top loading. However, in 2001, front loading clothes washers were still relatively unusual in the United States, representing only 3.0 percent of all clothes washers in the RECS sample. See DOE, "2001 Residential Energy Consumption Survey: Housing Characteristics Tables," table HC5-4a.

6. The fraction Energy Star in the RECS corresponds poorly to fraction Energy Star in appliance sales data from DOE. For example, in the RECS among households with appliances less than four years old, the percentage of households who report owning an Energy Star appliance is 58 percent for refrigerators, 63 percent for dishwashers, 30 percent for room air conditioners, and 59 percent for clothes washers. In contrast, the DOE reports that the percentage Energy Star among appliances sold in 2005 was 33 percent for refrigerators, 82 percent for dishwashers, 52 percent for room air conditioners, and 36 percent for clothes washers. These percentages are based on sales data reported to DOE by retail partners. The DOE warns users that the set of retail partners changes from year to year and urges caution in using these data, particularly for making comparisons across years.

generally similar for these alternative measures, suggesting that the results are not entirely driven by misreporting.

First, in addition to asking whether a household's clothes washer is Energy Star, RECS asks if the clothes washer is "front loading" or "top loading." As described in detail in Davis (2008), front-loading clothes washers tumble clothes on a horizontal axis through a pool of water at the bottom of the tub, using about 50 percent less energy per cycle than conventional washers. Thus, "front loading" is an excellent proxy for energy efficiency and, importantly, whether the clothes washer is front loading is likely to be salient to both homeowners and renters.

Second, results are reported for energy efficient lighting. After asking how many lights the household typically uses, the survey asks, "How many of these lights use energy efficient bulbs? An energy efficient bulb is a fluorescent tube or a compact fluorescent bulb that costs more than a regular bulb but is one that lasts much longer." The measure used in the analysis is whether the household reports having *any* energy efficient light bulbs though results are similar for the *percentage* of light bulbs that are energy efficient.

19.3   Results

19.3.1   Descriptive Statistics

Table 19.1 presents descriptive statistics. The first two columns report mean household characteristics for homeowners and renters. The final column reports p-values from tests that the means in the subsamples are equal. The table reveals pronounced differences between homeowners and renters. Homeowners have substantially higher annual household income, are less likely to receive welfare benefits, are older, are less likely to be nonwhite, and are more likely to live in suburban and rural areas. In addition, appliance saturation levels differ substantially with homeowners more likely to have clothes washers and dishwashers but less likely to have room air conditioners.

Energy efficient technologies are described near the bottom of table 19.1. Homeowners are significantly more likely to report having energy efficient refrigerators, dishwashers, clothes washers, and lighting. Differences range from 7 percentage points for refrigerators to 11 percentage points for clothes washers. Particularly striking are the means for front loading clothes washers. Nine percent of homeowners report having a front loading washer compared to only 2 percent for renters. For room air conditioners, the pattern is reversed, with more renters reporting Energy Star units. This primarily reflects the higher saturation levels of room air conditioners among renters. In addition, room air conditioners are somewhat different because they are often owned by renters. Whereas it would be unusual for a tenant to install

Table 19.1   **Comparing mean household characteristics of homeowners and renters**

	Homeowners	Renters	p-value
Household economic characteristics			
Household annual income (1000s)	55.7	34.2	.00
Proportion household head employed	0.90	0.88	.08
Proportion welfare	0.06	0.24	.00
Household demographics			
Household size (persons)	2.60	2.57	.69
Age of household head	52.7	42.2	.00
Proportion with children	0.34	0.38	.10
Proportion household head nonwhite	0.21	0.44	.00
Type of neighborhood			
Urban	0.36	0.57	.00
Town	0.16	0.19	.14
Suburban	0.23	0.14	.00
Rural	0.25	0.10	.00
Climate and electricity prices			
Annual cooling degree days (1000s)	1.58	1.61	.64
Annual heating degree days (1000s)	4.15	3.82	.09
Electricity prices (cents per kwh)	10.3	11.1	.09
Appliance saturation			
Refrigerator	1.00	1.00	.95
Dishwasher	0.67	0.39	.00
Room air conditioner	0.21	0.38	.01
Clothes washer	0.95	0.57	.00
Energy efficient technologies			
Energy Star refrigerator	0.24	0.17	.00
Energy Star dishwasher	0.18	0.07	.00
Energy Star room air conditioner	0.04	0.05	.01
Energy Star clothes washer	0.23	0.12	.00
Front loading clothes washer	0.09	0.02	.00
Energy efficient lighting (any)	0.41	0.33	.01
Sample size	2,979	1,219	
Implied number of households (millions)	77.8	28.6	

Notes: This table describes households in the 2005 Residential Energy Consumption Survey (RECS). Means are computed using RECS sampling weights. The final column reports p-values (clustering by census division) from tests that the means in the subsamples are equal. Some households have more than one refrigerator or room air conditioner, and the table reports whether the most used unit is Energy Star. The survey questions about clothes washers are careful to exclude community clothes washers located in, for example, the basement or laundry room of an apartment building.

his or her own refrigerator or clothes washer in a rental unit, room air conditioners are relatively portable and can be easily installed.

Comparison of means provides an important baseline. However, it is difficult to draw strong conclusions on the basis of the evidence in table 19.1. Although the differences for energy efficient technologies are consistent with the landlord-tenant problem, this pattern could also be driven by other factors such as household income that are correlated with homeownership. The analysis that follows adopts a regression framework, comparing the saturation of energy efficient technologies between homeowners and renters while controlling for household income and other household characteristics. It is worth emphasizing that although the means for many of the characteristics are very different, there is a fair degree of overlap between homeowners and renters. Consider household income, for example. Although mean annual household income is very different ($55,700 for homeowners compared to $34,200 for renters) there are a reasonable number of renters (291 out of 1,219) with household income higher than the median household income for homeowners, and a reasonable number of homeowners (895 out of 2979) with household income lower than the median household income for renters. This lends credibility to the regression framework and its ability to effectively control for the observable differences between groups.

19.3.2   Regression Results

Table 19.2 presents estimates from a linear probability model of the following form,

$$y_i = \beta_0 + \beta_1 1(\text{renter}) + \beta_2 X_i + \varepsilon_i.$$

The dependent variable y_i is an indicator variable equal to one if the household reports having a particular energy efficient technology. For example, in the first row, the dependent variable is an indicator variable for households with an Energy Star refrigerator. The table reports the estimated coefficient and standard error corresponding to 1 (renter), an indicator variable for renters. The coefficient of interest β_1 is the difference in Energy Star appliance saturation between renters and homeowners with a negative coefficient indicating that renters are less likely to have an energy efficient model. Households who do not have a particular technology type are excluded from the regression, so the sample size varies across rows from 4,198 (all households) for lighting to 1,184 for room air conditioners.

Table 19.2 reports estimates of β_1 from four different specifications ranging from no controls in column (1) to the complete vector of covariates X_i in column (4) including household income (cubic), household demographics including indicators for whether the household head is employed and whether the household receives welfare benefits, indicator variables for 1, 2, 3, 4, 5, and 6+ household members, the age of the household head, and indicators for whether the household has children and whether the household head is nonwhite, as well as electricity prices (cubic), heating and cooling degree days (cubics), census division, and available state indicators. One of the important reasons why it is important to control for these household characteristics is that homeowners and renters may differ in the level of uti-

Table 19.2   **Are renters less likely to have energy efficient appliances?**

	(1)	(2)	(3)	(4)
Energy Star refrigerator [sample mean = .22]	−.067	−.034	−.056	−.067
	(.014)	(.017)	(.015)	(.015)
Energy Star dishwasher [sample mean = .25]	−.100	−.073	−.086	−.095
	(.024)	(.024)	(.033)	(.036)
Energy Star room air conditioner [sample mean = .16]	−.032	−.016	−.018	−.009
	(.011)	(.016)	(.016)	(.023)
Energy Star clothes washer [sample mean = .23]	−.030	−.002	−.027	−.033
	(.014)	(.016)	(.017)	(.014)
Front loading clothes washer [sample mean = .08]	−.054	−.032	−.028	−.031
	(.007)	(.004)	(.005)	(.005)
Energy efficient lighting [sample mean = .39]	−.075	−.038	−.046	−.049
	(.023)	(.026)	(.031)	(.024)
Household income (cubic)	No	Yes	Yes	Yes
Household demographics	No	No	Yes	Yes
Electricity prices (cubic)	No	No	No	Yes
Heating and cooling degree days (cubics)	No	No	No	Yes
Census division and available state indicators	No	No	No	Yes

Notes: This table reports estimated coefficients corresponding to an indicator for renter from twenty-four separate regressions, all estimated using least squares with RECS sampling weights. For each regression, the dependent variable is an indicator variable equal to one if the household has the energy efficient technology indicated in the row heading. Standard errors (in parentheses) are robust to heteroskedasticity and arbitrary correlation within census divisions.

lization of appliances. Households with high utilization levels have more to gain from adoption of energy efficient technologies (Hausman and Joskow 1982) because the savings are larger.

Consider first the estimates for refrigerators. In column (1) without controls, renters are 6.7 percentage points less likely to report having energy efficient refrigerators. This difference is identical to the difference in sample means in table 19.1. Controlling for household income decreases the point estimate corresponding to 1 (renter), consistent with high-income households being both more likely to be homeowners and more likely to own energy efficient refrigerators. Adding additional controls in columns (3) and (4) increases the point estimates to 5.6 and then back to 6.7 percentage points.

For dishwashers without controls, the difference is 10.0 percentage points. This is relatively large compared to the sample mean of 25 percent. As with refrigerators, the point estimate decreases after adding income and then increases again after adding additional controls. Homeowners tend to be older, face lower electricity prices, and live in rural and suburban areas, all characteristics that tend to decrease the probability that a household reports having energy efficient appliances.

Estimates for room air conditioners and clothes washers are also negative though consistently smaller than the coefficients for refrigerators and dishwashers. As mentioned in the preceding, room air conditioners are relatively portable, potentially mitigating the landlord-tenant problem. Point estimates for front loading clothes washers are negative, precisely estimated, and large relative to the sample mean of 8 percent. Finally, the estimate for lighting in the full specification is 4.9 percentage points, compared to the somewhat larger sample mean of 39 percent. With lighting, it is relatively easy for a tenant to move into a rental unit and replace incandescent light bulbs with energy efficient light bulbs. On the other hand, the cost savings from energy efficient lighting are accrued over many years, and there may be moving costs or other factors that prevent renters from taking energy efficient light bulbs with them when they move.

19.3.3   Discussion of Alternative Possible Explanations

These results demonstrate a consistent pattern of renters being less likely to report having energy efficient technologies. Although these results are consistent with the landlord-tenant problem, it is important to consider possible alternative explanations.

First, the differences could reflect landlords choosing not to invest in energy efficient technologies because appliances may have a shorter lifespan in renter occupied units. Because they do not own the appliances, renters may treat appliances more roughly (e.g., slamming doors, breaking refrigerator shelves), increasing the wear and tear on appliances eventually leading to them needing to be replaced. If this behavior is prevalent, landlords would then efficiently choose less expensive appliances. Similarly, landlords may be concerned about possible theft of appliances. This might be particularly problematic for lighting, with expensive light bulbs likely to disappear when renters move out.

Second, the differences could reflect unobserved differences between homeowners and renters in taste for green products. Suppose that, controlling for observables, homeowners receive a warm glow from using an energy efficient technology but renters do not. Alternatively, it could be that controlling for observables, homeowners have stronger tastes for certain appliance characteristics that are correlated with energy efficiency. These differences in taste could lead landlords to efficiently invest less in energy efficient technologies. For tastes to explain these findings, this preference for "green" would need to be imperfectly correlated with household income and other control variables and positively correlated with home ownership.

The following subsection reports the results from alternative specifications aimed at evaluating these and other possible alternative explanations. Many of these specifications add additional controls, and, for the most part, the basic pattern of renters being less likely to have energy efficient technologies is not sensitive to the addition of these controls. Although it is impossible to definitively rule out possible alternative explanations, the fact that the results are robust across alternative specifications lends support to the interpretation of these estimates as evidence of the landlord-tenant problem.

19.3.4   Alternative Specifications

Table 19.3 reports results from the baseline specification and thirteen alternative specifications. The dependent variable is indicated in the top of each column. For example, in column (1), the dependent variable is an indicator variable equal to one if the household has an Energy Star refrigerator. All specifications control for household income (cubic) and other household demographics, as well as electricity prices (cubic), heating and cooling degrees (cubics), census division, and available state indicators as in column (4) of table 19.2.

Row (A) reports the baseline specification. For row (B), the model is estimated using a logit model. Average marginal effects are reported and are very similar to the baseline estimates. Row (C) excludes households that "don't know" if their appliance is Energy Star. In the baseline specification, these households are treated as not having Energy Star appliances, and this choice does not seem to be driving the results. Relatively few households answer "don't know," and the fraction is similar for homeowners and renters. For example, for refrigerators, 4.0 percent of homeowners and 5.3 percent of renters answer "don't know."

Rows (D–F) restrict the sample to households with relatively new appliances. Again, results are similar to the baseline specification, suggesting that the results are not driven by differences in appliance age between homeowners and renters. If anything, the point estimates tend to grow larger (in absolute value) as one restricts the sample to relatively newer appliances.

Rows (G) and (H) report estimates separately for renters below and above the mean level of annual household income for renters. Estimated coefficients are similar for both groups and overwhelmingly negative, providing mild evidence against the "green tastes" explanation. If we thought that the results were driven by taste for green products that is imperfectly correlated with household income, one would have expected smaller estimated coefficients for high-income renters.

Row (I) reports estimates for renters whose utilities are included in the rent. Point estimates are negative and statistically significant for refrigerators, room air conditioners, and clothes washers. This is somewhat surprising because landlords in these units are paying utilities and, thus, have incentive to invest in energy efficiency. Still, it is important to keep in mind that these households are a somewhat unusual and unrepresentative group, overwhelmingly living in smaller apartments in older multiunit buildings. Those that do have refrigerators and clothes washers are more likely to have smaller apartment-sized models where energy efficiency options are more limited.

Row (J) restricts the sample to multiunit buildings and row (K) controls

Table 19.3   Are renters less likely to have energy efficient technologies? Alternative specifications

	Energy Star refrigerator [mean = .22] (1)	Energy Star dishwasher [mean = .25] (2)	Energy Star room air conditioner [mean = .16] (3)	Energy Star clothes washer [mean = .23] (4)	Front loading clothes washer [mean = .08] (5)	Energy efficient lighting [mean = .39] (6)
(A) Baseline specification	−.067 (.015)	−.095 (.036)	−.009 (.023)	−.033 (.014)	−.031 (.005)	−.049 (.024)
(B) Logit model	−.071 (.015)	−.103 (.038)	−.011 (.022)	−.033 (.016)	−.044 (.006)	−.050 (.025)
(C) Excluding "don't know"	−.071 (.016)	−.106 (.037)	−.014 (.025)	−.040 (.016)	n/a	n/a
(D) Among households with appliances < 10 years old	−.080 (.020)	−.094 (.037)	−.009 (.037)	−.021 (.017)	−.039 (.011)	n/a
(E) Among households with appliances < 5 years old	−.140 (.029)	−.124 (.047)	−.020 (.049)	−.071 (.032)	−.066 (.017)	n/a
(F) Among households with appliances < 2 years old	−.120 (.037)	−.100 (.062)	−.012 (.019)	−.018 (.050)	−.079 (.033)	n/a
(G) Low-income renters only	−.066 (.021)	−.054 (.037)	−.013 (.021)	−.027 (.027)	−.001 (.013)	−.027 (.014)
(H) High-income renters only	−.047 (.027)	−.102 (.038)	.001 (.027)	−.032 (.014)	−.055 (.013)	−.060 (.037)
(I) Renters with utilities included	−.074 (.018)	.004 (.044)	−.100 (.036)	−.150 (.044)	−.050 (.011)	−.001 (.041)
(J) Among households living in multiunit buildings	−.064 (.027)	−.041 (.083)	−.006 (.046)	−.074 (.108)	−.032 (.019)	−.026 (.055)
(K) Including housing characteristics	−.038 (.015)	−.071 (.042)	.005 (.013)	−.033 (.012)	−.031 (.008)	−.040 (.028)
(L) Including self-reported utilization	n/a	−.095 (.036)	−.009 (.023)	−.031 (.014)	−.031 (.005)	−.045 (.024)
(M) Excluding households who receive energy assistance	−.077 (.017)	−.101 (.035)	−.017 (.023)	−.030 (.015)	−.033 (.005)	−.047 (.020)
(N) Excluding cities with rent control (NY, CA)	−.066 (.017)	−.086 (.033)	−.026 (.031)	−.032 (.014)	−.026 (.005)	−.059 (.015)

Notes: This table reports estimated coefficients corresponding to an indicator for renter from seventy-nine separate regressions, all estimated using least squares with RECS sampling weights. For each regression, the dependent variable is an indicator variable equal to one if the household has the energy efficient technology indicated in the column heading. All specifications control for household income (cubic) and other household demographics, as well as electricity prices (cubic), heating and cooling degrees (cubics), census division, and available state indicators as in column (4) of table 19.2. Standard errors (in parentheses) are robust to heteroskedasticity and arbitrary correlation within census divisions.

for housing characteristics including the age of the housing unit, an indicator variable for multiunit, number of bedrooms, number of total rooms, and total square feet. These characteristics help proxy for lifetime wealth. Brueckner and Rosenthal (2009), for example, point out that newer houses tend to be owned by high-income households and that, over time, neighborhoods with an older housing stock tend to attract lower-income households. The point estimates are similar with these additional controls.

Row (L) controls for self-reported measures of utilization. For dishwashers and clothes washers, RECS asks households to report the number of loads a household typically does in a week. For air-conditioning and lighting, utilization is assessed by asking about the number of hours typically used per day. Adding the self-reported measures of utilization does little to the estimates. This is perhaps not surprising because the household characteristics already included in the regressions are important determinants of utilization levels. For the baseline specification, it is better to exclude these self-reported measures because utilization is a function of energy efficiency. As discussed in Davis (2008), energy efficient technologies lower the cost of utilization, potentially leading to increased utilization.

Row (M) excludes households who receive energy assistance. In the RECS, 4.4 percent of households receive some public aid. The largest such program, the Low Income Home Energy Assistance Program (LIHEAP) has been in operation in the United States since 1982 and operates in all fifty states with a $4.5 billion dollar budget in 2009. Eligible household must meet income requirements, and, typically, assistance is awarded on a first come, first served basis. For households facing subsidized electricity rates, it makes sense that landlords would not make costly investments in energy efficiency, and it is reassuring that the results do not change when excluding these households.

Finally, row (N) excludes households in urban areas in California and New York. Where the rental housing market is subject to rent control, landlords are constrained from making costly investments in energy efficiency because there is no scope for these investments to be capitalized into rents. Rent control is relatively uncommon in the United States, though several urban areas in California and New York have rent controls for some units, and it is interesting to see that the results do not change when households in these areas are excluded.

19.4   Evaluating the Implied Total Cost

An appealing feature of the estimates in section 19.3 is that they provide some of the information necessary to evaluate the overall magnitude of the landlord-tenant problem for an important group of household technologies. This section illustrates how these estimates can be applied, under simplified assumptions, to infer the implied total energy consumption, expenditure, and carbon emissions from the landlord-tenant problem. This preliminary assessment indicates that the total cost of this market failure is not negligible but that it is small relative to total energy consumption in rental housing units.

Table 19.4 reports the total cost of the landlord-tenant problem as implied by the estimates in the baseline specification. These results are calculated using average annual energy consumption and energy expenditure for Energy Star appliances from Sanchez et al. (2008).[7] The thought experiment is to consider how many additional energy efficient appliances there would be in the United States if renters were equally likely as homeowners to have these technologies.

The estimates imply that if renters were equally likely to have energy efficient appliances, in the United States there would be 2.2 million more Energy Star refrigerators, 3.1 million more Energy Star dishwashers, and 6.3 million more energy efficient light bulbs.[8] The estimates imply smaller impacts for room air conditioners and clothes washers. Nationwide, this would reduce annual energy consumption by 9.4 trillion btus, reduce annual energy expenditures by 93 million, and reduce annual carbon emissions by 166,000 tons.

To put this in perspective, this is about 1/2 of 1 percent of total energy consumption in rental housing units.[9] There are several reasons why this is not a larger fraction. First, in this thought experiment, the saturation of energy efficient technologies is increasing by only between 1 and 9 percentage points. Although not negligible, this is very different from assuming, for example, comprehensive replacement of all conventional appliances with energy efficient appliances. Second, these end-uses represent only about one-fourth of total energy expenditure in rental housing units.[10] Third, these calculations assume that energy efficient technologies use between 10 percent

7. Sanchez et al. (2008, table 5) reports annual energy savings per Energy Star unit of 0.85 Mbtu ($7.59) for refrigerators (15 percent), 1.17 Mbtu ($11.45) for dishwashers (29 percent), 0.68 Mbtu ($6.05) for room air conditioners (10 percent), and 1.32 Mbtu ($12.23) for clothes washers (20 percent). Sanchez et al. (2008, table 6) report that these appliances generate between .015 and .018 tons of carbon per Mbtu depending on the types of energy (electricity, natural gas, etc.) used by each appliance. Energy efficient light bulbs are assumed to use fifteen watts, compared to sixty watts for conventional incandescent bulbs.

8. In related work, Murtishaw and Sathaye (2006) use data from the American Housing Survey to evaluate the scope for principal-agent problems in residential refrigeration, water heating, space heating and lighting, concluding that 24 percent of residential energy consumption in the United States is potentially subject to principal-agent problems. This study was part of an international project whose results are described in International Energy Agency (2007).

9. According to DOE, "2005 Residential Energy Consumption Survey, Total Energy Consumption, Expenditures, and Intensities," table US1, rental housing units in the United States used 2.39 quadrillion btus of energy in 2005.

10. According to DOE, "2005 Residential Energy Consumption Survey, Total Energy Consumption, Expenditures, and Intensities," table US12, air conditioners, refrigerators, lighting, and other appliances together represent 36 percent of total energy consumption in rental housing units. Space and water heating represent the other 64 percent.

Table 19.4   The implied total cost of the landlord-tenant problem

	Refrigerators	Dishwashers	Room air conditioners	Clothes washers	Light bulbs	All technologies combined
Total units in millions	2.2	3.1	0.3	1.1	6.3	13.1
	(0.5)	(1.2)	(0.8)	(0.5)	(3.2)	(3.6)
Annual energy consumption in btus, trillions	1.9	3.7	0.2	1.4	2.1	9.4
	(0.4)	(1.4)	(0.5)	(0.6)	(1.1)	(2.1)
Annual expenditure on energy in 2009 dollars, millions	17.8	37.9	1.9	13.9	20.1	92.9
	(4.1)	(14.4)	(4.8)	(6.1)	(10.3)	(20.0)
Annual carbon emissions in metric tons, thousands	34.0	65.9	3.6	21.3	38.3	165.8
	(7.7)	(25.0)	(9.2)	(9.3)	(19.6)	(35.7)

Notes: This table reports the total cost of the landlord-tenant problem as implied by the estimated coefficients in column (4) of table 19.2. Standard errors (in parentheses) are robust to heteroskedasticity and arbitrary correlation within census divisions.

and 30 percent less energy than conventional technologies. The one exception is lighting, for which savings are larger.

19.5   Concluding Remarks

This chapter provides one of the first empirical analyses of the landlord-tenant problem. Across specifications, the estimates indicate that renters are significantly less likely to have energy efficient refrigerators, clothes washers, dishwashers, and lighting. Taken literally, the estimates imply nine trillion btus of excess energy consumption annually in the United States. More research and better data are needed to fully evaluate this problem. The new questions in the RECS are a step in the right direction, but more information is needed including results from professional energy audits to assess potential problems about the accuracy of the self-reported measures of energy efficiency. In future work, it would also be valuable to extend the analysis to a broader class of residential energy efficiency investments including building insulation, windows, and heating equipment.

References

Blumstein, Carl, Betsy Krieg, and Lee Schipper. 1980. "Overcoming Social and Institutional Barriers to Energy Conservation." *Energy* 5:355–71.

Brueckner, Jan, and Stuart Rosenthal. 2009. "Gentrification and Neighborhood Housing Cycles: Will America's Future Downtowns Be Rich?" *Review of Economics and Statistics* 91 (4): 725–43.

Davis, Lucas W. 2008. "Durable Goods and Residential Demand for Energy and Water: Evidence from a Field Trial." *RAND Journal of Economics* 39 (2): 530–46.

Fisher, Anthony C., and Michael H. Rothkopf. 1989. "Market Failure and Energy Policy: A Rationale for Selective Conservation." *Energy Policy* 17:397–406.

Gillingham, Kenneth, Richard Newell, and Karen Palmer. 2009. "Energy Efficiency Economics and Policy." *Annual Review of Resource Economics* 1:597–619.

Hausman, Jerry A., and Paul L. Joskow. 1982. "Evaluating the Costs and Benefits of Appliance Efficiency Standards." *American Economic Review* 72 (2): 220–25.

International Energy Agency. 2007. *Mind the Gap: Quantifying Principal-Agent Problems in Energy Efficiency.* Paris: OECD Publishing.

Jaffe, Adam B., and Robert N. Stavins. 1994. "The Energy Paradox and the Diffusion of Conservation Technology." *Resource and Energy Economics* 16:91–122.

Levinson, Arik, and Scott Niemann. 2004. "Energy Use By Apartment Tenants When Landlords Pay for Utilities." *Resource and Energy Economics* 26:51–75.

Murtishaw, Scott, and Jayant Sathaye. 2006. "Quantifying the Effect of the Principal-Agent Problem on U.S. Residential Energy Use." Environmental Energy Technologies Division, Lawrence Berkeley National Laboratory Working Paper no. 59773.

Nadel, Steven. 2002. "Appliance and Equipment Efficiency Standards." *Annual Review of Energy and the Environment* 27:159–92.

Sanchez, Marla, Richard E. Brown, Gregory K. Homan, and Carrie A. Webber. 2008. "2008 Status Report: Savings Estimates for the Energy Star Voluntary Labeling Program." Environmental Energy Technologies Division, Lawrence Berkeley National Laboratory Working Paper no. 56380.

Comment   Olivier Deschênes

Most proposed climate legislations are centered on the establishment of a market-based mechanism to price the externality caused by carbon emissions. In many cases, these proposals also include other provisions such as industry-specific subsidies, standards, and other forms of regulations or incentives. The chapter by Lucas Davis begins by making the key observation that in settings where asymmetric information or principal-agent problems arise, carbon pricing alone may not be sufficient to solve the environmental externality problem. Such settings would justify combining standards and market-based approaches to address the externality, as is the case for example in H.R. 2454.

One example where a market failure still arises in the presence of a market price on carbon emissions is the "landlord-tenant" problem. Because information about the energy efficiency of certain appliances might be difficult to credibly convey to tenants, landlords will tend to furnish their rental units with cheaper, energy inefficient appliances. In that case, and to the extent that tenants cannot change their appliances in response to the higher energy costs, carbon pricing will lead to inefficient energy consumption amongst tenants. Lucas Davis's chapter fills an important gap in the literature by presenting the first comprehensive empirical analysis of the landlord-tenant problem using data from the 2005 Residential Energy Consumption Survey (RECS).

The evidence in this chapter clearly supports the notion of a landlord-tenant problem. First and foremost, Davis's analysis convincingly shows that renters are significantly less likely to have energy efficient appliances (defined as appliances with the "Energy Star" certification) than homeowners. This is especially notable for refrigerators and dishwashers, where the homeowner-renter energy efficiency gaps are 7 and 10 percentage points, respectively. The baseline coverage rate of these energy efficient appliances is roughly 25 percent so the estimated gaps are large. Importantly, most of the regression estimates reported are insensitive to the inclusion or exclusion of a rich set of control variables such as household income, demographic variables, energy prices, and weather variables. As such, concerns about omitted variables bias plaguing the estimates are unlikely to be important. Davis also

Olivier Deschênes is associate professor of economics at the University of California, Santa Barbara, and a research associate of the National Bureau of Economic Research.

For acknowledgments, sources of research support, and disclosure of the author's material financial relationships, if any, please see http://www.nber.org/chapters/c12131.ack.

considers a series of alternative explanations for the main findings and presents a thorough specification analysis aimed at evaluating them. By and large, the main results are robust to the alternative specifications considered, lending further support to the landlord-tenant hypothesis.

I have a few suggestions for the author and others doing research in this area. One interesting finding that needs further attention is the fact the homeowner-renter energy efficiency gap does not appear to interact with household income, as shown in rows (G) and (H) of table 19.3. Davis interprets this as suggestive evidence against the notion that the documented homeowner-renter gap reflects a difference in unobserved preferences for "green" products. Whether this is the case should be more carefully analyzed in future research. Also, while I share Davis's view that taken as a whole the empirical evidence supports the notion of a landlord-tenant problem, there are a few empirical irregularities in table 19.3 that might require further attention. One such issue is sign reversals and sizable changes in point estimates across some of the specifications. While some of these fluctuations may reflect nothing more than sampling variation, it will be important to continue probing these estimates as more data become available.

More generally, I think future work in this area should focus on the important appliances such as central heating and cooling units that are not analyzed in this chapter because of data limitations. As Davis notes, central heating and cooling demand accounts for roughly 75 percent of renters' energy consumption, and these types of appliances are possibly even more subject to creating perverse agency issues. Future installments of the RECS should consider expanding their questionnaires to include these appliances. Finally, another possible area for future research is to analyze differences in actual energy consumption between homeowners and renters conditional on income, household size, and house/unit square footage rather than simply analyzing the "coverage" of energy efficiency unit. The landlord-tenant problem is not as much of an issue if renters use their variable-usage appliances such as dishwasher and air conditioners less intensively than homeowners.

Contributors

Severin Borenstein
Haas School of Business
University of California, Berkeley
Berkeley, CA 94720-1900

James B. Bushnell
Department of Economics
University of California, Davis
One Shields Avenue
Davis, CA 95616

Lucas W. Davis
Haas School of Business
University of California, Berkeley
Berkeley, CA 94720-1900

Olivier Deschênes
Department of Economics
2127 North Hall
University of California, Santa
    Barbara
Santa Barbara, CA 93106

Meredith Fowlie
Department of Agricultural and
    Resource Economics
301 Giannini Hall
University of California, Berkeley
Berkeley, CA 94720-3310

Don Fullerton
Department of Finance
BIF Box #30 (MC520)
University of Illinois
515 East Gregory Drive
Champaign, IL 61820

Lawrence H. Goulder
Department of Economics
Landau Economics Building 328
Stanford University
Stanford, CA 94305

Kevin A. Hassett
American Enterprise Institute for
    Public Policy Research
1150 Seventeenth Street, NW
Washington, DC 20036

Stephen P. Holland
Department of Economics
Bryan School of Business and
    Economics
University of North Carolina at
    Greensboro
Greensboro, NC 27402-6165

Matthew E. Kahn
UCLA Institute of the Environment
Departments of Economics and of
    Public Policy
La Kretz Hall, Suite 300
Box 951496
Los Angeles, CA 90095-1496

Christopher R. Knittel
MIT Sloan School of Management
100 Main Street, E62-513
Cambridge, MA 02142

Charles D. Kolstad
Department of Economics
University of California
Santa Barbara, CA 93106

Matthew J. Kotchen
School of Forestry and Environmental
    Studies
Yale University
195 Prospect Street
New Haven, CT 06511

Kala Krishna
Department of Economics
523 Kern Graduate Building
The Pennsylvania State University
University Park, PA 16802

Arik Levinson
Department of Economics
ICC 571
Georgetown University
3700 O Street, NW
Washington, DC 20057

Erin T. Mansur
Dartmouth College
6106 Rockefeller Hall
Hanover, NH 03755

Aparna Mathur
American Enterprise Institute for
    Public Policy Research
1150 Seventeenth Street, NW
Washington, DC 20036

Gilbert E. Metcalf
Department of Economics
Tufts University
8 Upper Campus Road
Medford, MA 02155-6722

Michael J. Roberts
Department of Agricultural and
    Resource Economics
#8109 North Carolina State University
Raleigh, NC 27695-8109

Ryan Sandler
Department of Economics
University of California, Davis
One Shields Avenue
Davis, CA 95616

Wolfram Schlenker
Department of Economics
School of International and Public
    Affairs
Columbia University
420 West 118th Street, MC 3323
New York, NY 10027

Hilary Sigman
Department of Economics
Rutgers University
75 Hamilton Street
New Brunswick, NJ 08901-1248

V. Kerry Smith
Department of Economics
W.P. Carey School of Business
P.O. Box 879801
Arizona State University
Tempe, AZ 85287-9801

Robert N. Stavins
John F. Kennedy School of
    Government
Harvard University
79 JFK Street
Cambridge, MA 02138

Roberton C. Williams III
Department of Agricultural and
    Resource Economics
Symons Hall
University of Maryland
College Park, MD 20742

Catherine Wolfram
Haas School of Business
University of California, Berkeley
Berkeley, CA 94720-1900

Author Index

Acemoglu, D., 135n6, 142
Ackerman, K. V., 218
Ahlgren, O., 132
Aldy, J. E., 3, 38, 127
Allingham, M. G., 222n13
Alston, J. M., 280
Ambrosi, P., 197, 220n10
Andreoni, J., 155, 217n6
Antinori, C., 219
Arbell, J., 131
Arrow, K. J., 66
Atkinson, G., 54
Atkinson, S. E., 132
Auffhammer, M., 104
Ayres, R. U., 77

Babiker, M. H., 58, 63
Babock, B., 284n1
Battisti, D. S., 273n5
Baum-Snow, N., 260, 262
Becker, G. S., 216
Bell, M. L., 79
Bennear, L. S., 128, 142
Bergstrom, T. C., 155
Berman, E., 37
Bernard, A. L., 158n2
Berry Cullen, J., 267
Blanchard, O. J., 46n8
Bloom, N., 50
Blume, Lawrence E., 155
Blumstein, C., 302
Bohringer, C., 54, 129, 158n2

Borenstein, S., 231
Bovenberg, A. L., 22, 114n5, 160
Bradford, D., 246n3
Brown, G. M., Jr., 232n2, 235
Brueckner, J., 312
Brunnermeier, S., 129
Buchanan, J., 190
Bui, L., 37
Bull, N., 28n6
Burrows, P., 181n3
Burtrauw, D., 75, 80, 83n8, 112n4, 114n5
Bushnell, J. B., 3, 186n6, 187, 202, 219
Busse, M., 288, 291, 294

Cahue, P., 40
Canadell, J. G., 94n1, 94n3, 95, 97n7
Cao, X., 262
Capoor, K., 220n10
Carlton, D. W., 181n3, 231, 232, 236, 242
Cassman, K. G., 273
Caulkins, J. P., 131
Chang, H. F., 3, 222, 222n14
Chavez, C. A., 214, 216n4, 217n5
Chen, Y., 186n6
Cherkashin, I., 58, 58n5
Chetty, R., 191
Chiu, S., 190, 190n10
Cifuentes, L., 80
Clark, C. F., 150
Clotfelter, C. T., 222n13
Coase, R., 159n4
Conte, M. N., 219n9

Subject Index

Page numbers followed by the letter *f* or *t* refer to figures or tables, respectively.